CLIMATE TREND ATLAS OF EUROPE BASED ON OBSERVATIONS 1891–1990

CLIMATE TREND ATLAS OF EUROPE BASED ON OBSERVATIONS 1891–1990

by

Christian-D. Schönwiese

and

Jörg Rapp

Johann Wolfgang Goethe University,
Institute for Meteorology and Geophysics,
Frankfurt am Main, Germany

SPRINGER-SCIENCE+BUSINESS MEDIA, B.V.

A C.I.P. Catalogue record for this book is available from the Library of Congress

ISBN 978-90-481-4822-6 ISBN 978-94-015-8818-8 (eBook)
DOI 10.1007/978-94-015-8818-8

Printed on acid-free paper

TABLE OF CONTENTS

PREFACE

Climate varies on all scales of time and space, by a large variety of reasons. However, before any discussion of reasons can be performed it is necessary to realize the very facts of climate variability by means of observations or reconstructions, respectively. In this book we focus on observed long-term trends of selected climate elements (temperature, precipitation, humidity, pressure) as revealed by direct measurements of the European station network within the recent 100 years.

Of course, there are a number of problems in detail: Reliability and accuracy of data, time series homogeneity, statistical confidence of trends and so on. We hope that these problems are addressed in an instructive and, as far as possible, exhausting way. The main purpose of this work, however, was to provide a collection of trend charts which specify the regional particularities of observed climate trends in different months or seasons of the year leading us to an 'Atlas of observed climate trends in Europe'. Keeping in mind the recent World Meteorological Organization (WMO) climate normal (CLINO) period, we have also calculated the 1961–1990 trends.

This Atlas was originally published as Report No. 20 of the Frankfurt University Environmental Research Center (Zentrum für Umweltforschung, ZUF) where Tobias Fuchs and Michael Denhard have also contributed to this original German edition. In preparing this considerably revised and extended English version the authors are grateful to a number of anonymeous reviewers as well as to Christine Lidzba, Angelika Neuse, Tim Staeger, and Martin Weyres for various technical assistance, whereas the lay-out is mainly based on the activity of one of us (Jörg Rapp). Last not least we are grateful to the publisher for his interest, cooperation and high-quality printing of our 'Atlas'.

We hope that not only climatologists but also all people who are interested in this and related fields of science (e.g. geography, hydrology, biology, and so on), including all possible impacts concerning ecology, economics and social affairs may benefit from the results summarized in this 'Atlas'. However, we would like to stress that any extrapolation of observed trends into the future may be misleading without the knowledge of reasons of climate change. In this context, especially concerning the human-induced enhancement of the 'greenhouse effect' we refer to the parallel 'Climate Change Atlas' by Ann Henderson-Sellers and Ann-Maree Hansen.

Frankfurt/Main, November 1996

Christian-D. Schönwiese and Jörg Rapp

CHAPTER 1

INTRODUCTION

In a usual climate atlas, the reader finds average charts of the various climate elements (temperature, precipitation, etc.) or charts of any climate classification which again represent average aspects. Thus, the impression arises that climate, in contrast to weather, is more or less constant. Even the reference period (time interval which is covered by the climatic data) is often not specified. In reality, climate varies on all scales of time and space (Mitchell *et al.*, 1966; Schönwiese, 1994). To realize this fact in detail, however, and to survey it clearly, are not easy. This is why all measurements imply errors or uncertainly ranges, respectively, and even without errors the variation behaviour of climate is so complicated that proper statistical methods are needed to take into account the various aspects of climate variability and to separate the various components of climate variations.

Measurement errors may occur in a systematic way which means deviations from the real (unknown) values in exactly or approximately constant deviations. This type of error affects averages but, for example, not trends (if the deviations from the real values are exactly constant). Random errors are always existent, and their magnitude is usually quantified via the Gaussian theory of randomness leading to an accuracy assessment of the best estimate of a series of measurement data. Let a_i be these data, n the number of data and $\bar{a}$ the weighted or non-weighted arithmetic mean of these data, then

$$\Delta a = \sqrt{\frac{1}{n(n-1)} \sum_{i=1}^{n} (a_i - \bar{a})^2} \qquad (1.1)$$

is such a quantification where $\Delta a = s/\sqrt{n}$ and s is the usual expression for the data standard deviation (see also formula (1.2); Spiegel, 1972, Schönwiese, 1992). That means, in this case, that the accuracy of any average value $\bar{a}$ is given by $\bar{a} \pm \Delta a$. A discussion of data accuracy problems is incorporated in Chapter 2.

From the climatological point of view, data time series $a_i(t_i)$, $t_{i+1} - t_i = \Delta t$, inhomogeneity is an important special case of errors where t is time and Δt is the equidistant time step of such a series (hour, day, month, year, etc.). Inhomogeneity means that any observational climate time series at any station may imply variations which have nothing to do with climate itself. Examples are a change of instruments (with different systematic errors), change of environment conditions (changing vegetation in the neighbourhood of the station, formation of buildings, etc.), and displacement of the station (e.g. from the centre to the surrounding of a city). Often such inhomogeneities appear in the form of sudden 'jumps' of the time series data values. In this case they are easily detectable, especially if neighbour stations do not show this 'jump' and if the reason is known (e.g. change of instrument). In other cases this detection is difficult, namely if the inhomogeneity effect is continous. An example is the continous growth of a town leading to a related growth of the urban heat island effect which should not be misinterpreted as a large-scale or even global warming. The problem of time series inhomogeneity is addressed by means of statistical tests (summarized in Chapter 3), the causal discussion of the superimposed variation components or the investigation of the station history, respectively.

Without measurement errors including time series inhomogeneity problems, at any selected station, the variation behaviour of any climate element can be classified in respect of the 'characteristic time' of the variation components whose superposition produces the observed total time series variance. An adequate statistical method which enables such a disintegration is spectral variance analysis where the time series variance

$$s^2 = \frac{1}{n-1}\sum_{i=1}^{n}(a_i(t_i) - \bar{a})^2 \tag{1.2}$$

($s = \sqrt{s^2}$ = standard deviation) is evaluated as a function of period (T) or frequency ($f = 1/T$), respectively. The algorithms used for this purpose are 'Fast Fourier Transform' (FFT), 'Autocorrelation Spectral Analysis' (ASA) or 'Maximum Entropy Spectral Analysis' (MESA), alternatively; for details see, for example, Båth (1974), Mitchell *et al.* (1966), Olberg and Rákóczi (1984) and Schönwiese (1992). Some spectral variance aspects concerning the time series used in this study are presented in Appendix B.

Here only the possibility of such an analysis and the fundamental results are important. This is illustrated by means of an example (see Figure 1.1). This example shows the annual mean surface air temperatures 1781–1994 at station Hohenpeissenberg (Southern Germany, 47.8°N, 11.0°E, station height 977 m NN), a pre-Alpine mountain station which is not affected by the urban heat island effect and which provides a very long continuous series of direct measurements. The variance spectrum (ASA method, lower part of Figure 1.1) can be interpreted as follows. The time series under consideration is dominated by

- relatively high-frequent variations of cycle lengths ('periods') of 2 – 3.5 yr (year-to year or interannual variability);
- relatively low-frequent variations of cycle lengths of > 10 yr (interdecadal variability, maximum 13 yr in this case);
- ultra low-frequent variations roughly > 100 yr (secular variability, low-frequent residuum of the variance spectrum shown in Figure 1.1).

Many other temperature records show very similar results (Schönwiese *et al.*, 1986). The relatively low-frequent variation component is indicated in Figure 1.1 (upper part), by a heavy line (time series smoothing, more exactly 10 yr Gaussian low-pass-filter; for details see Mitchell *et al.*, 1966, or Schönwiese, 1992).

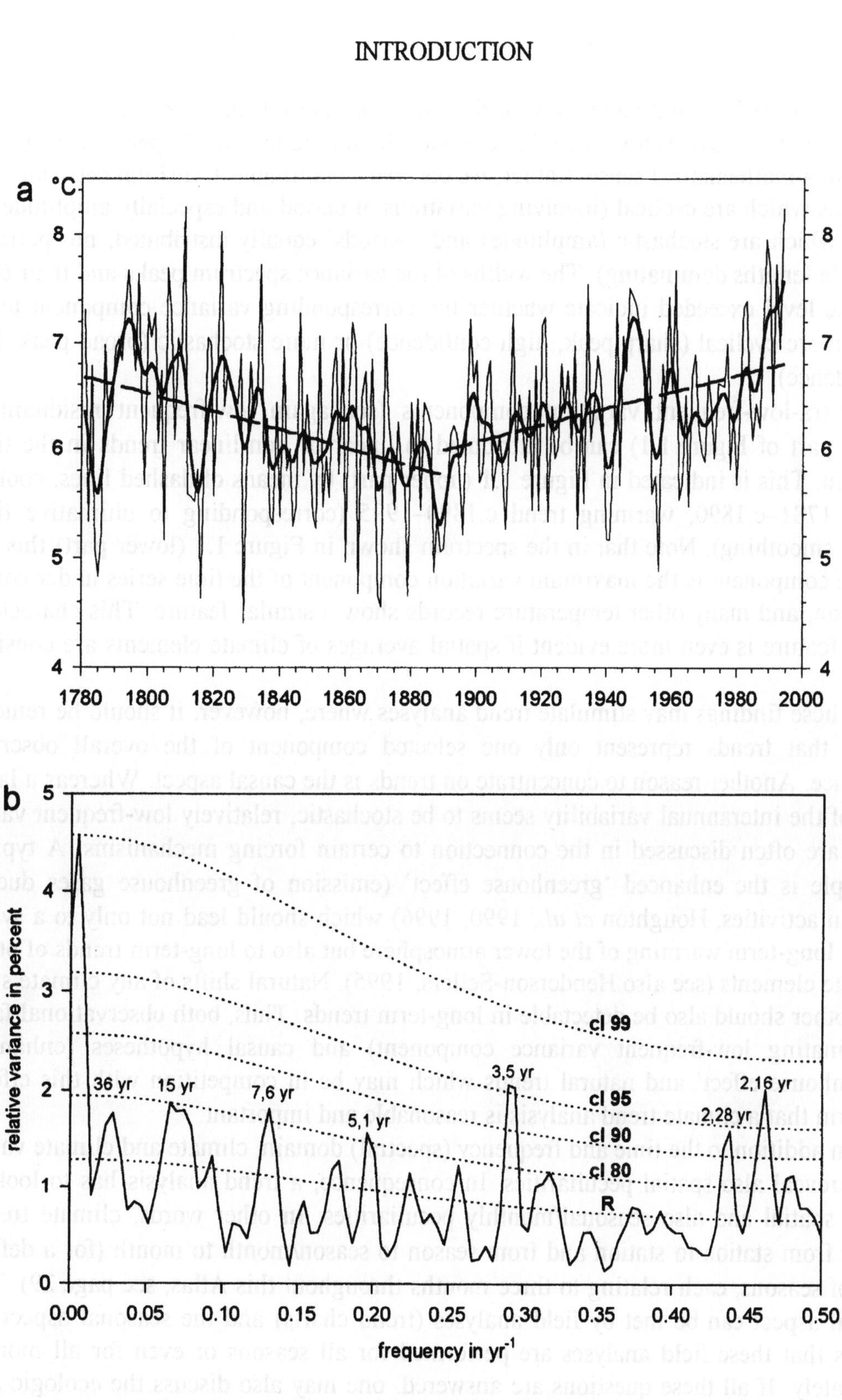

Figure 1.1. **(a) Observed annual means of surface air temperature (2 m) 1781–1995 at station Hohenpeissenberg (Southern Germany, see station list in Appendix A), including 10 yr Gaussian low-pass filtered data and linear trends 1781–1890 or 1890–1995, respectively.**
(b) Corresponding power spectrum (autocorrelation spectral variance analysis; from Schönwiese, 1992), where R is the 'red noise' background and cl is confidence levels.

Moreover it should be mentioned that the term 'period' in the context of variance spectrum analysis does not mean that climate would produce exactly periodical fluctuations in a mathematical sense. Rather, we observe a continuous transition between fluctuations which are cyclical (involving variations of period and especially amplitude) to those which are stochastic (amplitudes and 'periods' equally distributed, no 'periods' or cycle lengths dominating). The widths of the variance spectrum peaks and their confidence level exceeded indicate whether the corresponding variance component tends to be more cyclical (sharp peak, high confidence) or more stochastic (broad peak, low confidence).

Ultra-low-frequent variation components (see again low-frequent residuum in lower part of Figure 1.1) can be attributed to linear or non-linear trends in the time domain. This is indicated in Figure 1.1 (upper part) by means of dashed lines: cooling trend 1781–c.1890, warming trend c.1890–1995 (corresponding to ultimative time series smoothing). Note that in the spectrum shown in Figure 1.1 (lower part) this variance component is the maximum variation component of the time series under consideration, and many other temperature records show a similar feature. This characteristical feature is even more evident if spatial averages of climate elements are considered.

These findings may stimulate trend analyses where, however, it should be remembered that trends represent only one selected component of the overall observed variance. Another reason to concentrate on trends is the causal aspect. Whereas a large part of the interannual variability seems to be stochastic, relatively low-frequent variations are often discussed in the connection to certain forcing mechanisms. A typical example is the enhanced 'greenhouse effect' (emission of greenhouse gases due to human activities; Houghton *et al.*, 1990, 1996) which should lead not only to a systematic long-term warming of the lower atmosphere but also to long-term trends of other climate elements (see also Henderson-Sellers, 1995). Natural shifts of any climate state to another should also be detectable in long-term trends. Thus, both observational facts (dominating low-frequent variance component) and causal hypotheses (enhanced 'greenhouse effect' and natural trends which may be in competition with this effect) confirm that a climate trend analysis is reasonable and important.

In addition to the time and frequency (spectral) domain, climate and climate variations reveal also spatial peculiarities. In consequence, a trend analysis has to look at these spatial and also seasonal/monthly peculiarities. In other words, climate trends differ from station to station and from season to season/month to month (for a definition of seasons, each relating to three months throughout this Atlas, see page 29). The spatial aspect can be met by field analyses (trend charts) and the seasonal aspect requires that these field analyses are performed for all seasons or even for all months separately. If all these questions are answered, one may also discuss the ecologic and socioeconomic consequences of such climate change more or less independent of their natural or anthropogenic forcing. Finally, statistical confidence tests have to be used in order to assess the reliability of the analysis results and – as far as possible – to discern between stochastic and deterministic forcing.

CHAPTER 2

PURPOSE OF ANALYSIS AND RELIABILITY PROBLEMS

The region under consideration in this Atlas is Europe. This is due to the satisfactory time series data availability on a secular time scale, in contrast to nearly all other regions in the world. The purpose is to describe as exactly as possible the observed long-term climate trends in this region including their regional peculiarities with a fair regional dissolution, for all months, seasons and the year. Due to the fluctuative behaviour of climate which is not of interest if the consideration is focussed on long-term trends (eliminating relatively short-term fluctuations) and some spectral characteristics (long-term residuum, see again Figure 1.1, time and frequency domain) it seems to be reasonable to look at time intervals of at least 100 yr. As far as the causal aspect is concerned, note that also the enhanced 'greenhouse effect' (already mentioned in Chapter 1) acts on this time scale and this may be also true for some other competitive natural forcings of climate.

However, trends may intensify so that they may become evident only within the recent decades. Figure 2.1 shows two examples of such a behaviour where both are related to precipitation: station De Bilt (Netherlands), month of August (Figure 2.1 a); station Bad Mergentheim (Southern Germany), winter season (Figure 2.1 b). These special examples are in line with general definitions of the UN World Meteorological Organization (WMO) which has defined CLINO periods (climatic normals) each of a 30 yr time interval, recently 1961–1990. It is recommended by WMO that climate tables and charts should refer to these CLINO periods. In consequence, in this Atlas, in addition to the secular period 1891–1990, also the trends of the recent CLINO period 1961–1990 are analysed and compared with related average charts.

There is an additional argument available from trend analysis moving in time. This is documented in Figure 2.2 where, for surface air temperature at station Frankfurt/Main the trend values 1891–1990, 1892–1990, 1893–1990, and so on are shown (100 yr, 99 yr, 98 yr, etc.). It arises that the 100- to roughly 30 yr (see dotted line 1961 in Figure 2.2) trends are relatively stable, especially in the case of annual data (year) where the warming amounts to c. 0.1 K yr^{-1} nearly independently from the time interval under consideration. Only in winter (see Figure 2.2) a trend increase up to 0.3 – 0.4 K yr^{-1} is detectable for the intervals c. > 1930 until 1990. Short-term trends c. > 1960 (that means reference period < 30 yr), however, become enormously unstable in all seasons, even changing their sign. Similar results arise if, instead of the start year of the reference period, the end year is moving or both so that the trend values for all possible time intervals are computed. All these findings confirm the use of reference periods no shorter than approximately 30 yr.

As can be seen in the examples Figures 1.1 and 2.1 (temperature and precipitation), the relation of trends and other variation components can be very different. Parti-

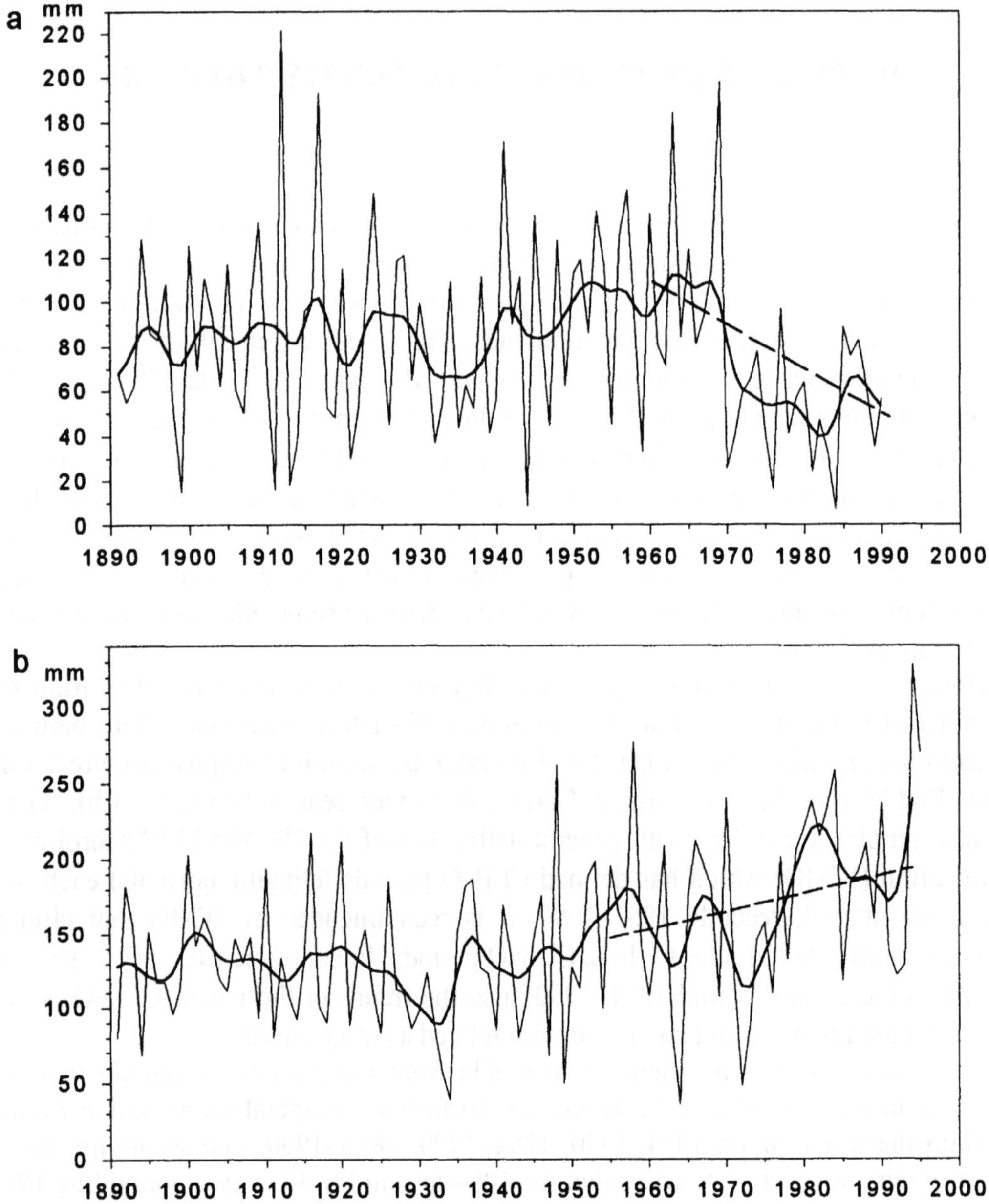

Figure 2.1. Precipitation annual totals, 10 yr Gaussian low-pass filtered data and recent linear trends at stations (a) De Bilt (Netherlands), month of August, or (b) Bad Mergentheim (Southern Germany), winter.

cularly in the case of precipitation (see Figure 2.1, secular time scale) year-to-year variations and fluctuations often dominate. On the one hand it is justified to express any time series $a_i(t_i)$ in terms of the series average $\bar{a}$ and the deviations a_i' from this average,

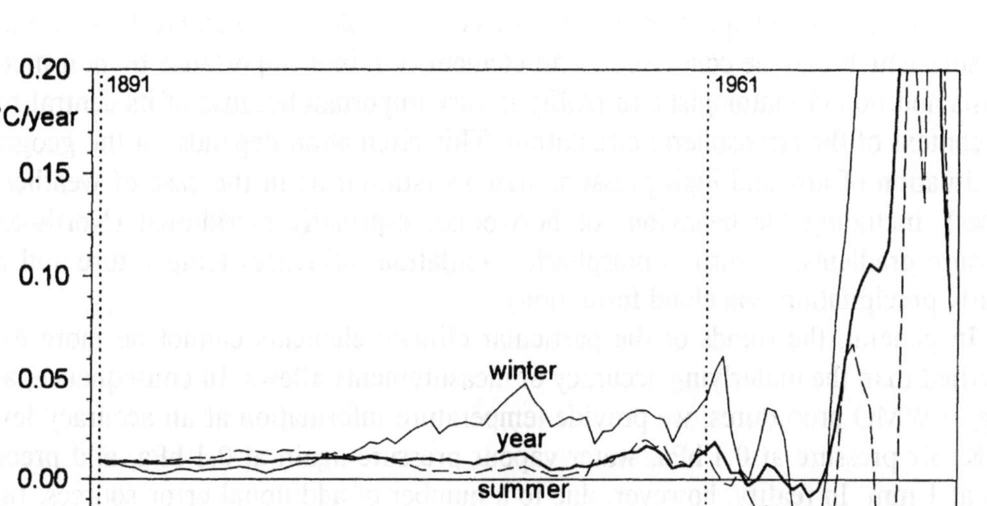

Figure 2.2. **Moving trend analysis. Standardized linear air temperature trends (in °C yr^{-1}) with respect to different analysis periods in Frankfurt/Main (Germany), for summer (dashed line), winter (solid line), and year (bold line). The first year of the time series is variable (1891, 1892, ..., 1989); the last one (1990), however, is constant. Trends become random for periods < 20–30 years.**

$$a_i(t_i) = \bar{a} + a_i'(t_i), \tag{2.1}$$

or, respectively in terms of the linear trend $\hat{a}(t)$ and the deviations $a_i^*(t_i)$ from this trend (with n = number of data = sample size)

$$a_i(t_i) = \hat{a}(t) + a_i^*(t_i), \quad Tr = \hat{a}(t_n) - \hat{a}(t_1). \tag{2.2}$$

On the other hand it is necessary to quantify the relation of the trend against the residuum variance. This is congruent with the statistical question of whether this trend is confident against randomness, and if yes, at which confidence level. A simple approach is the trend-to-noise ratio Tr/N where often the data time series standard deviation s (see formula (1.2)) is used as a proxy for 'noise' (total observed variability). A more sophisticated method is the Mann-Kendall test, described in the next section.

The aim to compute long-term trends of as many climate elements as possible is restricted by the data availability. Although, as already mentioned, the situation in Europe is favourable – particularly in Central Europe, England, and Scandinavia –

only surface air temperature (reference height 2 m above ground level), mean sea level air pressure, and precipitation remain to be available at a secular time scale. Air pressure which may be conceived as an element of minor importance by non-meteorologists and non-climatologists, in reality is very important because of its central role in the context of the atmospheric circulation. This circulation depends on the geographical situation of low and high pressure systems (similar as in the case of weather processes), including the behaviour of horizontal, especially meridional (North–South) pressure gradients. In turn, atmospheric circulation influences temperature and particularly precipitation (via cloud formation).

In general, the trends of the particular climate elements cannot be more exactly specified than the underlying accuracy of measurements allows. In consequence, according to WMO procedures, we provide temperature information at an accuracy level of 0.1 K, air pressure at 0.1 hPa, water vapour pressure again at 0.1 hPa, and precipitation at 1 mm. In reality, however, due to a number of additional error sources, the accuracy is often considerably more restricted.

One of these sources is the urbanization effect already mentioned in Chapter 1, especially the "heat island effect" of temperature. In detail, there are different pecularities of this effect in relation to space and time. Thus, at any particular time, a temperature difference between the central city and the green surrounding of several K may be observed (for example London nearly 10 K, or Karlsruhe, Germany, nearly 7 K) if the weather is sunny and windless (Oke, 1973). However, on a secular time scale, this temperature difference amounts to only 0.1 – 0.2 K (ECSN, 1995) and is, therefore, considerably smaller than most of the related long-term temperature trends. Different methods to calculate daily averages of temperature should not influence significantly monthly averages (Kalb and Noll, 1980; Kalb and Hänle, 1981; Kalb and Baucus, 1982) as used in this Atlas.

Whereas in the case of temperature and pressure the usual measurement techniques are relatively simple so that earlier measurements, say 100 yr ago, are comparable with modern ones, this condition is not fulfilled in the case of precipitation so that this problem needs a detailed discussion:

First, in many European countries discontinuities of precipitation measurements appeared which were extensively described by Groisman and Easterling (1994) and which may disturb the results of a trend analysis. Often, the change to shielded instruments (in order to diminish the wind influence) is most important, especially if a considerable part of precipitation is observed in the form of snow. Thus, in Finland within the years 1909 and 1981 new instruments were introduced which may have lead to an increase of 25–30 % of measured precipitation in case of snow whereas the measured precipitation in liquid form was only marginally influenced. In Sweden around 1900, in Norway between 1904 and 1908 a similar effect due to change of instruments may have occurred. To summarize, the Scandinavian precipitation measurements in the early interval of the secular period under consideration in this Atlas may be underestimations so that the increase resulting from a trend analysis may be overestimated. However, it is very hard to quantify this error which may be in the order of magnitude of roughly 20–30 % in winter.

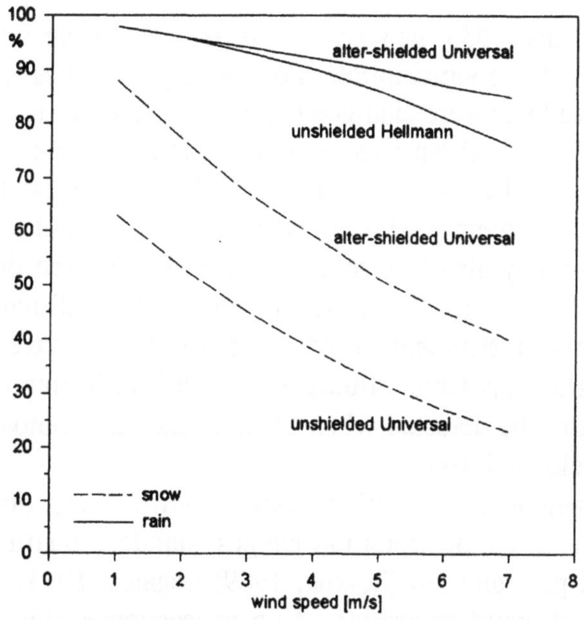

Figure 2.3. **Percent of measured in relation to 'real' precipitation using the Hellmann or Universal gauge with (alter-shielded) and without (unshielded) wind protection arrangement (from Legates, 1993).**

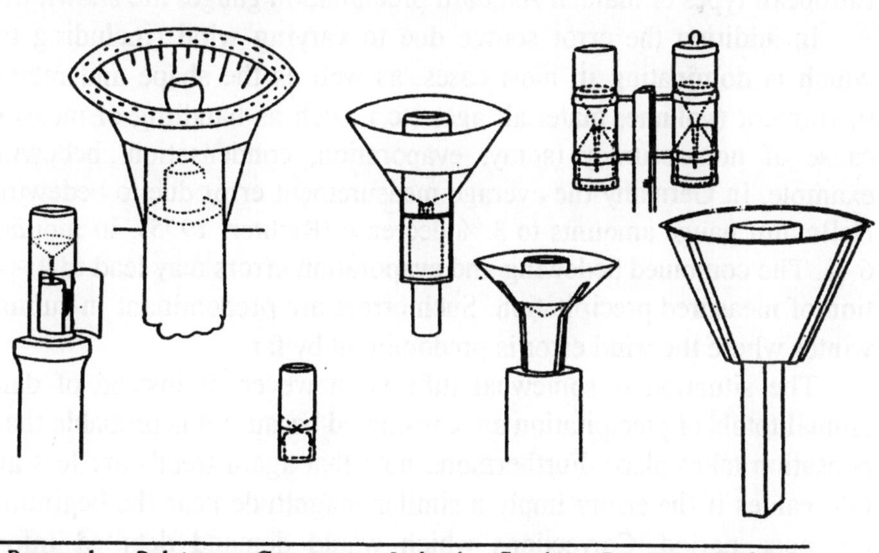

	Portugal	*Belgium*	*Germany*	*Iceland*	*Sweden*	*Romanie*	*Finland*
Type	Portuguese	RRM P50	Hellmann	Icelandic	SMHI	IMC	Wild
Area of orifice [cm²]	200	100	200	200	200	200	500
Height of gauge [cm]	43	35	43	56	35	45	35

Figure 2.4. **Types of manual standard precipitation gauges in some European countries (from Sevruk and Klemm, 1989)**

In the Netherlands in 1945 the altitude of the precipitation gauges was changed, leading to an increase of measured precipitation of approximately 5 %. In the former Soviet Union, in 1948 and 1953, new types of instruments were introduced, leading to an increase of measured snow precipitation of roughly 5 – 40 %. In addition, in 1967 the measurement procedures were changed leading again to an increase, in this case, of liquid precipitation. Because these reasons and due to the poor density of stations, the precipitation trends revealed for the region of the former Soviet Union are very unreliable and should be interpreted with extreme care. It should be mentioned that a change of instruments may also lead to a decrease of measured precipitation. This is reported concerning Poland in 1945 (introduction of the Hellmann type gauge) and Switzerland where new instruments used since 1980 may have led to a 5 – 15 % reduction. Although this may have influenced the 1961-1990 trend calculations, a corresponding jump cannot be detected when these trends are compared by those of the adjacent countries (field analysis).

Second, experimental investigations have shown that, due to the different orientation and height of the precipitation gauges, standard instruments measure only 50 – 95 % of the real precipitation (Sevruk, 1989; Legates, 1993) (see Figure 2.3) negatively correlated with wind exposition. As a consequence, this error is large if the station is situated in a flat area without mountains, outstanding buildings, vegetation, etc., otherwise it is decreased. In case of snowfall this error may be even larger than 50 % (Sevruk, 1994). Note that such measurement errors affect much more the (questionable) absolute values whereas relative trends are much less disturbed. Some European types of manual standard precipitation gauges are shown in Figure 2.4.

In addition the error source due to varying wind (including turbulence effects) which is dominating in most cases, as well as the shape and related features of the instrument (volume, material, age, etc.), such as handling of measurement, may be a cause of non-accuracy (spray, evaporation, condensation, bedewing, leaking). For example, in Germany the average measurement error due to bedewing in the case of a Hellmann gauge amounts to 3 % decrease (Richter, 1995), in mountain regions up to 6 %. The combined bedewing and evaporation errors may lead to 4 – 8 % underestimation of measured precipitation. Such errors are predominant in summer, in contrast to winter where the wind error is predominant by far.

The situation is somewhat difused, however, if instead of daily, monthly, and annual totals of precipitation are considered because it is probable that some error compensation takes place. Furthermore, note that again trends are less affected than absolute values if the errors imply a similar magnitude near the beginning and end of the reference period. Corrections which would demand detailed information from all stations over all the time interval considered are impossible.

A special case worth discussing separately is a warming trend which leads to an increasing attribution of rain to precipitation, snow vice versa (decreasing). This effect is important in winter and an assessment for the German station Karlsruhe (Rapp and Schönwiese, 1996) leads to the suggestion that approximately one–tenth of the observed winter trend (1891–1990) of +65 mm may be due to the indirect contribution

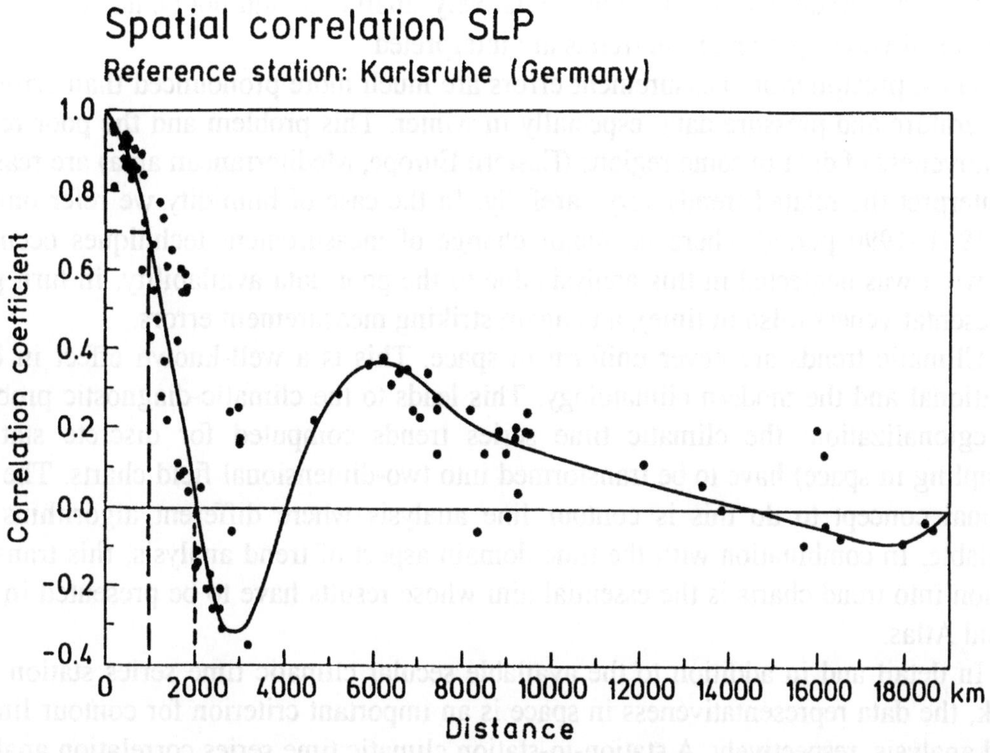

Figure 2.5. **Spatial correlation (dots) of monthly mean MSL air pressure data, reference station Karlsruhe (Germany) with respect to the distance of the other stations under consideration, including non-linear regression function.**

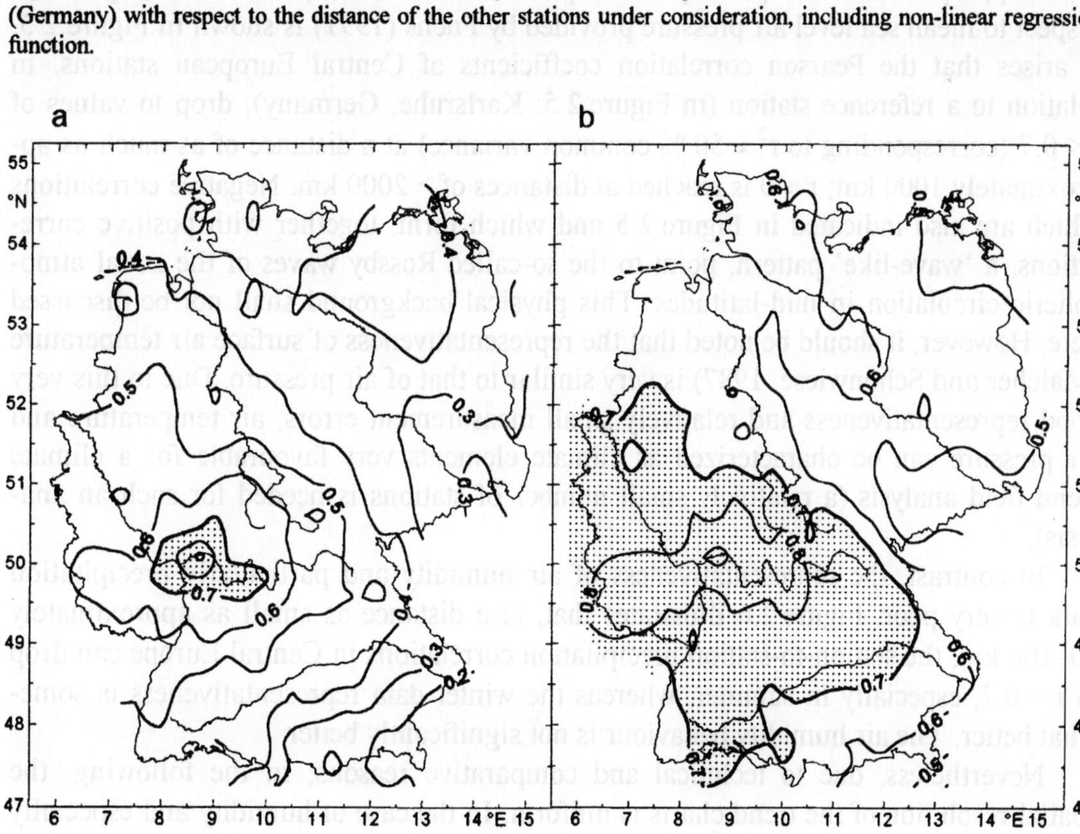

Figure 2.6. **Spatial correlation analysis of (a) summer and (b) winter precipitation in Germany in relation to reference station Frankfurt/M. in terms of correlation contour lines (field analysis).**

of warming. Although this deviation is relatively small it should not be neglected if the amounts of winter precipitation trends are interpreted.

Thus, precipitation measurement errors are much more pronounced than errors of temperature and pressure data, especially in winter. This problem and the poor representativeness of data in some regions (Eastern Europe, Mediterranean area) are reasons to interpret the related trends very carefully. In the case of humidity we refer only to the 1961–1990 period where no major change of measurement techniques occurred and wind was neglected in this analysis due to the poor data availability, in turn poor representativeness (also in time), and again striking measurement errors.

Climatic trends are never uniform in space. This is a well-known effect in both traditional and the modern climatology. This leads to the climatic-diagnostic problem of regionalization: the climatic time series trends computed for discrete stations (sampling in space) have to be transformed into two-dimensional field charts. The traditional concept to do this is contour line analysis where different algorithms are available. In combination with the time domain aspect of trend analysis, this transformation into trend charts is the essential aim whose results have to be presented in this Trend Atlas.

In detail and in addition to the available secular climatic time series station network, the data representativeness in space is an important criterion for contour line or field analysis, respectively. A station-to-station climatic time series correlation analysis is the appropriate tool to quantify data representativeness in space. An example with respect to mean sea level air pressure provided by Fuchs (1991) is shown in Figure 2.5. It arises that the Pearson correlation coefficients of Central European stations, in relation to a reference station (in Figure 2.5: Karlsruhe, Germany), drop to values of $r < 0.7$ (corresponding to $r^2 \approx 50$ % common variance) at a distance of as much as approximately 1000 km; $r \approx 0$ is reached at distances of > 2000 km. Negative correlations which are also indicated in Figure 2.5 and which form, together with positive correlations, a 'wave-like' pattern, point to the so-called Rossby waves of the zonal atmospheric circulation in mid-latitudes. This physical background shall not be discussed here. However, it should be noted that the representativeness of surface air temperature (Malcher and Schönwiese, 1987) is very similar to that of air pressure. Due to this very good representativeness and relatively small measurement errors, air temperature and air pressure can be characterized as climate elements very favourable for a climate trend field analysis (a relatively small number of stations is needed for such an analysis).

In contrast, the representativeness of air humidity and particularly precipitation data is very poor. Figure 2.6 illustrates that, at a distance as small as approximately 50–100 km, the station-to-station precipitation correlations in Central Europe can drop to $r < 0.7$, especially in summer, whereas the winter data representativeness is somewhat better. The air humidity behaviour is not significantly better.

Nevertheless, due to technical and comparative reasons, in the following, the spatial resolution of the trend charts is uniform. In the case of humidity and especially precipitation, however, one should be aware of both the representativeness and measurement restrictions. This requires a careful interpretation of the corresponding trend

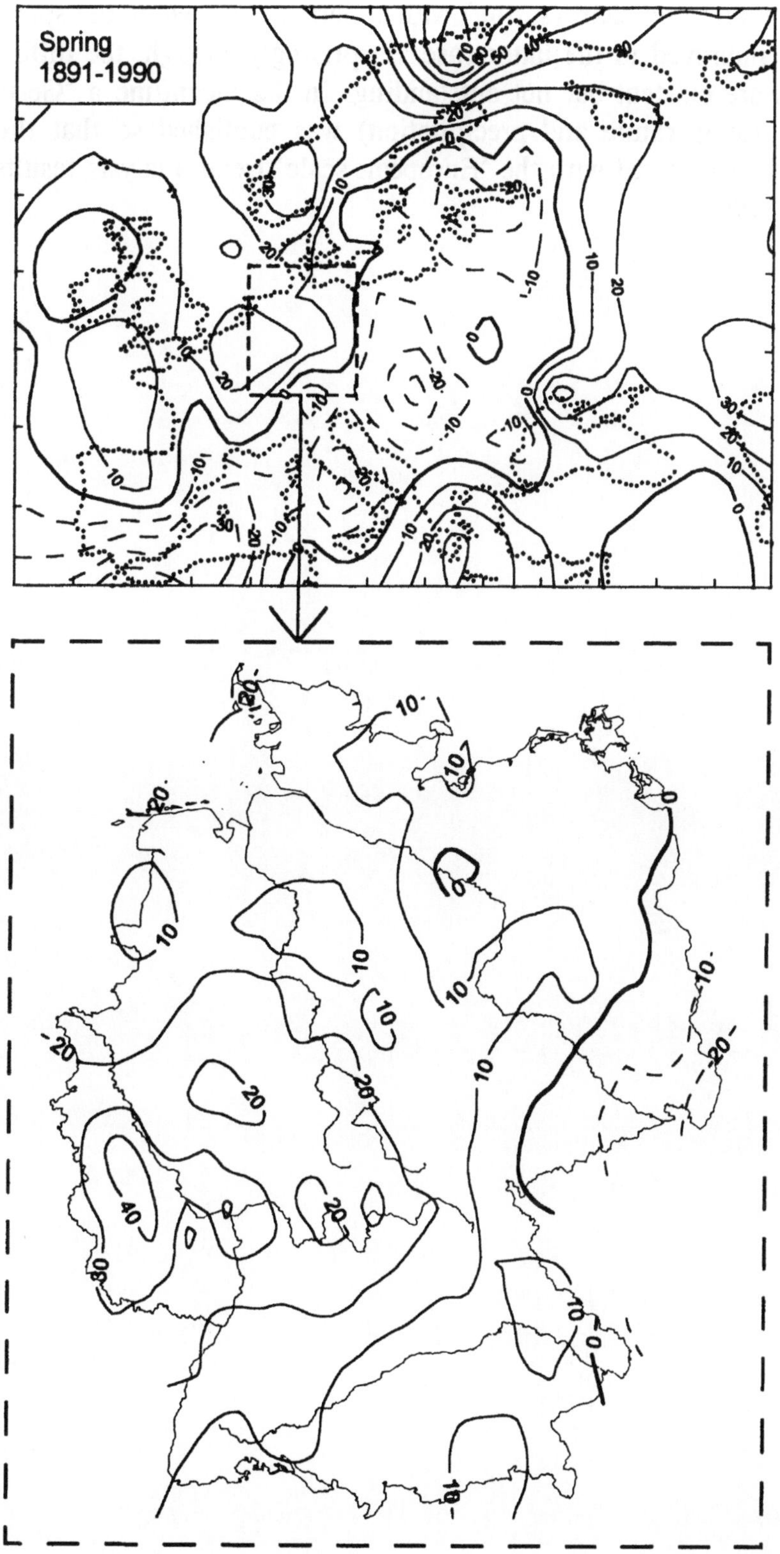

Figure 2.7. Linear precipitation trends during 1891–1990, spring (field analysis, in mm), comparing an analysis for Europe (this Atlas) and Germany (Rapp and Schönwiese, 1996); see text.

charts, particularly outside the Central European region. Figure 2.7 compares an European precipitation trend chart (spring) with a German one where the spatial resolution is essentially improved in the latter case. Surprisingly enough, the differences of the contour lines are existent but not outstanding. In the meantime a 'German Climate Trend Atlas' (temperature and precipitation) was published so that this 'subscale' analysis can be compared with the 'European scale' trend analysis results (Rapp and Schönwiese, 1996).

CHAPTER 3

METHOD OF ANALYSIS

In orientation to the general climatological background and the purpose of analysis, the following step-scheme of methods has arisen:

- Registration and supplementation of data.
- Error corrections.
- Homogeneity tests.
- Analysis of spatial representativeness.
- Computation of averages and particularly trends for each data time series.
- Testing of trend confidence.
- Field analysis by means of contour line computation with respect to averages and particularly trends, such as trend confidence estimates.
- Arrangement and production of charts for all trend and trend confidence results.

In the following, the methods used for this purpose are briefly summarized.

3.1. Data

In the context of some previous work (e.g. Schönwiese *et al.*, 1986; Malcher and Schönwiese, 1987; Schönwiese and Birrong, 1990; Ahmadzadeh, 1991; Fuchs, 1991) our data sources for secular time series of surface air temperature, air pressure, precipitation, and air humidity (in the latter case only the last 30 yr) are as follows: NCAR (National Center for Atmospheric Research, USA), data collections on magnetic tapes or diskettes; CDIAC (Carbon Dioxide Information Analysis Center, USA), data collections on diskettes and CD-ROMs; GISS (Goddard Institute for Space Studies, USA), data collections on CD-ROMs; periodical journals, especially 'Monthly Climatic Data for the World' edited by WMO (World Meteorological Organization); publications such as year-books and special information (on request) from national Weather Services; and private data collections, particularly by von Rudloff (1967, 1985).

Data gaps of no more than 3 yr were interpolated by averages (based on the previous and following 10 yr), and time series which contained longer gaps were omitted. Evident errors were corrected. After all combinations, supplementations and corrections, we obtained a system of 90 temperature, 117 precipitation, and 59 air pressure time series all covering the 1891–1990 period on a monthly mean data basis; in addition, 55 air humidity series for 1961–1990 were provided. The related detailed information is available from the station list as given in Appendix A. Figure 3.1 shows the geographical distribution of the stations used (temperature and precipitation time series).

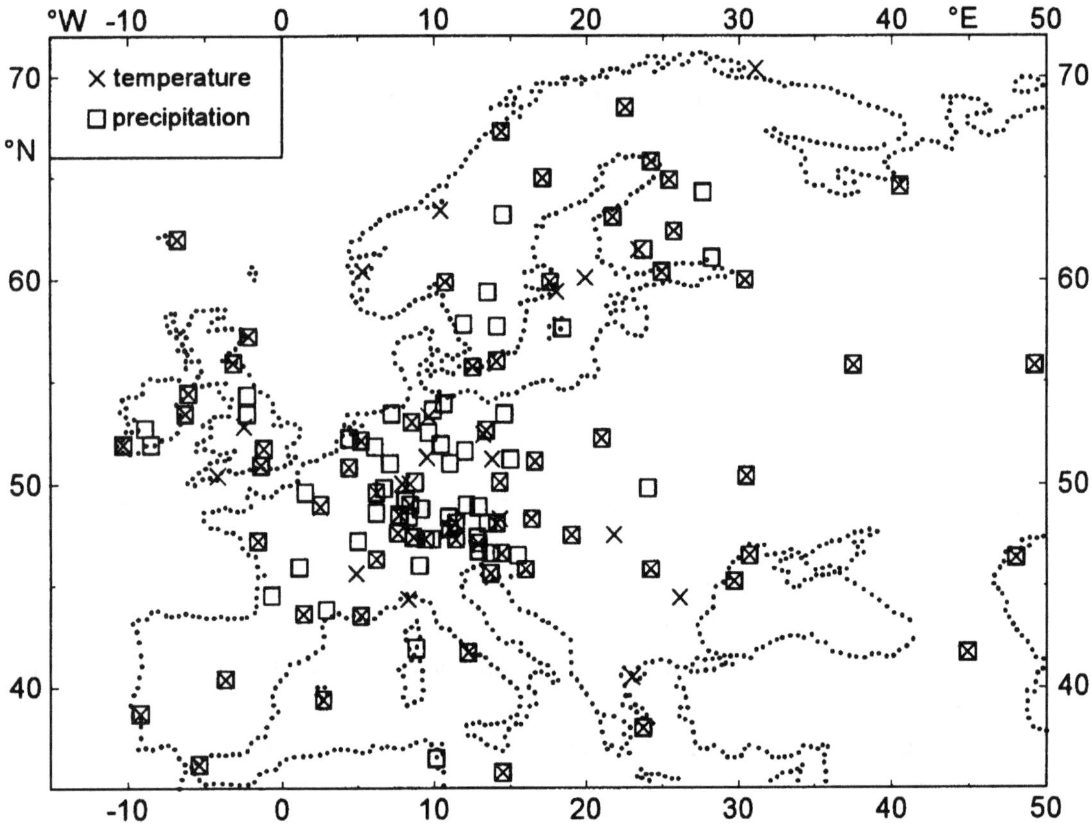

Figure 3.1. Geographical distribution of stations from where temperature and precipitation time series are used.

3.2. Homogeneity tests

With respect to homogeneity tests, one has to discern between absolute and relative methods. In the case of absolute homogeneity tests, only one time series (from one station) is considered and usually one is looking for 'jumps' (abrupt changes) which do not seem to be of a climatological nature. Based on the discussion provided in the previous chapter, Table 3.1. summarizes the possible influences which may lead to such inhomogeneities (abrupt and continuous) in the case of precipitation. In the following, first absolute homogeneity tests are addressed which are appropriate for detecting non-climatological 'jumps'.

One example for this type of testing is the method from Abbe (Conrad and Pollack, 1962) used in a modified form (Schönwiese *et al.*, 1986) where the time series under consideration is high-pass filtered so that relatively low-frequent (long-term) oscillations, often occuring in the real climate, cannot be addressed as 'inhomogeneities'. The Abbe test takes into account both the change of sign and magnitude of deviations from a time series average. In the case of homogeneity of a time series a_i the following relation must be met:

$$1-\frac{1}{\sqrt{n-1}} \leq \frac{2A}{B} \leq 1+\frac{1}{\sqrt{n-1}} \tag{3.1}$$

where

$$A = a_1'^2 + a_2'^2 + \ldots + a_n'^2 - \tfrac{1}{2}\left(a_1'^2 + a_n'^2\right), B = \left(a_1' - a_2'\right)^2 - \left(a_2' - a_3'\right)^2 + \ldots + \left(a_{n-1}' - a_n'\right)^2;$$

a_i values are the deviations from the arithmetic mean $\bar{a}$, and n is the sample size (number of data, length of time series). A disadvantage of this test procedure is that a long-term climate trend, which may be real, is misinterpreted as an inhomogeneity. It is possible, however, to avoid this disadvantage by using related high-pass filtered data (Schönwiese, 1990), i.e. by subtracting long-term trends and fluctuations (related low-pass-filtered are smoothed data) which are supposed to have a climatic nature from the original time series. The Abbe test modified in this way, then, is applied on the high-pass filtered data time series.

In relative homogeneity tests, the time series under consideration is compared with a related time series averaged from several stations in the surroundings. The presuppositions of such methods are that there is a significant correlation of all these stations and that the time series of the surrounding stations are at least approximately homogeneous. (In general, therefore, one has to start with absolute tests and/or to use relative tests twice repeated.) The relative homogeneity tests used here are from Buishand (1982), Craddock (1979) and WMO (reported by Mitchell *et al.*, 1966). For more details see the referenced literature.

The relative homogeneity test provided by Craddock (1979) is used in a graphical form and implies, therefore, some subjectivity. It uses the expressions

$$s_1 = \frac{\bar{b}}{\bar{a}} a_1 - b_1,\ s_2 = \frac{\bar{b}}{\bar{a}} a_1 - b_1 + \frac{\bar{b}}{\bar{a}} a_2 - b_2, \ldots, \tag{3.2}$$

TABLE 3.1. **Influence mechanisms effecting errors in precipitation measurements which may lead to inhomogeneities and, in turn, may lead to incorrect or inaccurate trend analysis results.**

Influence (error)	Type	Effect in time
Wind velocity trend	Natural *	Continuous
Temperature trend	Natural *	Continuous
Trend of precipitation type	Natural *	Continuous
Trend of vegetation	Natural/anthropogenic	Often continuous
Change of instrument	Anthropogenic	Abrupt
Change of observation time	Anthropogenic	Abrupt
Change of station site	Anthropogenic	Abrupt
Change of surrounding buildings (urbanization etc.)	Anthropogenic	Abrupt or continuous (urbanization continuous)

* *Note that these trends may be also due to large-scale anthropogenic influence, e.g. by the enhanced 'greenhouse effect'.*

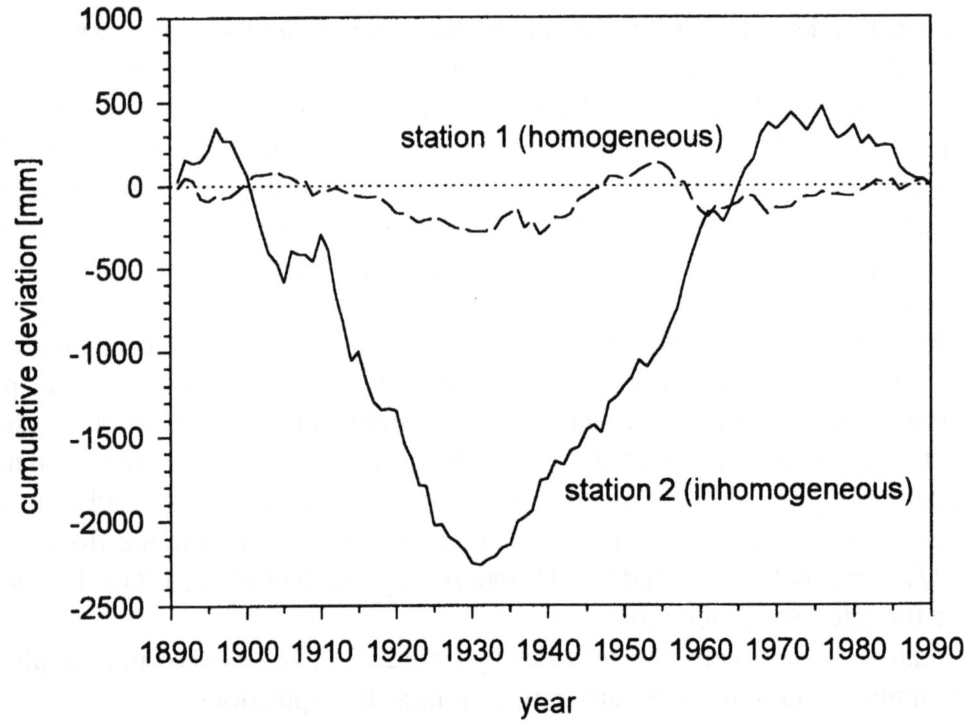

Figure 3.2. Craddock test. Cumulative deviations s_1, s_2, ... of a homogeneous and an inhomogeneous precipitation time series.

where a_i values are the data taken from a (hopefully) homogeneous area average time series (reference time series) and b_i values are the data of the time series for which homogeneity has to be checked; $\bar{a}$ and $\bar{b}$ are the related arithmetic means. If the terms s_i are plotted in relation to time, evident extreme values (v-shaped or inverse v-shaped, i.e. changing trends, so-called change points) are criteria for possible inhomogeneities in the corresponding year. In the case that s_i values fluctuate only marginally around an average, homogeneity can be supposed (examples are shown in Figure 3.2.).

The disadvantage of some subjectivity in the case of the Craddock test is circumvented by the relative homogeneity test as proposed by Buishand (1982) which, in addition, is able to relate to definitive confidence levels (error probabilities). This test is a multiple procedure again computing cumulative deviations. However, in this case, four test criteria Q, R, U, and A are defined. These criteria are values compared with those from special tables to decide on either homogeneity or non-homogeneity on a specified confidence level. These criteria are

$$Q = \text{Max} \left| S_{k^*} \right| \quad \text{mit } 0 \leq k \leq n, \tag{3.3}$$

$$R = \text{Max}\left(S_{k*}\right) - \text{Min}\left(S_{k*}\right) \quad \text{mit } 0 \le k \le n, \tag{3.4}$$

$$U = \frac{1}{n(n+1)} \sum_{k=1}^{n} \left(S_{k*}\right)^2, \tag{3.5}$$

$$A = \sum_{k=1}^{n-1} \left\{ \left(k(n-k)\right)^{-1/2} S_{k*} \right\}. \tag{3.6}$$

where

$$S_{k*} = \frac{\sum_{i=1}^{k} \left(d_i - \bar{d}\right)}{\sqrt{\frac{\sum_{i=1}^{n} \left(d_i - \bar{d}\right)^2}{n}}} \quad \text{and} \quad S_{0*} = 0. \tag{3.7}$$

$k = 1, 2, \ldots, n$; $d_i = a_i - b_i$ are the differences between the tested and the reference time series, and $\bar{d}$ is the related average. This test which was evaluated preferably in respect to precipitation series is, however, parametric, i.e. Gaussian distributed data are necessary. Moreover, the data should not be autocorrelated (no significant persistence effects).

Another relative homogeneity test recommended by the World Meteorological Organization (WMO, Mitchell *et al.*, 1966) is based on the same difference values, d_i, as described before. For the time lag $\tau = 1\Delta t$ (Δt = observation or time series data interval) the autocorrelation coefficient is computed and compared with

$$r(p) = \frac{-1 + z(p)\sqrt{n-2}}{n-1}, \tag{3.8}$$

where $z(p)$ is the value of the standardized Gaussian distribution with probability p ($p = 1 - \alpha$, where α = error probability). In the case where $r(p) > z(p)$ at the confidence level p, an inhomogeneity at this level is supposed.

All these tests have their advantages and disadvantages, so it is important to use not only one selected method which may lead to a wrong test decision. Instead, we summarized and interpreted the results from the various test methods as shown in Table 3.2. (see also Malcher and Schönwiese, 1987).

Note that one of the relative homogeneity tests is a multiple method leading to four test quantities (Buishand, 1982). We applied all tests to the annual temperature and pressure data time series, whereas in the case of precipitation, due to the poor station-to-station correlations, only the absolute test was applicable. In the case of humidity,

TABLE 3.2. Homogeneity classes as defined in this study (see also Malcher and Schönwiese, 1987).

All tests say 'homogeneous'.	Very probably homogeneous	Class 1
Nearly all tests (with mostly one exception) say 'homogeneous'.	Probably homogeneous	Class 2
Several tests say 'homogeneous', several say 'non-homogeneous'.	Undicided, pending	Class 3
Nearly all tests (with mostly one exception) say 'non-homogeneous'.	Probably non-homogeneous	Class 4
All tests say 'non-homogeneous'.	Very probably non-homogeneous	Class 5

no tests were applied (problematical due to the very short time interval). The results of all homogeneity tests are implied in the station list (see Appendix A). An example for an inhomogeneous time series is shown in Figure 3.3 where the sharp decrease in air pressure 1906–1915 is probably not real (leading to a probably incorrect trend statement: decreasing instead of increasing, but, since c. 1945, actually decreasing). One additional method for homogeneity testing arises from field analysis and is described later (near the end of this section).

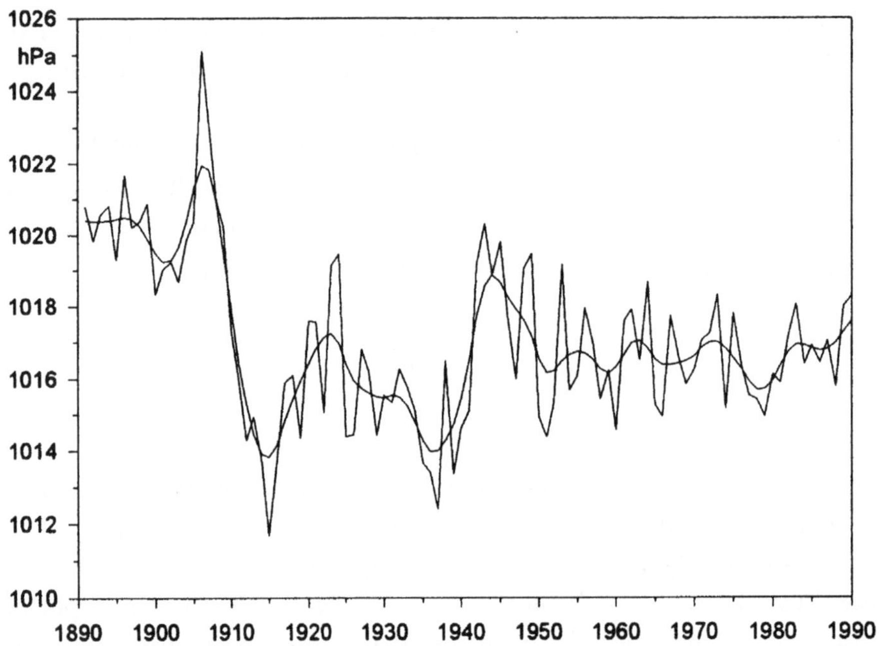

Figure 3.3. Mean Sea Level (MSL) air pressure (SLP) annual means at the Luxembourg station for 1891–1990. The trend analysis is incorrect in this case due to an inhomogeneity indicated c. 1910.

3.3. Trend analysis

We present all trends, calculated linearly as illustrated in Figure 3.4, in the units of the climate element under consideration (K ≙ °C in the case of air temperature, hPa = mbar in the case of surface air pressure and humidity, i.e. water vapour partial pressure, mm = lm^{-2} in case of precipitation) where the trend value is the difference

$$Tr \equiv \Delta a = \hat{a}_n(t_n) - \hat{a}_1(t_1) \tag{3.9}$$

of the values of the linear regression equation $\hat{a}(t)$ at time t_1 and t_n (n = number of data; see Figure 3.4) based on a Pearson correlation of the climatic time series under consideration with time (+Tr in case of an increasing, –Tr in case of a decreasing trend). In order to enable a discussion of the relative trends, we calculated also the time series averages, $\bar{a}$. Note, however, that thereafter, the relative trends may be calculated as

$$Tr/\bar{a} \quad \text{or} \quad Tr/a_1(t_1) \quad \text{or} \quad Tr/a_n(t_n), \tag{3.10 a–c}$$

where a_1 is the start and a_n the end value of the time series under consideration. Say, for example, that $\bar{a}$ = 80 mm, $a_1(t_1)$ = 110 mm and $a_n(t_n)$ = 50 mm (similar to Figure 2.1a, rounded), then one obtains Tr = –60 mm and

(a) $Tr/\bar{a}$ = –60/80 = –0.75 ≙ –75%, or

(b) $Tr/a_1(t_1)$ = –60/110 = –0.545 ≙ –54.5%, or

(c) $Tr/a_n(t_n)$ = –60/50 = –1.2 ≙ –120% (!).

This means that the second method (b) leads to smaller trend values in the case of decreasing trends and to larger trend values in the case of increasing trends; the third method (c) gives the opposite sign. In general, the first method ($Tr/\bar{a}$) is more reasonable. In the case of precipitation we have also added the results of a relative trend analysis following formula 3.10 a (see Chapter 8.3).

The term Tr/s (s = standard deviation) leads to the assessment of the so-called trend-to-noise ratio (where s is used as a proxy for noise N, Tr/s ≡ Tr/N). If the data time series are Gaussian (normally) distributed, this simple assessment would allow confidence statements. Otherwise, where precipitation data especially are very seldom Gaussian distributed, the Mann–Kendall test (Mann, 1945; Kendall 1970) is a much more adequate method (Sneyers, 1990; Denhard and Schönwiese, 1992) because neither a Gaussian frequency distribution (non-parametric test) nor a linearity of the trend is required.

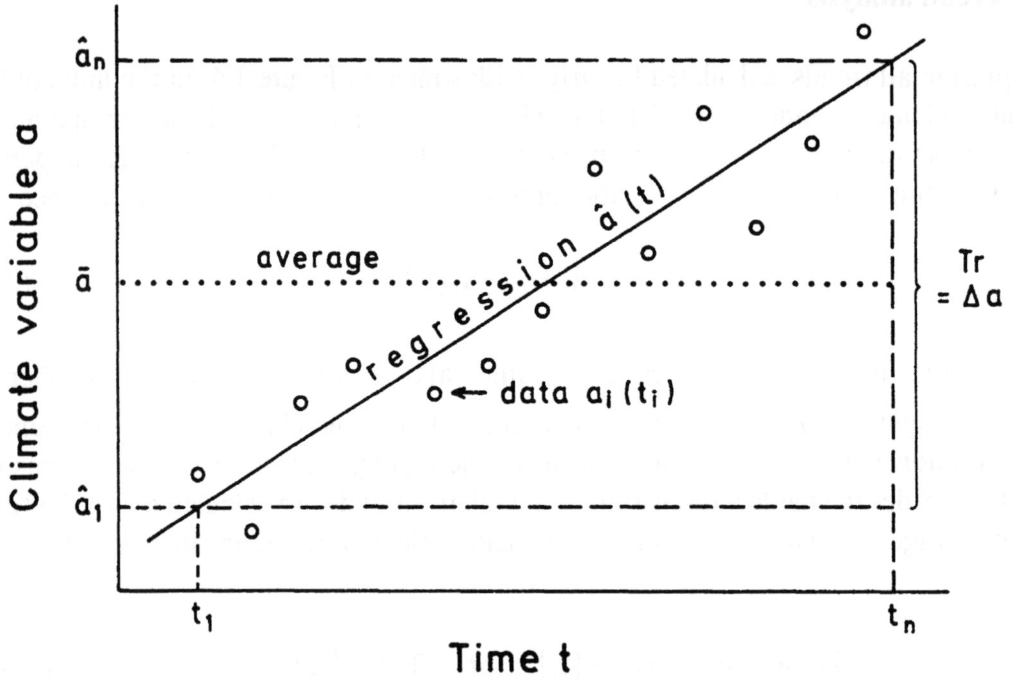

Figure 3.4. Schematical explanation of the term 'trend' and different possibilities to compute it.

In its original form (Mann, 1945) the raw test quantity is

$$Q_s = \frac{\sum_{i=1}^{n-1} \sum_{J=i+1}^{n} \operatorname{sgn}(a_j - a_i)}{\frac{1}{2} n(n-1)} = \frac{s}{\frac{1}{2} n(n-1)} , \tag{3.11}$$

where n = record length, sgn = sign and all possible time series data pairs (a_i, a_j) are summed up for $i < j$. (In a series of n data exist ½ [n(n-1)] pairs.) For $n > 10$ the raw test quantity Q_s is approximately Gaussian distributed with average $\mu = 0$ and variance

$$\sigma^2 = \frac{2(2n+5)}{9n(n-1)} . \tag{3.12}$$

Then, the final standardized test quantity is

$$Q = \frac{Q_s - \mu}{\sigma} = \frac{s}{\sqrt{\frac{1}{18} n(n-1)(2n+5)}} . \tag{3.13}$$

TABLE 3.3. Confidence levels CL and error probabilities α for the Mann–Kendall trend test (and all other trend test quantities following a Gaussian distribution).

Q	CL	α	CL	α	Q
> 1	> 68.3 %	< 0.317	> 80 %	< 0.2	> 1.282
> 1.5	> 86.6 %	< 0.134	> 90 %	< 0.1	> 1.645
> 2	> 95.4 %	< 0.046	> 95 %	< 0.05	> 1.960
> 3	> 99.7 %	< 0.003	> 99 %	< 0.01	> 2.576
> 4	> 99.999 %	< 0.00001	> 99.9 %	< 0.001	> 3.290

Kendall (1970) has modified this test for the case of possible identical time series data and proposed

$$Q = \frac{S}{\sqrt{\frac{1}{18}\left[n(n-1)(2n+5) - \sum_{l} b_l(b_l-1)(2b_l+5)\right]}}, \tag{3.14}$$

where b_l is the number of identical data values (for each value). Similar to other test quantities which are Gaussian distributed $Q > 2$ means approximately > 95% confidence (< 5% error probability) and $Q > 3$ means approximately > 99 % confidence (< 1% error probability); for exact values see Table 3.2 (from Spiegel, 1972, or Schönwiese, 1992).

As will be outlined later (Chapter 4) there is a pronounced secular winter warming detectable in Europe (see related trend chart, page 91). If one computes the corresponding trend-to-noise ratios Tr/N (see Figure 3.5, upper part (a)), this trend appears to be weakly significant only within Central Spain (Tr/N > 1.5, i.e. confidence level CL approximately > 85%). The same data analysed by means of the Mann–Kendall test indicate much higher confidence levels: $Q > 4$ within Spain, i.e. CL > 99.9%, $Q > 2$ within parts of SE Europe around the Black Sea, i.e. CL > 95% (error probabilities $\alpha < 0.001$ or < 0.05, respectively).

If the trend analysis is performed correctly, the residuals ($a_i^*(t_i)$ in (2.2)) should represent a random process with properties of stationarity (no change of mean, variance, etc. with time), no autocorrelation, and Gaussian distributed data. Residuals of a trend analysis are, of course, detrended and therefore stationary with respect to the average (from subinterval to subinterval). In the case of variance, we observe a very irregular behaviour instead of systematic trends (Ahmadzadeh, 1991; Rapp, 1991). An example of a residual frequency distribution is shown in Figure 3.6 (precipitation, annual totals, at station Frankfurt/M., Germany). It appears that the condition of a Gaussian distribution is fulfilled (despite some curtosis) which is confirmed by the application of the Kolmogorov–Smirnov test: the null hypothesis of Gaussian distribution is accepted at a 99 % confidence level. We did not check the autocorrelation functions but suppose that there are no problems because a Gaussian process is consistent with a non–autocorrelated one (except some persistence which may remain, leading to a red instead of a white noise; Mitchell *et al.*, 1966).

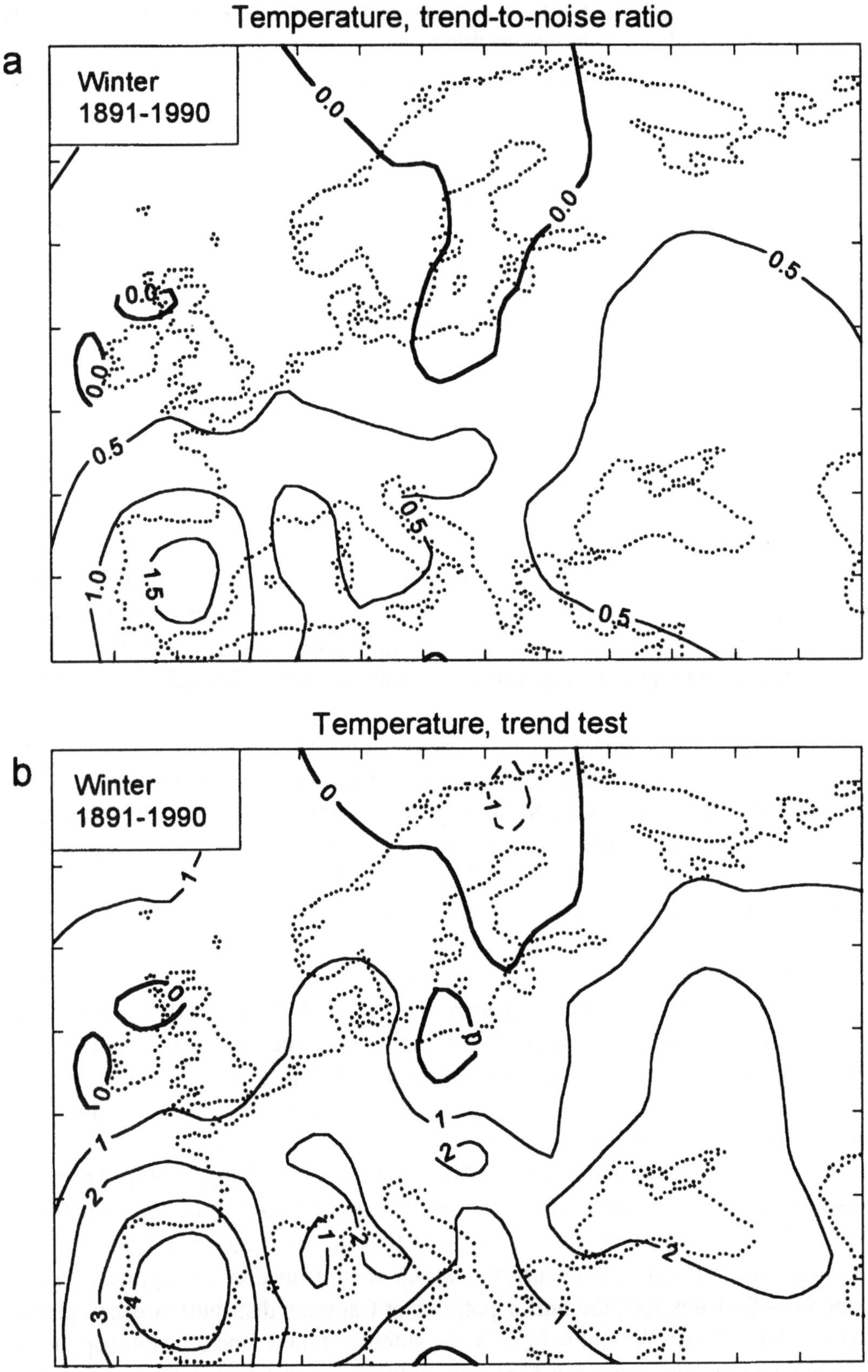

Figure 3.5. Comparison of (a) trend-to-noise ratios and (b) Mann–Kendall test values, winter temperature trends for 1891–1990 (field analysis).

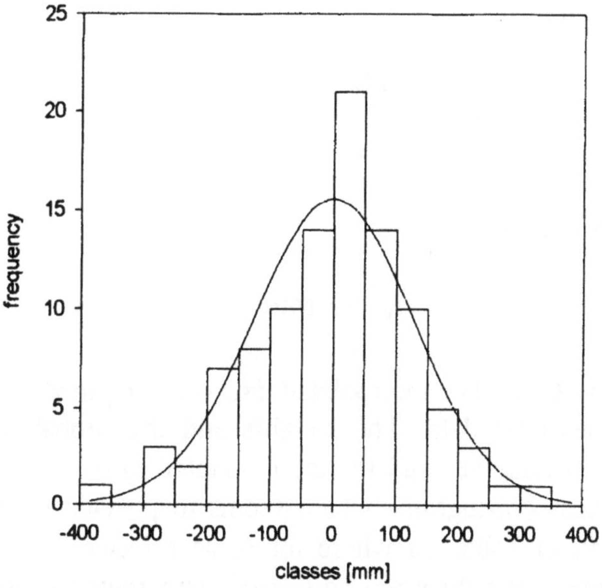

Figure 3.6. Example of a residual frequency distribution analysis following the trend computation (precipitation, annual data, 1891–1990, station Frankfurt/M., Germany).

3.4. Field analysis

The geographical relations needed for the field analysis are available by adequate projections. Similar to the presentation of many climate model results we used an equidistant rectangular coordinate system based on the cylinder projection (see Figure 2.1). The area under consideration is 35°N – 72°N, 15°W – 50°E. The reduction as used in this Atlas corresponds to a scale of approximately 1:50 Mill. (at 50°N).

The aim of field analysis, in general, is to transform data (in space) which are available from irregularly distributed stations into a regular grid system. After that, contour lines or box values may be estimated as correctly as possible. For this surpose, an interpolation scheme is necessary. An important presupposition of interpolations, in turn, is a significant correlation of the data of two or more neighbouring stations to estimate the interpolated data values between them. Therefore, a representativeness analysis as outlined in Chapter 2 (see Figures 2.4 and 2.5) is necessary before a reasonable field analysis can be performed. As a rough guideline, a minimum station-to-station correlation of $r > 0.7$ ($r^2 > 50$ %) may be required. This is well fulfilled in the cases of temperature and pressure, but the related precipitation analysis remains problematical (see Figure 2.5).

The interpolation scheme used here is the 'Ordinary Kriging' as recommended by Brown and Eischeid (1992), Bárdossy and Muster (1993), and many other authors.

This scheme enables the average difference between interpolated $\hat{z}_i$ and real (measured) data z_i to be reduced to zero:

$$m_R = \frac{1}{K} \sum_{i=1}^{K} (\hat{z}_i - z_i) = 0 \tag{3.15}$$

and minimizes the error variance:

$$\sigma_R^2 \rightarrow \text{Min}, \tag{3.16}$$

where K = number of stations. The interpolated data are computed as a weighted linear combination of the measured data. The weights and the probability model can be chosen among several alternatives and therefore adapted to the interpolation problem. We have chosen a linear 'variogram' which seems to provide optimal interpolation values as demonstrated in Table 3.4 where for some selected (and then suppressed) stations, the real (measured) values are compared with their estimated interpolation values (temperature, maximum error in this case 2.1 K, average error 0.54 K).

The main advantages of the Kriging interpolation scheme are, in comparison with other point-to-point interpolations, a (relatively) very small error probability, an (again relatively) optimal smoothed contour line picture and a small influence of very irregularly distributed stations (station 'clustering') on the results. Moreover, it is very easy to 'smooth away' errors which are detected during the field analysis. For further information, see Isaaks and Srivastava (1989) or Journel and Huijbregts (1981).

We used a 3° × 3° grid (exactly 3.095° × 3.083°) (see also Figure 3.7), so that we obtained 22 × 13 = 286 grid-points, corresponding to a spatial dissolution of c. 320 × 320 km at 50° N. For the interpolation of any grid point, at least 10 stations were used with a 'search radius' of 69° (degree of longitude and latitude) around every grid point. In general, we tried to find an optimum between results as accurately as possible with minimal computation time for all charts. Based on the interpolated grid-point data, an usual plot programme was used to produce the trend– or trend test contour lines, respectively (horizontal field analysis). No doubt, in Central Europe a trend field analysis using a smaller grid-point distance would have been possible. We decided, however, to use the same dissolution for the whole of Europe.

A particular problem is the suppression of the station data time series which – due to testing as described above – appear to be probably (classes 4 and 5) non-homogeneous. A very strict procedure would have led to a very sparse data basis. Therefore, it was determined graphically which consequences are affected by the use of possible non-homogeneous data time series. A dramatic example is shown in Figure 3.8 where the air pressure trend field analysis (1891–1990, spring) is performed with and without some very probably non-homogeneous records (including the Luxembourg station, see also Figure 3.3). These consequences ('isle' effect of contour lines) are very clearly detectable in the region of Benelux and SW Europe. In this way we decided, step by step, especially with respect to the secular period 1891–1990, which station time series

TABLE 3.4. Comparison of some observed temperature data (spring, average 1961–1990) with related estimations if the observations are suppressed and estimated by means of the Kriging interpolation scheme.

Station	Observed °C	Estimated °C
Vardo	-0.7	-0.8
Kristianstad	6.0	6.2
St. Petersburg	4.3	3.9
Wien	11.7	11.8
Geneve	8.9	11.0
Luxembourg	7.7	9.6
Madrid	12.7	13.1
Potsdam	8.3	8.0
Thessaloniki	14.5	14.1
Odessa	8.9	9.2
Kazan	4.1	4.3
Moskva	5.5	5.2
Thorshavn	5.2	5.7
Tbilisi	12.4	12.1

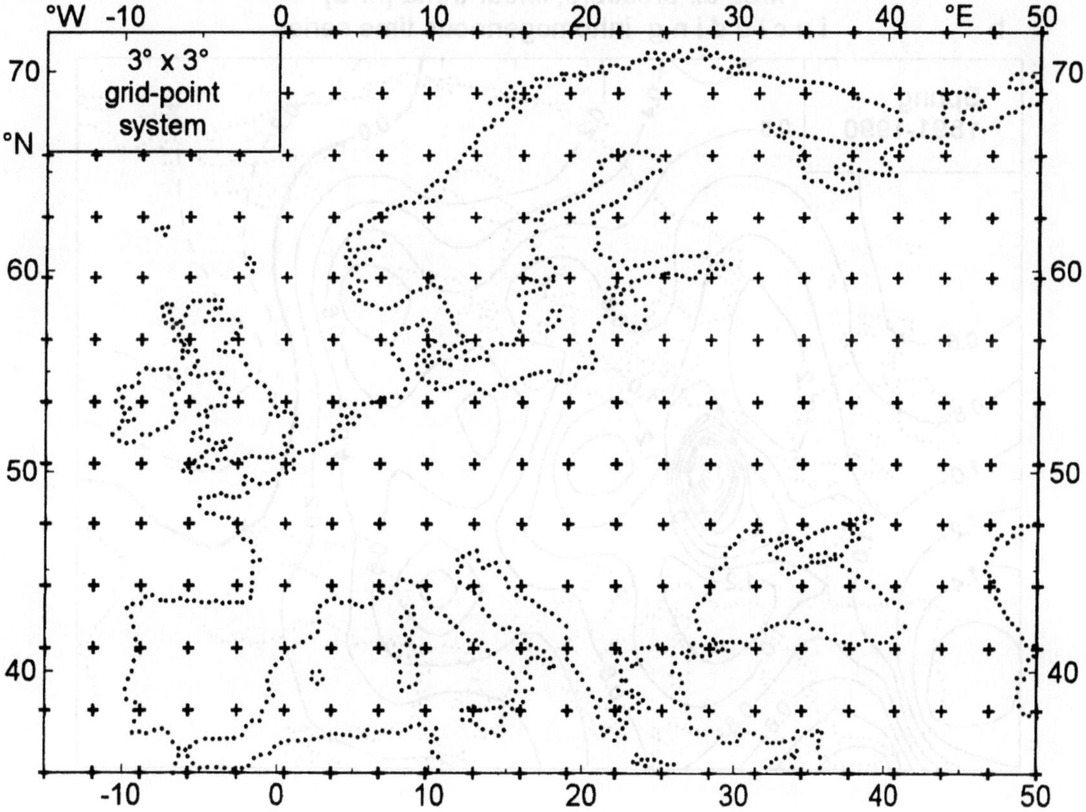

Figure 3.7. Geographical projection and grid-point system used for the 'Kriging' interpolation scheme (see text).

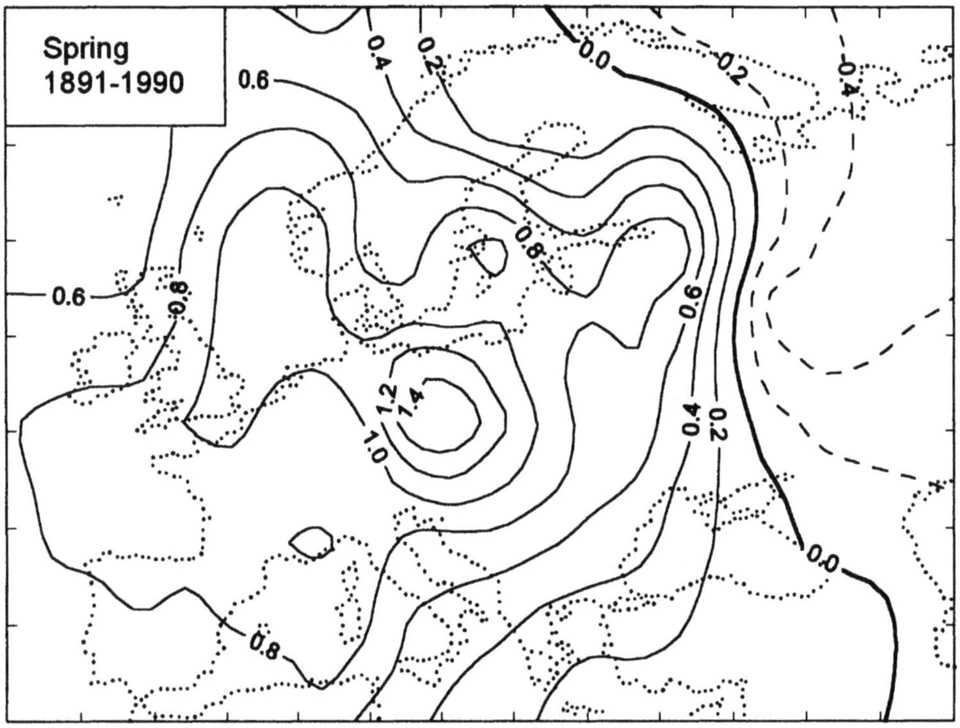

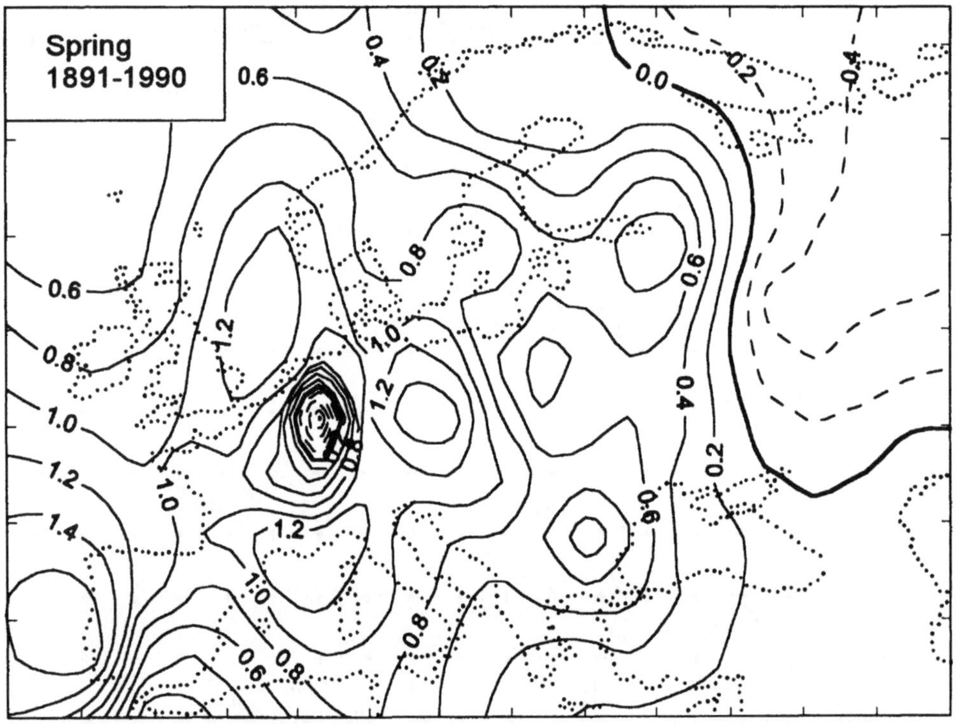

Figure 3.8. MSL air pressure trend (SLP; 1891–1990, field analysis in hPa), spring, (a) excluding and (b) including inhomogeneous time series (class '5', see text; compare also Figure 3.3, Luxembourg).

may be implied in the trend field analysis and which may not. Fortunately, in the period 1961–1990 there are not so many homogeneity problems.

There are also homogeneity problems due to different station altitudes. Except air pressure, we used the 'real' data as observed near the surface (exactly 2 m above ground level as measured). Given that, for example, the air temperature lapse rate (vertical differences) did not change, these different altitudes should not affect the trends (same systematic altitude errors which are eliminated when the difference values in time, i.e. the trends, are calculated). However, this error affects the average values so that, in this case, we omitted mountain stations. The mean sea level (MSL) pressure data (SLP), average and trend charts, are without mountain stations > 600 m above MSL (due to reduction problems). Precipitation, average and trend charts, include mountain stations so that all orographic effects (luv and lee, etc.) are reflected.

The choice of the distance of the contour lines was oriented to optimum principles of conformity and readability. Therefore, in the case of temperature, we used a distance of 2 K (°C) for averages and 0.5 K (°C) for trends uniformly. In the case of pressure and humidity, these distances are 1 hPa (averages) and 0.5 hPa (trends), respectively. For precipitation, the situation is more complicated because these data are totals, i.e. the values are accumulated (summed up) for a day, month, season, year and so on. In this case we used the following distances:

- monthly data → 20 mm (averages) or 10 mm (trends);
- seasonal data → 50 mm (averages) or 20 mm (trends);
- annual data → 100 mm (averages) or 50 mm (trends).

Based on these totals the data are averaged for any period (e.g. for 1961–1990, based on totals in each season). Note that the meteorological seasons are December of the previous year, January and February (winter), March – May (spring), June – August (summer), and September – November (autumn). All trend test charts use a contour line distance of 0.5. Generally, all positive contour lines are solid, and all negative contour lines are dashed. Moreover, the 0 (zero) contour line is printed in bold.

CHAPTER 4

OVERVIEW OF RESULTS

In this section we present an overview of the results. The corresponding charts are compiled in Chapter 8: 8.1 air temperature, 8.2 air pressure, 8.3 precipitation, and 8.4 humidity. Every chart implies information on the variable under consideration (as listed above), type of computation (averages, trends, or trend tests), period (1891–1990 or 1961–1990, respectively), particular month (e.g. March), season (e.g. summer) or annual display. Obviously we cannot discuss at length all details which are available from the charts. Rather, we try to concentrate on the most important aspects.

Generally it is worth nothing that all trend charts show both positive and negative values and often the spatial trend pattern is rather complicated, especially in the case of precipitation. Different reference periods or months/seasons can reveal a different trend behaviour. In the latter case this different behaviour is often masked if annual instead of monthly or seasonal data are used. The statistical confidence of the trends is often rather weak; however, there are also some very significant results available. Finally, note that due to the changing station coverage the reliability of interpolation also changes (relatively reliable in Central Europe, problematical especially in the Mediterranean, East European and adjacent Atlantic areas).

4.1. Temperature

4.1.1. AVERAGES

The average charts of air temperature (Chapter 8.1a) provide only a rough overview for the recent CLINO period 1961–1990. This is why mountain stations are neglected and the station coverage along the coasts is too poor to reflect satisfactorily the land–ocean contrasts (e.g. a 0°C isotherm should be situated along the Norwegian coast).

The coldest month is January with a temperature range of c. –12 °C (NE Europe) to c. +12 °C (North African coast). In July the corresponding range amounts to +11 °C (near Iceland) and +26 °C (Eastern Mediterranean area). Temperatures in spring are generally 1 – 2 K lower than in autumn. In summer the contour lines are oriented approximately from NE to SW, in spring and autumn from NW to SE. In winter, two ‘cold pools’ exist in N Norway and in the outermost NE region (continental climate) contrasted by the Mediterranean reservoir of warm air (0 °C boundary N to S until the Alpine area and W to E eastward of the Alps, other contour lines similar). The annual mean temperature covers a range from c. 0 °C (Northern Norway) to c. 12 °C (North African coast) with a remarkably sharp contrast near the Alpine region.

4.1.2. TRENDS

The temperature trends (see 8.1b) show different regional patterns from month to month. Most outstanding is a sharp January warming with the exception of Northern Scandinavia where a remarkable cooling (maximum c. 2 K) took place. The maximum warming was observed in Eastern Europe (1891–1990: > 2 K) which has intensified in recent decades (1961–1990: > 3 K) whereas in Spain the secular warming trend (> 1.5 K) has recently weakened. In the uppermost SE Mediterranean area the secular warming was recently (1961–1990) substituted by a weak cooling. Also in February and March warming trends have dominated but with a distinctly smaller magnitude than in January (expect a very pronounced March warming in NE Europe 1961–1990, > 4.5 K). In these months, again in contrast to January, Northern Scandinavia has also warmed. In April, a Central European cooling in recent decades is observed, whereas other regions have warmed weakly (except a remarkable secular April warming of c. 3 K in Eastern Europe). Spring (since April/May) and summer months all together show weak trends where, in Central European areas, coolings are often detectable although in other parts, again warming prevails. The autumn trends are rather complicated with increasing temperatures prevailing but also pronounced cooling, e.g. c. 1 K in September in a small part of the East Sea area and especially > 3 K in November in the Bulgarian area, 1961–1990 in both cases. Note that in November, 1961–1990, in the whole of Eastern Europe a cooling took place, whereas the 1891–1990 period reveals a warming nearly all over Europe with a maximum north of the Black Sea. The December trends are very impressive showing a very pronounced warming with a 1891–1990 maximum (> 2.5 K) in Eastern Europe and a 1961–1990 maximum (> 3 K) in Central Europe.

On a seasonal scale, warming is the dominant feature with maximum values in spring (> 2 K) and especially in winter (c. 2.5 K), in Eastern Europe in both cases. A secular-season cooling is indicated in the Central European summer (c. 0.5 K); on a decadal time scale, the maximum cooling appeared in autumn (Turkey region, > 1.5 K; note also the winter cooling in the uppermost SE area). On an annual time scale, warming clearly prevails with a secular maximum of c. 1.5 K in Eastern Europe, whereas in the recent decades these annual average trends have been very weak.

As far as the confidence tests are concerned (see 8.1 c), on a monthly scale, the most confident trends (CL > 95 %, Q > 2) appeared in the following months and regions, all during the 1891–1990 period. January: Spain (see also discussion in Chapter 3) and SE Europe; February: again Spain; March: NE and SW; April: E; June: small parts in the W Mediterranean area; July – September: near Spain; October: larger regions of W Europe; November: Spain; December: Spain and E Europe. In all other areas and generally on a decadal time scale (1961–1990, except May, Southern Scandinavia warming; July, warming in NW Europe; September – December, some varying parts of SW Europe) the confidence of the monthly data trends is very weak. On a seasonal scale a few very confident (CL > 99 %) trends are detectable: spring, Eastern Europe and Spain; summer, near Spain; autumn, Spain and Southern Germany; winter, again Spain (all secular 1891–1990 reference period). The trend test based on secular (1891–1990) annual mean temperatures reveals the most significant results: only warming is observed, and in relatively large areas, a confidence level of 95 % is exceeded; the maxima (CL > 99 %) appear in Spain, Southern France, the whole of Germany, Alpine countries, parts of Central England, and a large region in Central Eastern Europe.

4.2. Pressure

4.2.1. AVERAGES

The average MSL air pressure charts 1961–1990 (8.2a) reveal for most times of the year, the well-known pattern of low pressure in the NW (especially in winter, as a consequence of the Icelandic Low pressure system) to the N and high pressure in the S (in summer with a well-marked ridge of high pressure in the SW as a consequence of the Azores High pressure system; but also high pressure in the SE in all other seasons). The largest horizontal pressure differences appear in December when a relative minimum of c. 1002 hPa is contrasted by relative maxima of > 1020 hPa in SW (near Spain) or > 1022 hPa in SE (near the Caspian Sea). The consequences are a meridional pressure gradient of c. 18 hPa and, in turn, a strong westerly airflow over North and Central Europe. The south is then dominated by high pressure in the SW (> 1020 hPa) and SE (> 1022 hPa) with a minor low (c. 1018 hPa) in between. This minor low can be traced until the month of April; in May it weakens and shifts eastward, and in summer it practically vanishes. In these months the Mediterranean area is influenced by a north-easterly flow. In the same season the ridge of high pressure which is connected with the Azores High pressure system (roughly 1016 hPa) extends to Central Europe. In September it forms a bridge of high pressure so that nearly the whole region south of roughly 50°N is characterized by high pressure (roughly > 1016 hPa).

These mean month-to-month changes in the air pressure distribution are, of course, also reflected in the mean seasonal pressure pattern charts leading to strong zonal airflow in Northern Europe in autumn and especially in winter where the high pressure influence in summer causes more or less long periods of fair weather in Central and, especially, Southern Europe, whereas in the northern parts a westerly to north-westerly airflow dominates. The annual mean air pressure chart again shows low pressure (< 1009 hPa) in the north and high pressure in the outermost SW or SE, respectively (c. 1018 hPa in each case).

4.2.2. TRENDS

The most important air pressure trends (8.2b) can be seen in winter and, less pronounced, in autumn characterized by a drastic increase of the zonal flow in the recent decades (1961–1990). This is why, in this period, the meridional pressure gradient has intensified by up to more than 10 hPa (decreasing pressure trend in the north, increasing in the south). Strangely, in the longer reference period (1891–1990) the opposite was true, however much less pronounced; this means that in this secular period the meridional pressure gradient has decreased by not more than 1.5 hPa. In summer, on a secular time scale, an air pressure increase was dominant (except NW and SW) whereas in recent decades in Eastern Europe, a decrease (up to c. 1.5 hPa) occurred. Averaged for the whole year, virtually no trends remain on the long-term scale, but during 1961–1990 the winter and autumn pressure changes recognizably; this means that a pressure decrease in the north (up to c. 2 hPa) and an increase in the south (roughly by 1.5 hPa) prevail (see also Flohn *et al.*, 1992).

The confidence of all air pressure trends is weak on a seasonal scale with the exception of increasing pressure in SW (roughly Spain) in spring, summer and especially autumn (CL > 95 %, autumn > 99 %), all during 1891–1990, and decreasing pressure in the summer, in Eastern Europe (CL > 85 %). The mean annual data trends are significant at the 95 % level in several parts of Southern Europe (especially Iberian peninsula, increasing pressure) and at the 85 % level in the outermost NW. On a monthly data basis the level of significance very rarely exceeds 95 %; examples are a decreasing trend in August in the Eastern Mediterranean area, and an increasing trend in October in a major part of SW Europe, 1891–1990 in both cases (and partly > 99 %).

4.3. Precipitation

All precipitation charts should be interpreted very carefully because of the very restricted spatial representativeness of precipitation data and, with respect to this restriction, the poor station coverage. In particular, the results concerning the Mediterranean area, still more important Eastern Europe and the Atlantic section, are uncertain.

4.3.1. AVERAGES

As far as the average charts (1961–1990) are concerned (8.3a) in general, precipitation decreases from NW to SE, often in a very pronounced way (for example in January from 160 mm near Island to less than 20 mm in the Caspian Sea region; mean monthly totals of precipitation). The regional Alpine precipitation maximum is well-established, especially in the summer months (convective precipitation prevailing) whereas corresponding relative maxima in other mountain regions (e.g. Pyrenees, Scandinavia, etc.) are not detectable, certainly an effect of the poor station coverage. However, there is a remarkable regional autumn maximum in Northern Italy (due to low pressure activity, the so-called Genua cyclones). The regional differences in precipitation are most pronounced in summer (> 250 mm in NW and the Alpine region, < 50 mm in the Mediterranean area). On an annual time scale, the precipitation covers a range of c. 1500 mm in the NW and roughly 1200 mm in the Alpine region compared to < 400 mm in the Eastern Mediterranean and Caspian Sea region. In the Western maritime and Mediterranean areas, winter precipitation prevails, whereas in the remaining continental areas, including mountain regions, summer precipitation dominates.

4.3.2. TRENDS

Much more diverse and complicated than in the case of temperature are the precipitation trends (8.3b, 8.3c), such that it is very hard to summarize the main results. If one concentrates on the seasonal scale, one finds in spring predominantly increasing precipitation in SW and NW Europe (more or less both reference periods), whereas in Central Europe, drier conditions prevail. In the summer, drier conditions are detectable in major parts of Central Europe compared with increasing precipitation in Eastern Europe (1961–1990 also W and SW parts). In autumn and especially winter a drastic decrease of precipitation in the Central Mediterranean region (e.g. Italy, Greece) is observed (winter only recent decades) contrasted by increasing precipitation in nearly all Western parts but also, for example, Southern Germany and the Alpine region. On an annual and secular time scale, precipitation has decreased in an area extending from SW (Southern Spain) via most of the Mediterranean and some of the Central European regions up to the East Sea region (but very weak in the latter case) and increased otherwise. However there are a lot of regional and monthly peculiarities.

Confidence (8.3d) is weak in most cases. On a seasonal scale, some of the exceptions are as follows: increasing precipitation 1891–1990 in autumn, Southern Scandinavia (CL > 99 %); again increasing precipitation 1891–1990 in parts of Southern Scandinavia in winter, France, NE Spain (> 95 %) and outermost NE Europe, decreasing 1961–1990 in winter in parts of Greece and W Turkey (> 95 %). On a monthly scale, corresponding trends are significant (> 95 %) in January (increasing 1891–1990, e.g. Southern Scandinavia, February (decreasing 1891–1990, Eastern East Sea region), March (increasing 1891–1990 Northern Scandinavia, simultaneously decreasing Southern Spain), June (increasing 1891–1990 Southern Scandinavia, 1961–1990 north of the Black Sea), September (increasing 1891–1990 again Southern Scandinavia), October (decreasing 1891–1990 former Yugoslavia), November (decreasing 1961–1990 parts of France and Northern Italy), and December (increasing 1891–1990 Northern Scandinavia and adjacent outermost NE Europe).

4.4. Humidity

In the case of humidity, only water vapour pressure charts are shown based on the 1961–1990 interval. Especially with respect to Eastern Europe, the data coverage is poor.

4.4.1. AVERAGES

As far as the average charts are concerned, in all months, humidity is decreasing from S/SW in direction to N/NE, an effect which is closely correlated with the related temperature patterns. The magnitude covers, in January, a range of c. 3 – 10 hPa, and in July, 11 – 21 hPa. The spatial differences are most established in summer and autumn when, in the Mediterranean region, the maximum values are reached (21 hPa or 19 hPa, respectively). On an annual average, the magnitude covers a range of 6 hPa (NE) to 15 hPa (S).

4.4.2. TRENDS

The humidity trends are quite similar to those of temperature because of the positive temperature–evaporation correlation. The dominant effect is an increase in humidity increase in all seasons, with the exceptions of SE in spring and winter. In April this decrease also covers major parts of Central and Western Europe (significant at the 95 % level in Southern England and Northern France). The maximum increase (c. 2 – 3 hPa) is observed in the summer months July – September in the Central Mediterranean region (near Sicily, CL > 99 %). However, it cannot be strictly ruled out that this phenomenon, at least its magnitude, may be an inhomogeneity effect of the related time series. In general, the significance of the humidity trends is quite poor.

CHAPTER 5

SOME RELATIONS TO UPPER AIR DATA

Although, in this Atlas, we refer to surface data only (see also summarizing paper by Schönwiese *et al.*, 1994) a few relations to upper air data (Bayer *et al.*, 1994; Meyhöfer *et al.*, 1996) should be mentioned. These data are available by radiosonde measurements performed in recent decades so that only the 1961–1990 reference period can be considered. Also, the upper air trend analysis is performed only for the area 30° W – 50° E / 60° – 72° N and 15° W – 50° E / 35° – 60° N. This is due to representativeness problems, especially in oceanic areas, although, in general, the representativeness of upper air data is considerably better than that of surface data. Note that Angell (1988) has used no more than 63 stations for a world-wide analysis of radiosonde data. For example, the 500 hPa (c. 5.5 km) geopotential height correlations drop to 0.9 at a distance of as much as c. 1500 km (Bayer *et al.*, 1994). In the case of temperature, the situation is similar. Near the tropopause the station-to-station correlations are somewhat smaller which may be due to increased variability in this level (c. 200–300 hPa).

Gaffen (1994) showed that different types of radiosondes or radiation correction methods may lead to inhomogeneities in space. However, in general, such uncertainties are not necessarily detectable in the time series. Changes in methods of radiosonde measurement or radiation correction were performed in Europe during various years but without any evident effect on the seasonal data. Similarly, the uncertainties of the radiosonde measurement methods in the beginning of the 1960s seem to not have considerably influenced the trends.

In our analysis, at first the time series of geopotential height passed a system of homogeneity tests, which is similar to our earlier studies (Schönwiese *et al.*, 1990, 1994). Thereafter linear trends were computed. We used the "Kriging" interpolation scheme (see Chapter 3.4) again for the transformation into a regular grid-system (3° × 3°). Trends detecting only a single station which evidently disturbed a relatively smoothed contour line field picture were removed at this stage of analysis due to a possible undetected inhomogeneity.

It seems that, for example, the large-scale continental winter warming is a phenomenon not only of the surface layer but also of the whole troposphere. This is demonstrated in Figure 5.1 for the 500 hPa level corresponding to an altitude of c. 5.5 km (approximately in the middle of the troposphere; 850 and 700 hPa trends similar). The corresponding results of the Mann–Kendall test (see Chapter 3.3) are incorporated by shading (> 90 %, confidence level, see figure caption). Even the magnitude of the winter warming is similar to that of the surface layer (roughly 0.5 – 1.5 K) with the exception of the Mediterranean area where this magnitude is exceeded in the upper air data. Thus, the winter warming cannot be simply an effect of the urban heat island (established up to a maximum altitude of about 1 km). In general, in the upper air tempera-

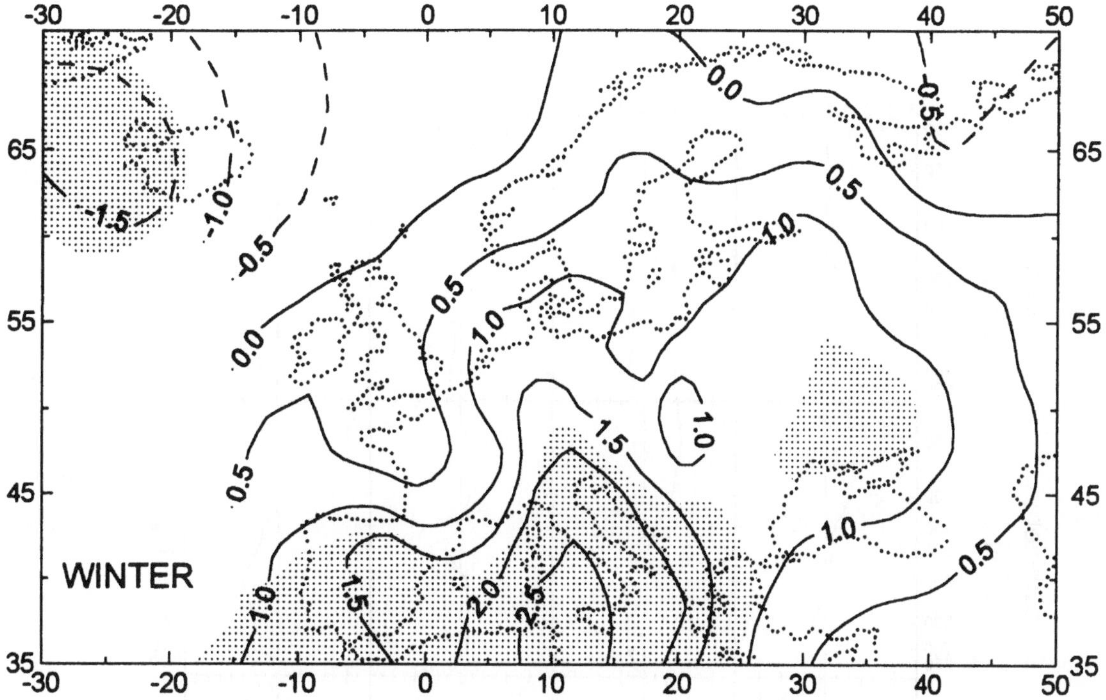

Figure 5.1. Temperature trend (1961–1990, field analysis in K), 500 hPa (c. 5.5 km altitude), winter (Dec., Jan., Feb.). Confidence level > 90 % (Mann–Kendall test) shaded.

ture data warming prevails in all levels of the troposphere, whereas in the stratosphere, cooling is predominant (Houghton *et al.*, 1990, 1996; Bayer *et al.*, 1994).

The 200 hPa (c. 13 km) level (near the tropopause) shows only weak statistically insignificant summer trends. However, a considerable cooling in parts of Western and Central Europe contrasted by a pronounced warming in the Eastern parts and the Atlantic area can be seen in winter (figures not shown). Cooling arises as a dominant feature in the 30 hPa (c. 24 km, figures not shown) level, expect the Mediterranean area and, more or less, the Atlantic area. Although all trends again vary in space, warming seems to be prevailing in the troposphere, and cooling mainly appears in the stratosphere.

This general behaviour is confirmed if trend profiles are produced (see examples presented in Figure 5.2) which also include spring (March, April, May) and autumn (September, October, November) and, additionally, the results of the Mann–Kendall test (dashed lines). The station Stuttgart (Germany) is very typical for Central Europe: increasing temperatures in the troposphere with a maximum in winter and a minimum in autumn near the ground layer, decreasing temperatures in the stratosphere, and Mann–Kendall test values > 2 (corresponding to > 95 % confidence) in all seasons expect spring near the 300 hPa level (and not near the surface). Only Oslo (weak) and Kiev (pronounced) indicate coolings in the troposphere, an effect which is, however, not detectable in the surface data set. In particular, the Kiev troposphere autumn trends are strange from this point of view.

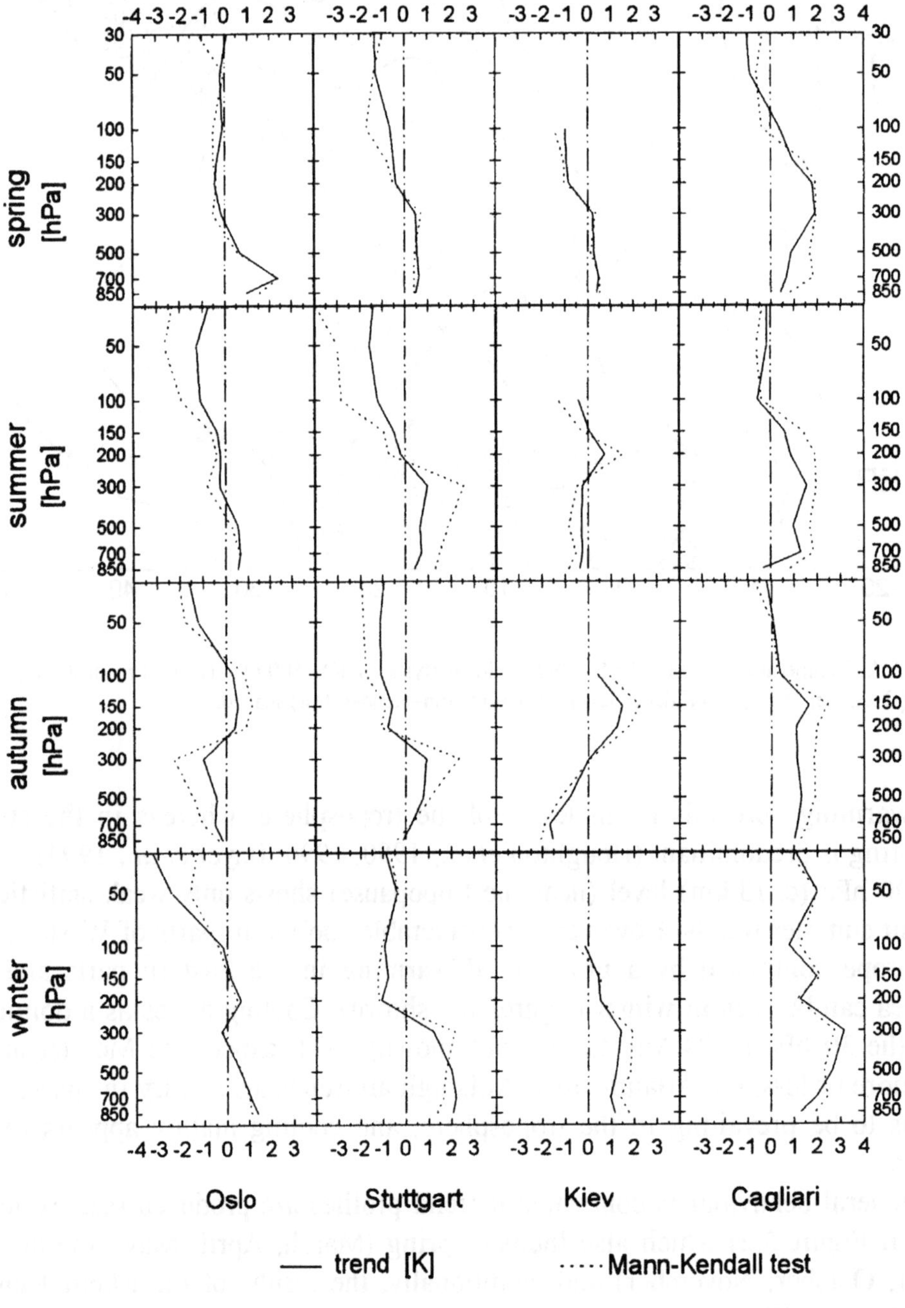

Figure 5.2. **Linear temperature trend profiles 850–30 hPa (1961–1990), solid lines, and related Mann–Kendall test values, dashed lines, all seasons, at selected stations (spring = March, April, May; summer = June, July, August; etc.)**

The relative topography 300/850 hPa which represents the corresponding layer thickness and therefore the mean virtual temperature of this layer (figures not shown, too) reveals that all over Europe a tropospheric warming has taken place (similar to Figure 5.1); that is an additional reason as to why urbanization effects are not the causes of this warming. It is more plausible that "greenhouse" warming (Houghton *et al.*, 1990, 1996) or another large-scale mechanism may have been the physical background of this particular European tropospheric warming. Concerning the same layer (300/800 hPa), but also 500/850 hPa, the lapse rate over most of the European regions (except the Mediterranean and near the Black Sea) has become more unstable (similar summer and winter results, except the Black Sea region). This could lead to more convective clouds and showers but is contradicted by increasing summer dryness in most of Western and Central Europe (Chapter 8.2) whereas the related winter precipitation has increased. The stabilization of the Mediterranean lapse rate coincides with decreasing autumn and especially winter precipitation in this area and with a pronounced increase in the geopotential height.

Our results coincide with corresponding global radiosonde analyses by Angell (1988) and Oort and Liu (1993); see also IPCC reports by Houghton *et al.* (1990, 1996). Angell's finding, based on the 1973–1987 interval, reveals the most pronounced 850–300 hPa warming trends in the northern hemisphere (c. + 0.4 K / 15 yr) and southern hemisphere subtropical (c. + 0.5 K / 15 yr) zones, however with no detailed regional pattern analysis (and based on an incoherent stations network as mentioned before). Oort and Liu use 700–800 stations and state an "excellent agreement" with Angell's and satellite-derived microwave sounding data. They found for the mean northern hemisphere a 1964–1988 warming trend, again at the 850–300 hPa level, of 0.2 K $\pm$0.1 K compared to 0.12 K $\pm$0.11 K from Angell (1988). However, these authors also find a cooling roughly within the 1952–1965 period so that in Angell's analysis which also discerns winter (December/January/February) and summer (June/July/August) data, related to the 850–300 hPa layer and the 1952–1957 period, in the northern hemisphere temperate zone a cooling trend prevails in winter contrasted by very little warming in summer. This cooling, following our analysis, cannot be typical for the European but is rather typical for the North Atlantic and North American sector. On the contrary, the stratospheric cooling seems to be a well-known global effect within all seasons (Angell, 1988; Labitzke *et al.*, 1986; Oort and Liu, 1993; see also IPCC reports).

In the case of pressure, again close similarities are detectable. Figure 5.3 shows the 1961–1990 trend of the geopotential height of the 500 hPa level in winter (which is a proxy for the pressure pattern). As in the case of MSL pressure (SLP) an intensification of the zonal flow component (westerlies) is outstanding, indicated by increasing pressure in the south and decreasing pressure in the north. This trend indicates an increased advection of relative warm and moist air masses from the North Atlantic to Europe consistent with warming and increasing winter precipitation in West and Central Europe. Mann–Kendall tests (see Figure 5.3) indicate that the winter trends are very significant (> 99 % confidence in areas of maximum trends; see also related results from Flohn *et al.*, 1990).

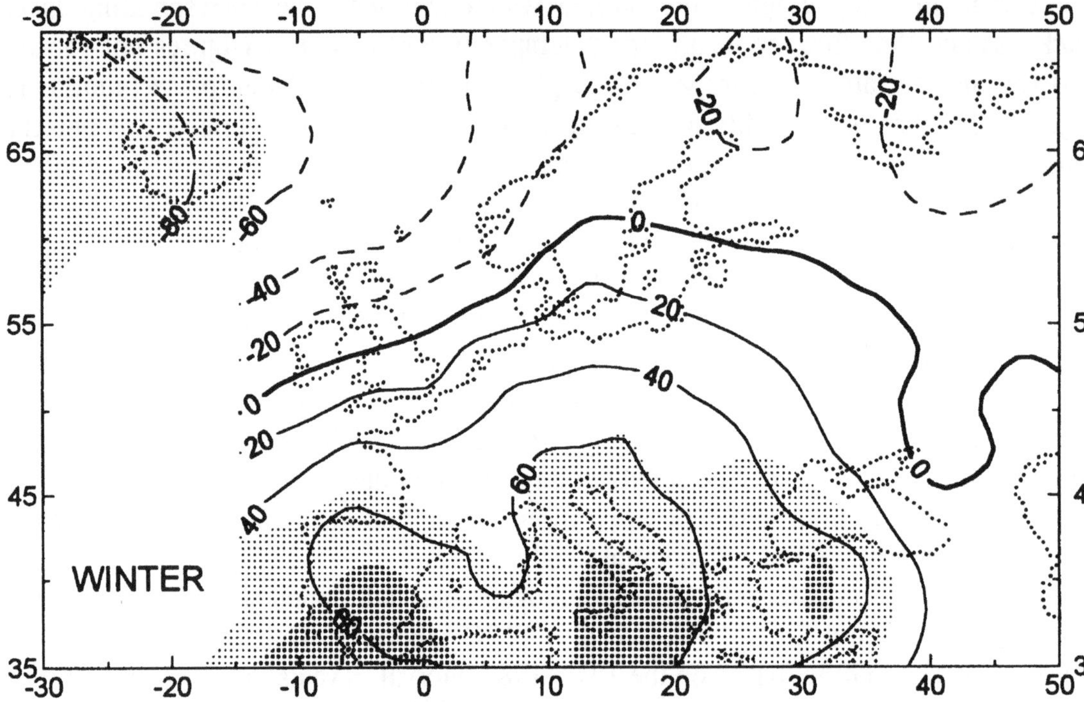

Figure 5.3. Geopotential height trend of 500 hPa level 1961–1990 (field analysis in gpm), winter (Dec., Jan., Feb.). Stippled shading: confidence level > 90 %; dark stippled shading: confidence level > 99 % (Mann–Kendall test).

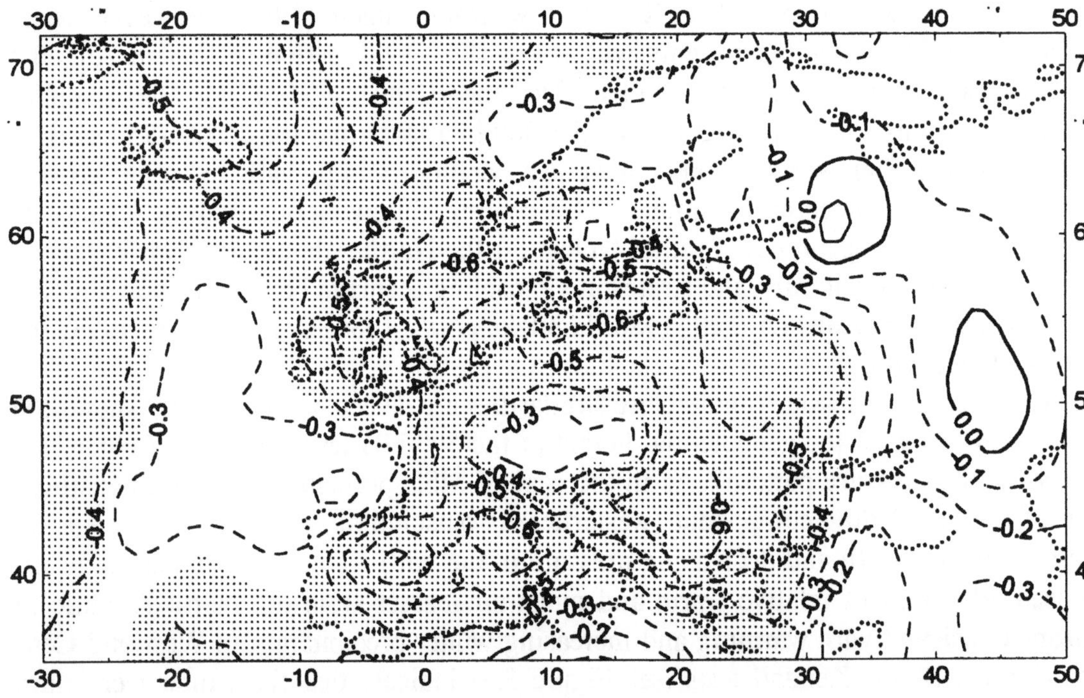

Figure 5.4. Correlation (field analysis) of geopotential height in 500 hPa and precipitation as observed at ground level, annual data 1961–1990, confidence level > 95 % shaded.

The Mediterranean precipitation decrease in autumn and winter (trend 1961–1990), furthermore, is consistent with increasing pressure in that area and those seasons. At the 500 hPa level, geopotential heights have increased up to 70 gpDm (geopotential Dekameters) in this area in winter (Figure 5.3). The correlation of 500 hPa geopotential height and precipitation as observed at the ground level (annual data) is specified in Figure 5.4 with maximum values in an area reaching from Southern Spain to former Yugoslavia and the North Sea/Southern Scandinavia region.

This outstanding change in winter circulation is, however, a characteristic feature of the 1961–1990 or similar time intervals and not a secular effect as far as can be seen from the 1891–1990 analysis of surface air pressure data (Chapter 8.2) and the behaviour of the North Atlantic Oscillation (NAO; see IPCC, Houghton *et al.*, 1996). Whereas NAO has intensified since approximately 1960 (consistent with Figure 3.3), both surface pressure data and NAO point to a weakening of the zonal air flow component on a 100 yr time scale. As a consequence, one has to be very careful to assign observed trends of a specified region and time interval to definite causes. Rather all aspects of climate variability have to be collected, including fluctuations and anomalies, in order to compare these observations and related statistics with the results from various types of climate models.

CHAPTER 6

CONCLUSIONS

The present scientific and public discussion indicates that climate change is one of the most important problems of mankind. In the context of this discussion the question of which way both observed and future climate change has or will influence ecosystems and human affairs is of particular interest.

The collection of charts compiled in this Atlas shows that secular or decadal observed climate change occurs in the form of complicated regional patterns or in other words: the quantitative observational facts of climate change vary with element, space, season or month, and reference period. These regional differences (where we hope that we present a maximum of reality and a minimum of inhomogeneities), with respect to the different climate elements and their seasonal or monthly relations, have to be taken into account when any conclusions concerning ecological, economic, and social consequences (climate impact analysis) shall be considered. Moreover, climate observations have to be approximated by climate models which try to understand the very reasons of climate change.

These reasons have to be known when future climate change is assessed. However, one of the most pronounced shortcomings of climate models used for the prognosis of large-scale climate change is their very restricted spatial dissolution (often c. 500 km gridpoint width, moving to a order of c. 200 km in the case of 'high-resolution' models). Another prominent shortcoming is the simulation of the hydrological cycle, including precipitation, leading to large quantitative uncertainties and therefore to another striking problem of climate modelling. The alternative possibility to extrapolate observed climate trends in the future such as, for example, the winter warming or the seasonally different precipitation trends as presented in this Atlas is hampered by the very fact that these trends are not stable in time. However, it may be useful to extrapolate observed trends one or two decades into the future in the sense of a possible scenario keeping in mind that this procedure is uncertain and that also other scenarios are possible.

On the other hand, in the context of the 'greenhouse' debate (IPCC, Houghton *et al.*, 1990, 1996; anthropogenic emissions of greenhouse gases, GHG), there is some rough parallelism of observed climate trends and predicted GHG forced climate change where again a pronounced winter warming and a seasonal change of the precipitation behaviour (drier summers and wetter winters in mid-latitude continents, Mediterranean generally drier) are predicted (e.g. Cubasch *et al.*, 1995).

As an alternative possibility various multiforced statistical models can be used to evaluate hypotheses on future climate change (Schönwiese and Stähler, 1991; Bayer *et al.*, 1994; Schönwiese and Bayer, 1995). As an example, Figure 6.1 presents future same observational climate summer and winter temperature trends based on such sta-

Table 6.1. Principal results on the future European climate for a doubling of the CO_2-content of the atmosphere (around 2030) based on climate models together with results from analysis of trends in European climate observations (according to ECSN, 1995, with additional results in *italic)*.

Climate element	Future climate change indications based on climate model experiments	Source of information	Inferences from trends indicated by observations – present study and ECSN (1995) (*italic:* only present study)
Temperature	Land surfaces warm more rapidly than the ocean	IPCC, 1996	Supported. The temperature rise in continental areas, or central part of Europe is, however, *slightly* higher than in the maritime, or coastal areas, *but* is still within the limits of natural variability.
	Winter: northern Europe will warm more than the global.	IPCC, 1996	Not supported; *it concerns a temperature decrease.*
	Summer: southern and central Europe will warm more.	Hadley, Emeraude	Supported after 1940 *(without East and South East Europe).*
	The rise of temperature in southern Europe is also expected to be higher than the global mean.	IPCC, 1996	Features uncertain. *Western Mediterranean region* has been warming.
Precipitation	Generally wetter conditions appear in the northern Europe and drier conditions in the southern part of Europe	MPI, Hadley, Emeraude	Supported. The changes in precipitation patterns are rather different in the various model simulations and the uncertainty obviously very high.
	Summer: wetter conditions in Northern Europe and drier in Southern Europe	Hadley, Emeraude	The precipitation conditions, though variable, generally show no definite trends during summer in Europe, *but drier in parts of Central and Western Europe.*
	Winter: it will be drier in the western Mediterranean area.	MPI, Hadley, Emeraude	Supported, but drying is also observed over Greece during winter.

tistical assessments where the data base enables the same spatial dissolution as in the charts collected in this Atlas. A remarkable similarity of the regional patterns of the predicted future) and observed (past, this Atlas) temperature trends is evident (not surprising because this type of statistical modelling is strictly based on observations).

However, there are also similiarities between observed trends and GHG forced general circulation models (GCMs, atmosphere and ocean), see Table 6.1. Details, including global charts of possible GHG-forced trends are available from Henderson-Sellers (1995). Such results should be interpreted very carefully due to shortcomings even of the most sophisticated GCM simulations concerning, for example, precipitation and, in general, the regional patterns of future climate change. Moreover, additional forcing mechanisms like the anthropogenic tropospheric sulphate aerosol formation (IPCC, Houghton *et al.*, 1996) and natural variability have to be taken into account. For a detailed discussion of climate model confidence problems see IPCC (Houghton *et al.*, 1996) and Henderson-Sellers (1995). An overview of recent European climate variability (observations and model simulations) is available from ECSN (1995).

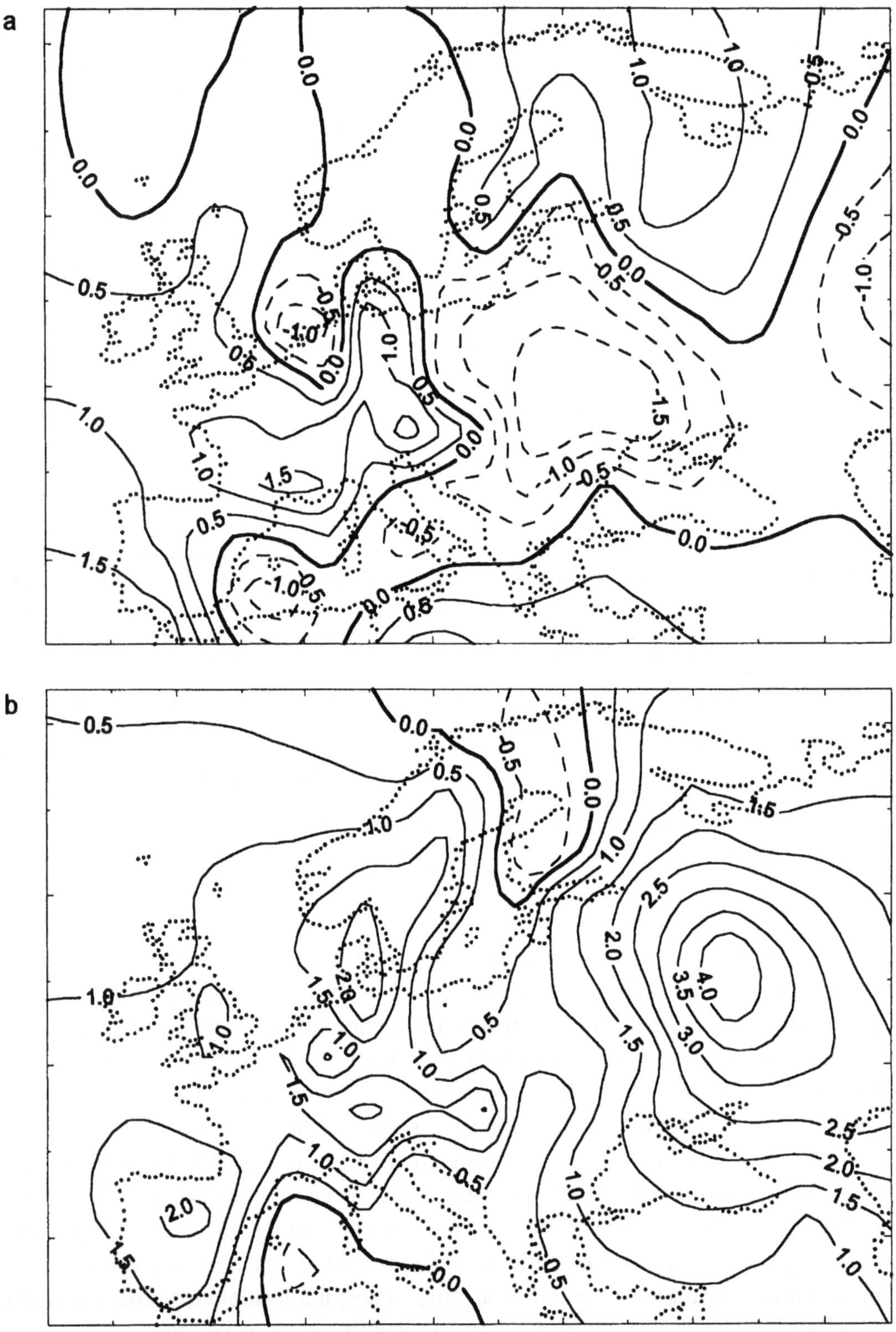

Figure 6.1. Assessment of the enhanced 'greenhouse' effect so far ('industrial signals' in K) due to GHG concentration increase since preindustrial time assessed by means of a multiforced statistical regression model, Europe, surface air temperature, (a) summer and (b) winter; from Bayer *et al.* (1994).

Even without addressing the problem of predicting future climate change we think it is reasonable and important to analyse and present ideas and results on observed climate change, on a restricted time (secular, decades) and space (Europe) scale but in a detailed way and in form of a field analysis, just to describe the very facts of observations. No doubt, the knowledge of observed past climate change, as accurate and comprehensive as possible including all seasonal and regional particularities, is a necessary presupposition to unterstand all climate variability both past and future.

CHAPTER 7

REFERENCES

Ahmadzadeh, S. (1991) *Klimavariabilität in Europa 1888–1988*, diploma thesis, Inst. Meteorol. Geophys. Univ. Frankfurt/Main.

Angell, J.K. (1988) Variations and trends in tropospheric and stratospheric global temperatures 1958–87, *J.Climate* **1**, 1296–1313.

Bárdossy, A. and Muster, H.(1993) *Comparison of interpolation methods of daily rainfall amounts under different meteorological conditions*, Report of a GEWEX workshop on analysis methods of precipitation on a global scale, WCRP-81, WMO/TD-No. **588**, Koblenz.

Båth, M. (1974) *Spectral Analysis in Geophysics*, Elsevier, Amsterdam.

Bayer, D., Denhard, M., Meyhöfer, S., Rapp, J., and Schönwiese, C.-D. (1994) Trend- und multiple Signalanalyse globaler bzw. europäischer Klimavariationen, Report No. **98**, Inst. Meteorol. Geophys. Univ. Frankfurt am Main.

Box, G.E.P. and Jenkins, G.M. (1970) *Time Series Analysis*, Holden-Day, San Francisco.

Brown, T.J. and Eischeid, J.K. (1992) An examination of spatial statistical techniques for interpolation of gridded climate data, in Am. Meteorol. Soc., Environment Canada (ed.), *12th Conf. Prob. Statist. Atm. Sci., 5th Int. Meeting Statist. Climatol., Joint Proc.*, J39–J42.

Buishand, J.A. (1982) Some methods for testing the homogeneity of rainfall records, *J. Hydrol.* **58**, 11–27.

Conrad, U. and Pollack, L.W. (1962) *Methods in Climatology*, Cambridge (USA).

Craddock, J.M. (1979) Methods for comparing annual rainfall records for climatic purposes, *Weather* **34**, 332–346.

Cubasch, U., Waskewitz, J., Hegerl, G.C., and Perlwitz, J. (1995) Regional climate changes as simulated in time-slice experiments, Max Planck Institute for Meteorology, Hamburg, report No. **153**.

Denhard, M. and Schönwiese, C.-D. (1992) Non-parametric trend statistics and rank correlations of long European sea level pressure time series, in Am. Meteorol. Soc., Environment Canada (ed.), *5th Internat. Meeting Statist. Climatol.*, 575–578.

ECSN, European Climate Support Network (1995) *Climate of Europe*, 1st European Climate Assessment, KNMI, De Bilt.

Flohn, H., Kapala, A., Knoche, H.R. and Mächel, H. (1992) Water vapour as an amplifier of the greenhouse effect: new aspects, *Meteorol. Z., N.F.* **1**, 122–138.

Fuchs, T. (1991) *Globale Statistik langer Luftdruckreihen (NN) 1889–1988*, diploma thesis, Inst. Meteorol. Geophys. Univ. Frankfurt/Main.

Gaffen, D.J. (1994) Temporal inhomogeneities in radiosonde temperature records, *6th Conf. Climate Variations, AmMet.Soc.*, 110–113.

Groisman, P.Y. and Easterling, D.R. (1994) Precipitation changes over the Northern hemispheric extratropics during the last hundred years, in Desbois M. and Désalmand F., *Global Precipitations and Climate Change*, Springer Verlag, Berlin, 119–133.

Henderson-Sellers, A. and Hansen, A.-M. (1995) *Climate Change Atlas*, Kluwer, Dordrecht, 159 pp.

Houghton, J.T., Jenkins, G.J., and Ephraums, J.J. (eds., 1990) *Climate Change. The IPCC Scientific Assessment*, Cambridge Univ. Press, Cambridge, 365 pp.

Houghton, J.T., Meira Filho, L.G., Callandar, B.A., Harris, N., Kattenberg, A., and Maskell, K. (eds., 1996) *Climate Change 1995. The Science of Climate Change.* Contribution of Working Group I to the 2nd Assessment Report of the IPCC, Cambridge Univ. Press, Cambridge, 572 pp.

Isaaks, E.H. and Srivastava, R.M. (1989) *An Introduction to Applied Geostatistics*, Oxford Univ. Press, New York.

Jenkins, G.M. and Watts, D.G. (1968) *Spectral Analysis and its Applications*, Holden-Day, San Francisco.

Journel, A.G. and Huijbregts, C.J. (1981) *Mining in geostatistics*, 2nd ed., Academic Press, London, 600 pp.

Kalb M. and Noll, H. (1980) *Klimadaten von Europa, Teil I: Nord-, West- und Mitteleuropa*, Deutscher Wetterdienst, Selbstverlag, Offenbach/Main.

Kalb M. and Hänle, B. (1981) *Klimadaten von Europa, Teil II: Südwesteuropa und Mittelmeerländer*, Deutscher Wetterdienst, Selbstverlag, Offenbach/Main.

Kalb M. and Baucus, M. (1982) *Klimadaten von Europa, Teil III: Südost- und Osteuropa*, Deutscher Wetterdienst, Selbstverlag, Offenbach/Main.

Labitzke, K., Naujokat, B., and Angell, J.K. (1986) Long-term temperature trends in the middle stratosphere of the northern hemisphere, *Adv.Space Res.* **6**, 7-16.

Legates, D.R. (1993) Biases in precipitation gage measurements, in World Climate Research Programme, *Global observations, analyses and simulation of precipitation*, WMO, Td-No. **544**, Geneva, 31-34.

Malcher, J. and Schönwiese, C.-D. (1987) Homogeneity, spatial correlation and spectral variance analysis of long European and North American air temperature records, *Theor. Appl. Climatol.* **38**, 157-166.

Mann, H.B. (1945) Nonparametric test against trend, *Econometrica* **13**, 245-259.

Kendall, M.G. (1970) *Rank correlation methods*, Griffin, London, 4th ed.

Meyhöfer, S., Rapp, J., and Schönwiese, C.-D. (1996) Observed three-dimensional climate trends in Europe 1961-1990, *Meteor. Zeitschrift*, N.F. **5**, 90-94.

Mitchell, J.M., Dzerdzeevskii, B., Flohn, H., Hofmeyr, W.L., Lamb, H.H., Rao, K.N., and Wallén, C.C. (1966) *Climatic Change*, WMO Publ. No. **195** (Tech. Note No. **79**), Geneva.

Oke, T.R. (1973) City size and the urban heat island, *Atm. Environment* **7**, 769-779.

Olberg, M. and Rákószik, F. (1984) *Informationstheorie in Meteorologie und Geophysik*, Akademie-Verlag, Berlin.

Oort, A.H. and Liu, H. (1993) Upper-air temperature trends over the globe, 1958-1989, *J.Climate* **6**, 292-307.

Rapp, J. and Schönwiese, C.-D. (1996) *Atlas der Niederschlags- und Temperaturtrends in Deutschland 1891-1990*, Frankfurter Geowiss. Arbeiten, Band **B5**, 2nd ed.

Rapp, J. (1991) Temperaturvariabilität in der Bundesrepublik Deutschland 1949-1985, diploma thesis, Inst. Meteorol. Geophys. Univ. Frankfurt/Main.

Richter, D. (1995) *Ergebnisse methodischer Untersuchungen zur Korrektur des systematischen Meßfehlers des Hellmann-Niederschlagsmessers*, Report of Deutscher Wetterdienst No. **194**, Offenbach/Main, 93 pp.

v. Rudloff, H. (1967) *Die Schwankungen und Pendelungen des Klimas in Europa seit dem Beginn der regelmäßigen Instrumenten-Beobachtungen (1670)*, Vieweg, Braunschweig.

v. Rudloff, H. (1985) *priv. comm.*

Schönwiese, C.-D, (1987) Moving spectral variance and coherence analysis and some applications on long air temperature series, *J.Clim.Appl.Meteor.* **26** (12), 1723-1730.

Schönwiese, C.-D. (1992) *Praktische Statistik für Meteorologen und Geowissenschaftler*, Borntraeger, Stuttgart, 2nd ed.

Schönwiese, C.-D. and Birrong, W. (1980) European precipitation trend statistics 1851-1980 including multivariate assessments of the anthropogenic CO_2 signal, *Z. Meteorol.* **40**, 92-98.

Schönwiese, C.-D. and Stähler, U. (1991) Multiforced statistical assessments of greenhouse-gas-induced surface air temperature 1890-1985, *Clim. Dyn.* **6**, 23-33.

Schönwiese, C.-D. and Bayer, D. (1995) Some statistical aspects of anthropogenic and natural forced global temperature change, *Atmósfera* **8**, 3-22.

Schönwiese, C.-D., Malcher, J. and Hartmann, C. (1986), *Globale Statistik langer Temperatur- und Niederschlagsreihen*, Report No. **65**, Inst. Meteorol. Geophys. Univ. Frankfurt/Main; 2nd ed. (1990).

Schönwiese, C.-D., Rapp, J., Fuchs, T. and Denhard, M. (1994) Observed climate trends in Europe 1891–1990, *Meteor. Zeitschr., N.F.*, **3**, 22–28.

Sevruk, B. (1989) Reliability of precipitation measurement, in Sevruk, B. (ed.) Precipitation measurement, *Workshop on precipitation measurement*, WMO/IAHS/ETH, Zurich, 13–19.

Sevruk, B. (1994) Spatial and temporal inhomogeneity of global precipiation data, in Debois, M. and Désalmand, F. (eds.), *Global precipitation and climate change*, Springer, Berlin, 231–246.

Sevruk, B. and Klemm, S. (1989) Types of standard precipitation gauges, in Sevruk, B. (ed.) Precipitation measurement, *Workshop on precipitation measurement*, WMO/IAHS/ETH, Zurich, 227–232.

Sneyers, R. (1990) *On the Statistical Analysis of Series of Observations*, WMO Publ. No. **415**, Geneva.

Spiegel, M.R. (1972) *Statistics*, McGraw-Hill, New York.

Abbreviations:

DVA = Deutsche Verlagsanstalt

ECSN = European Climate Support Network

ETH = Eidgenössische Technische Hochschule Zurich

GEWEX = Global Energy and Water Cycle Experiment

IAHS = International Association of Hydrological Science

IPCC = Intergovernmental Panel on Climate Change

WCRP = World Climate Research Programme

WMO = World Meteorological Organization, U.N. Geneva

CHAPTER 8

PRESENTATION OF CHARTS

8.1. Temperature

Temperature, average [°C]

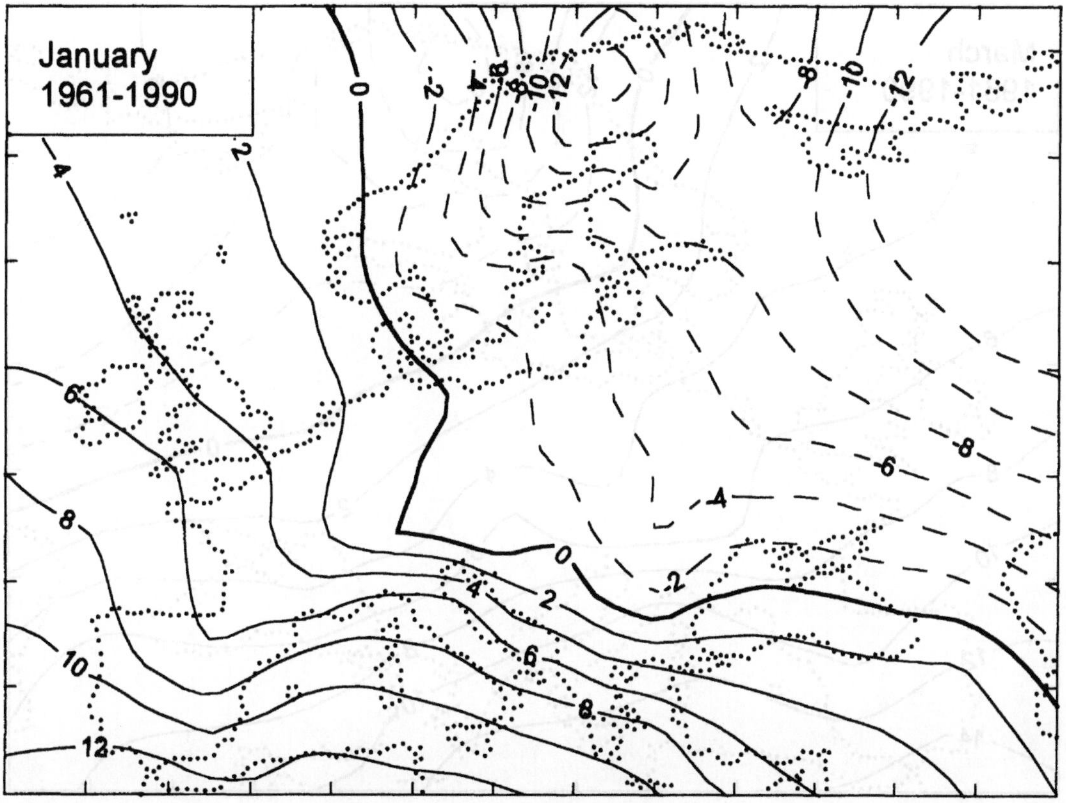

Temperature, average [°C]

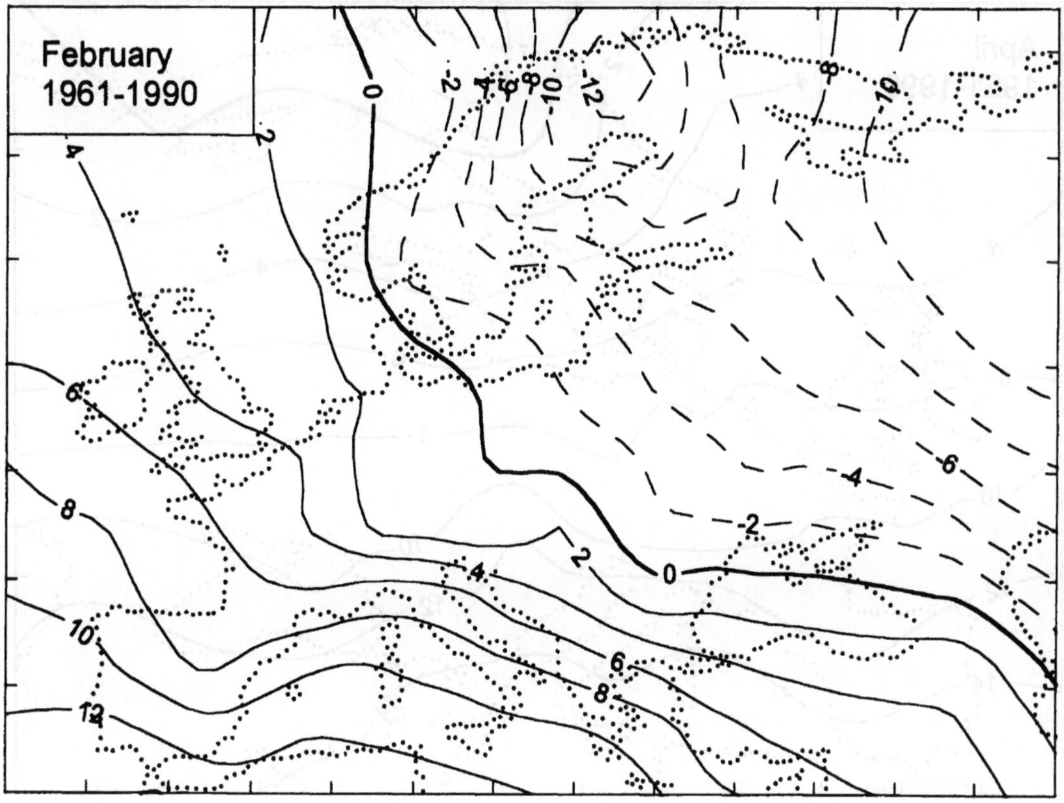

Temperature, average [°C]

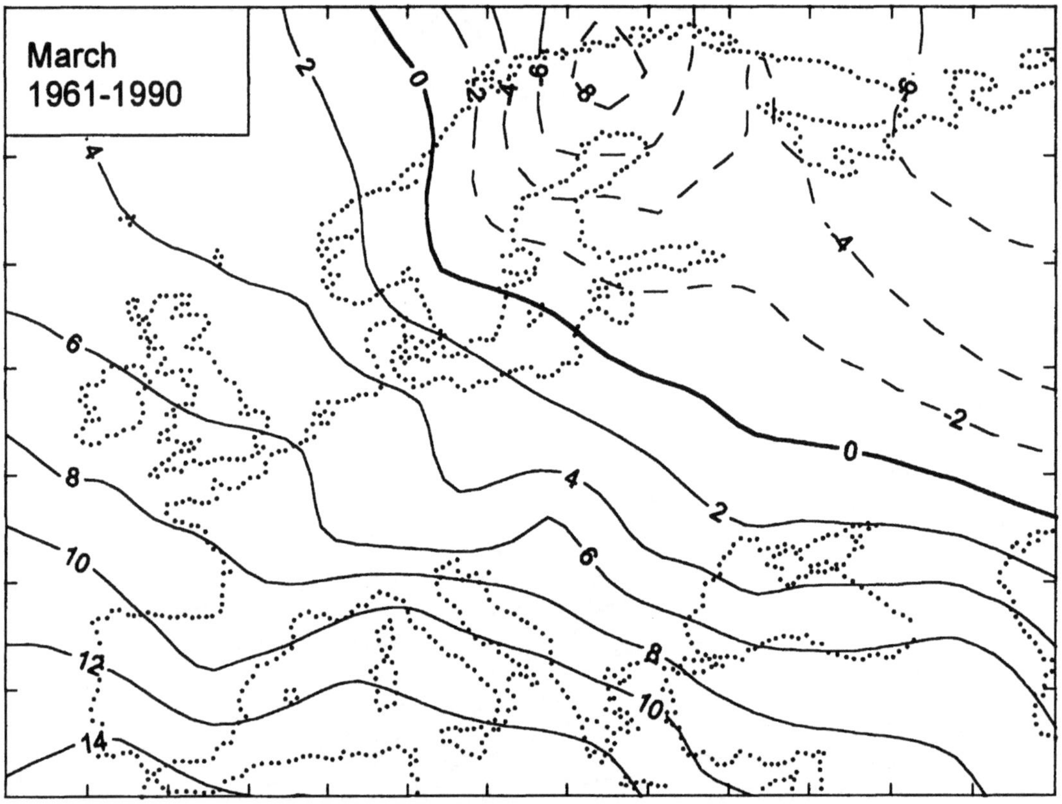

Temperature, average [°C]

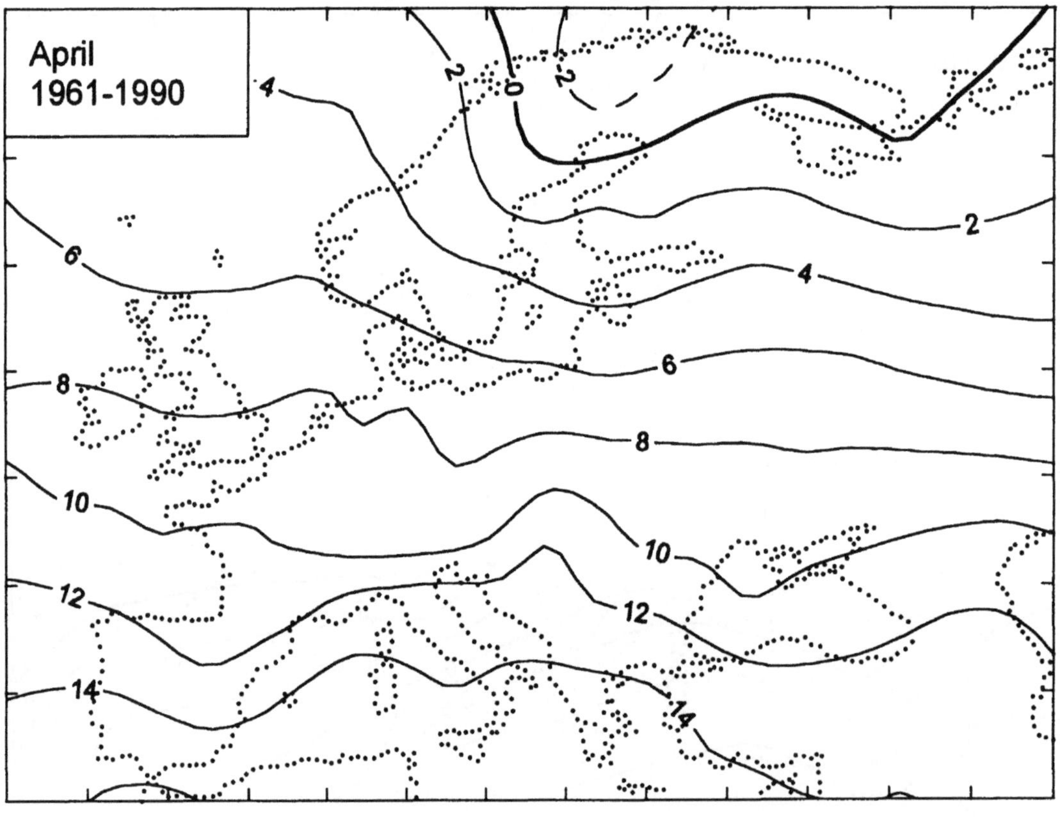

Temperature, average [°C]

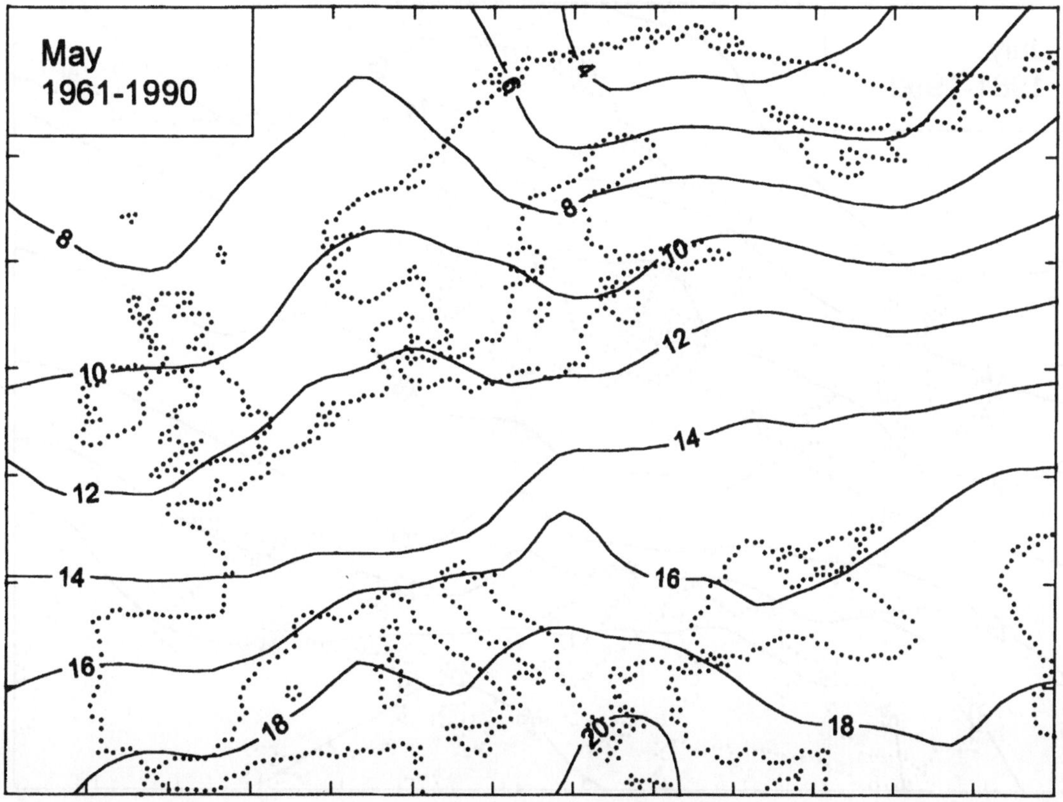

Temperature, average [°C]

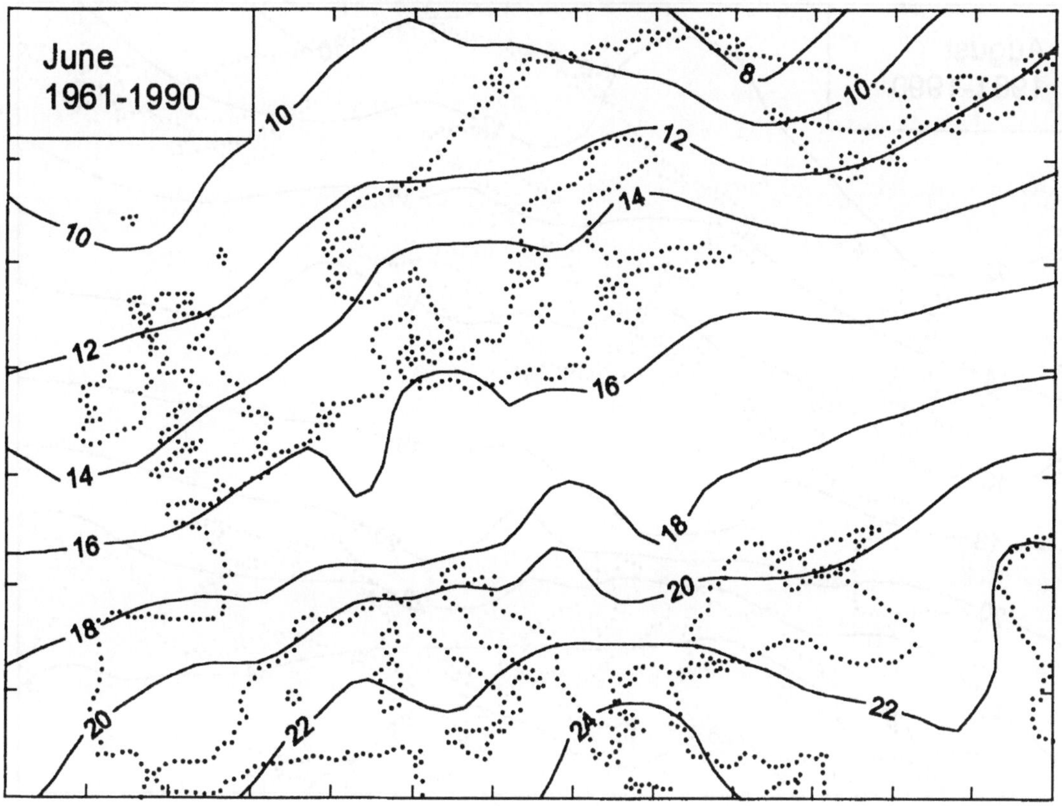

Temperature, average [°C]

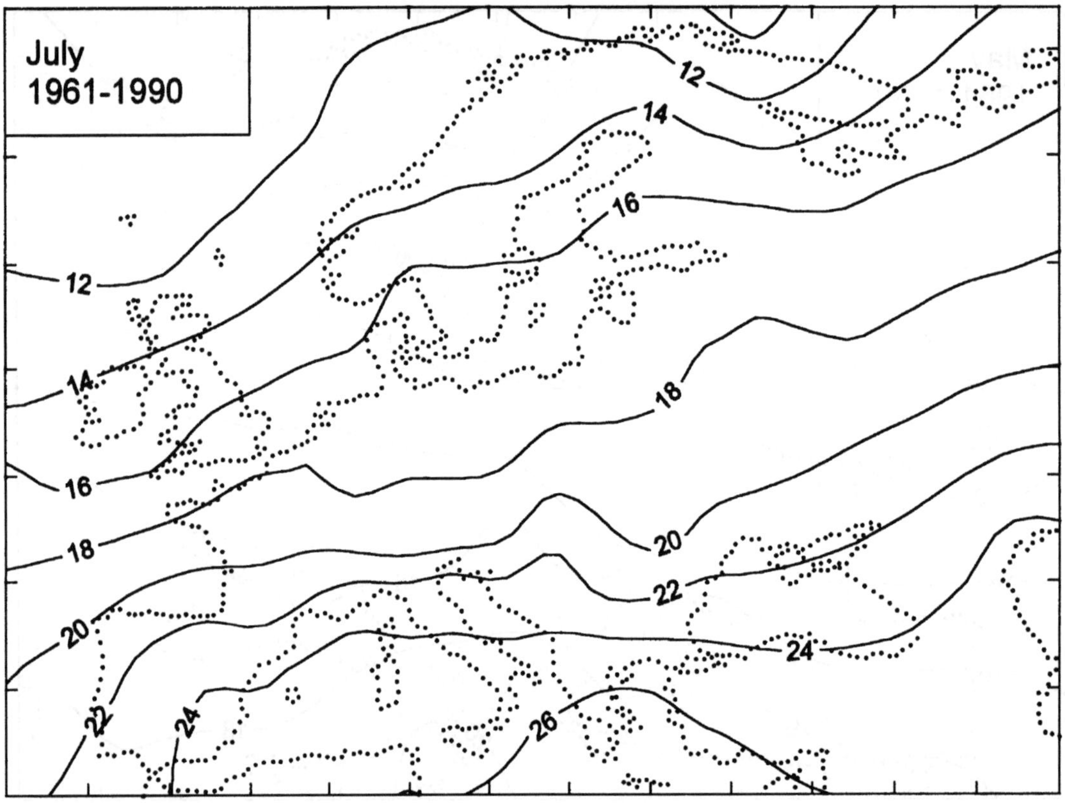

Temperature, average [°C]

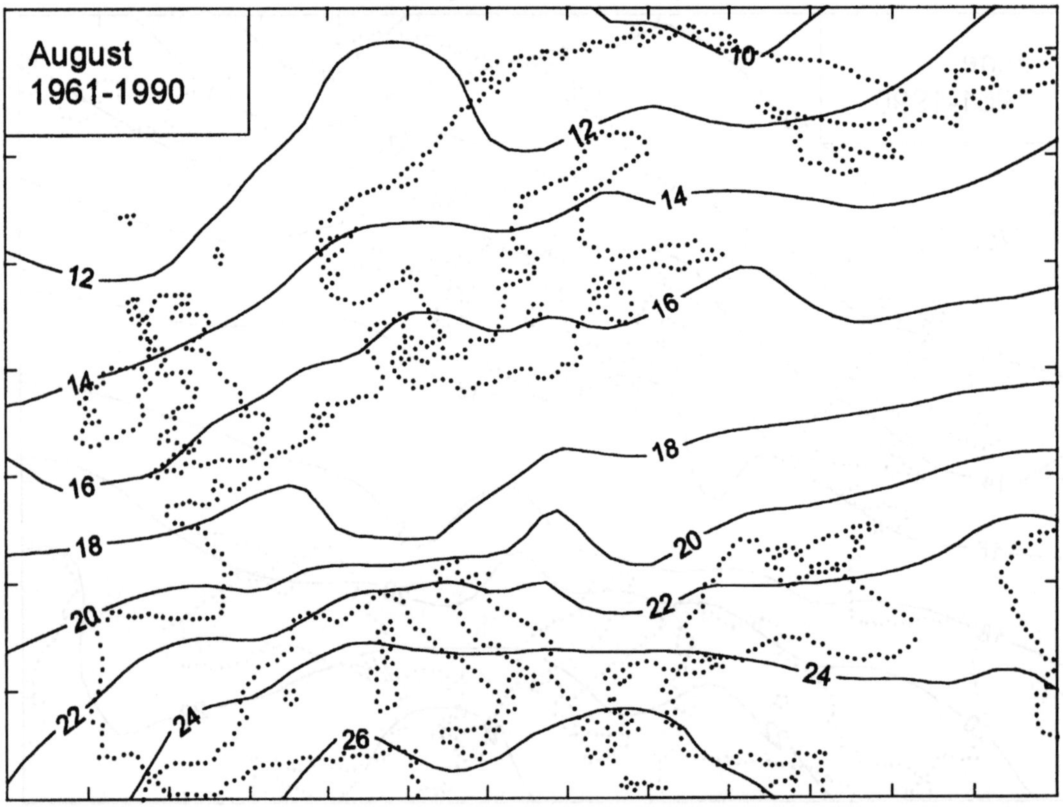

Temperature, average [°C]

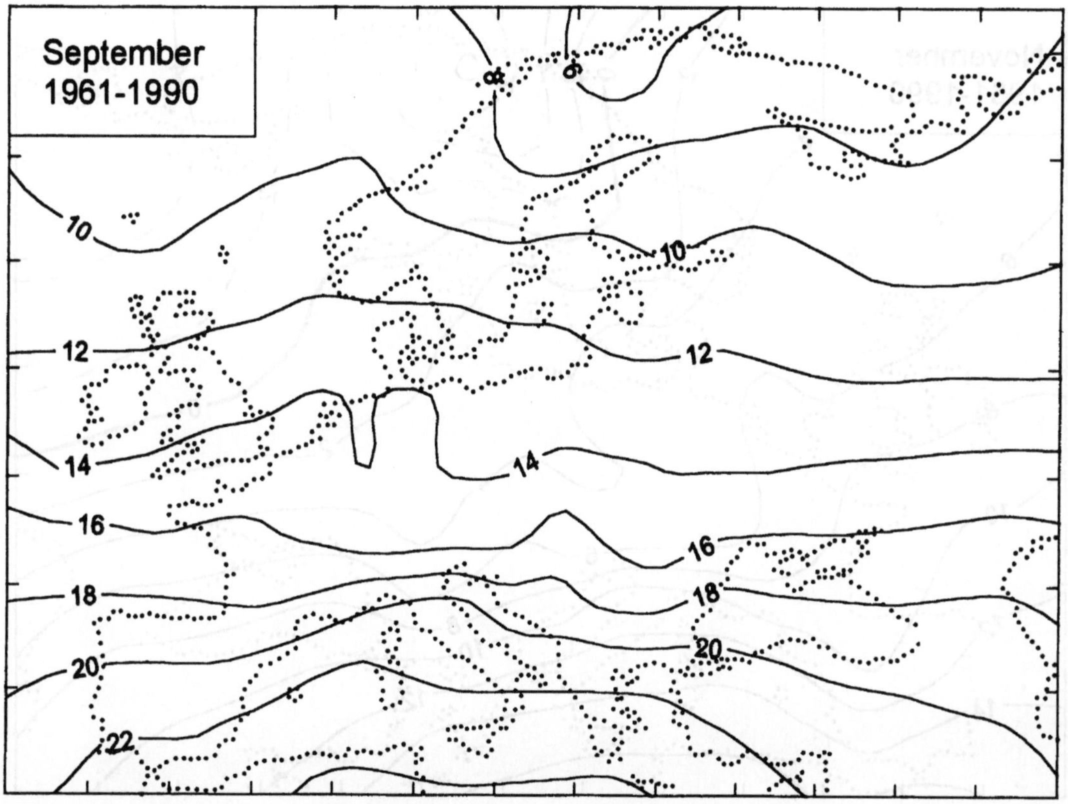

Temperature, average [°C]

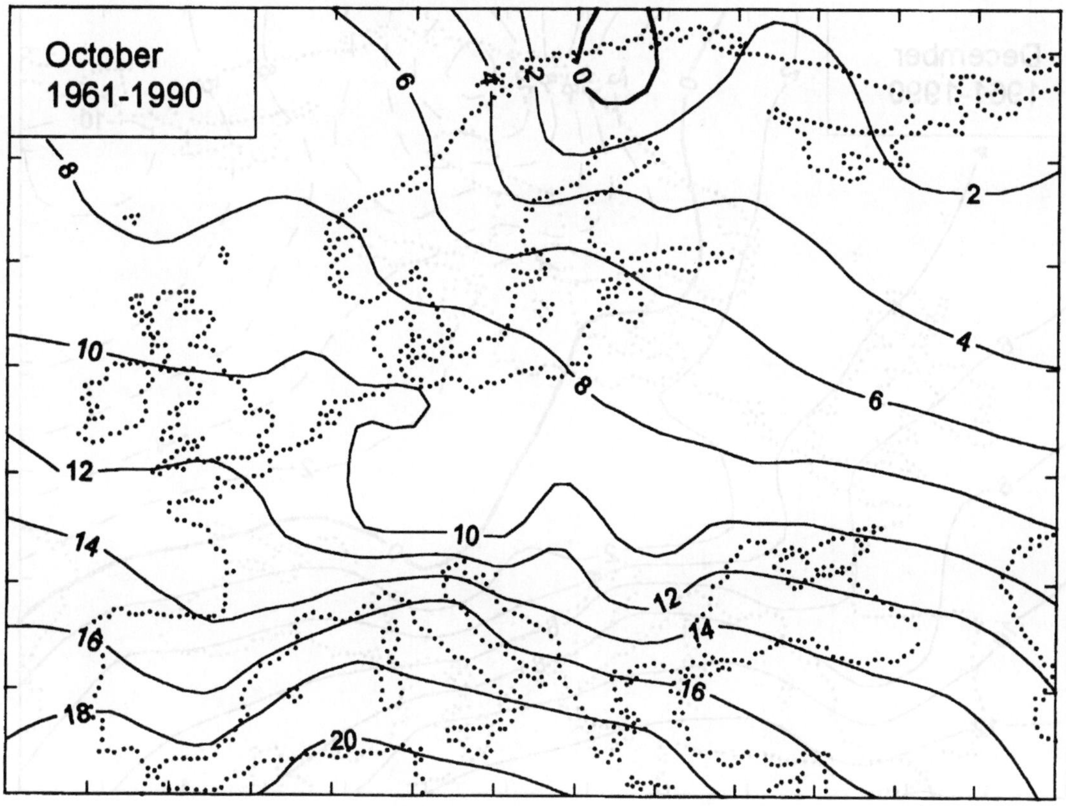

Temperature, average [°C]

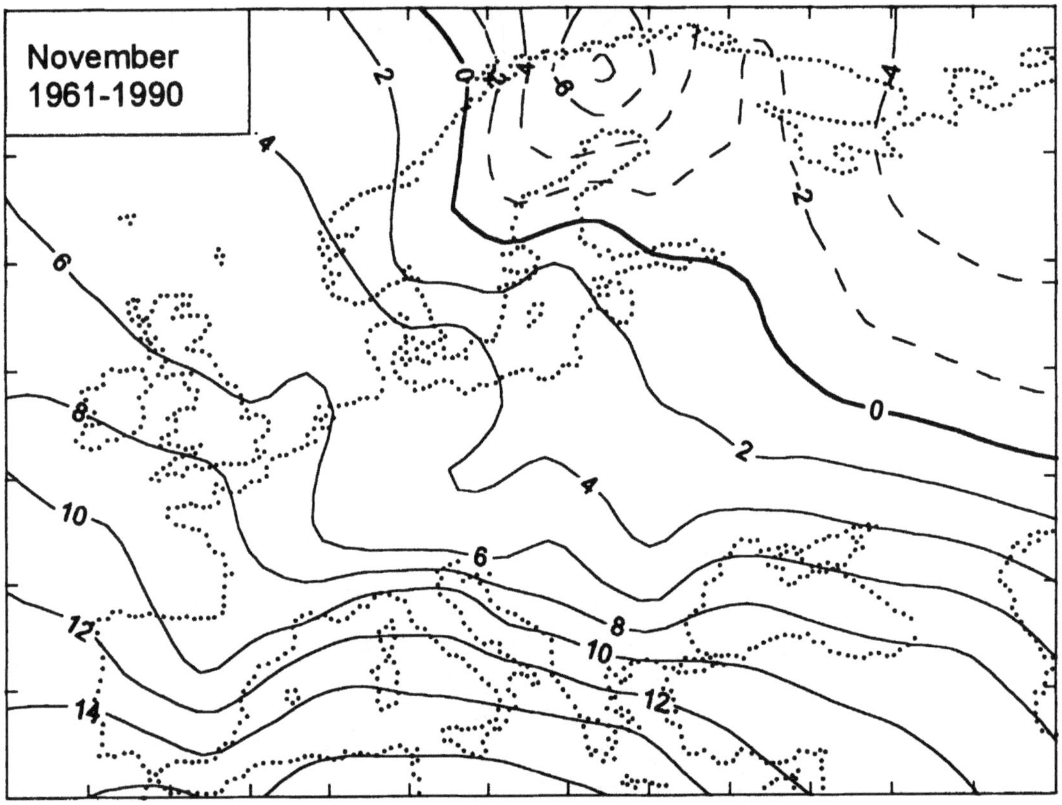

Temperature, average [°C]

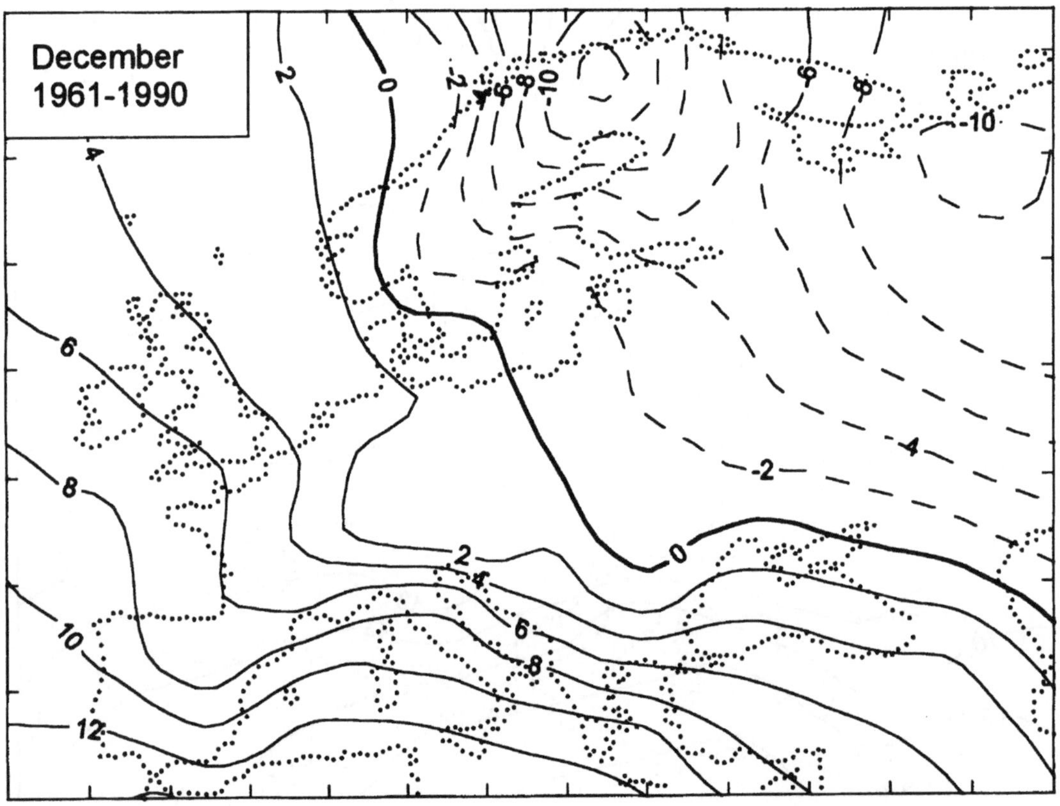

Temperature, average [°C]

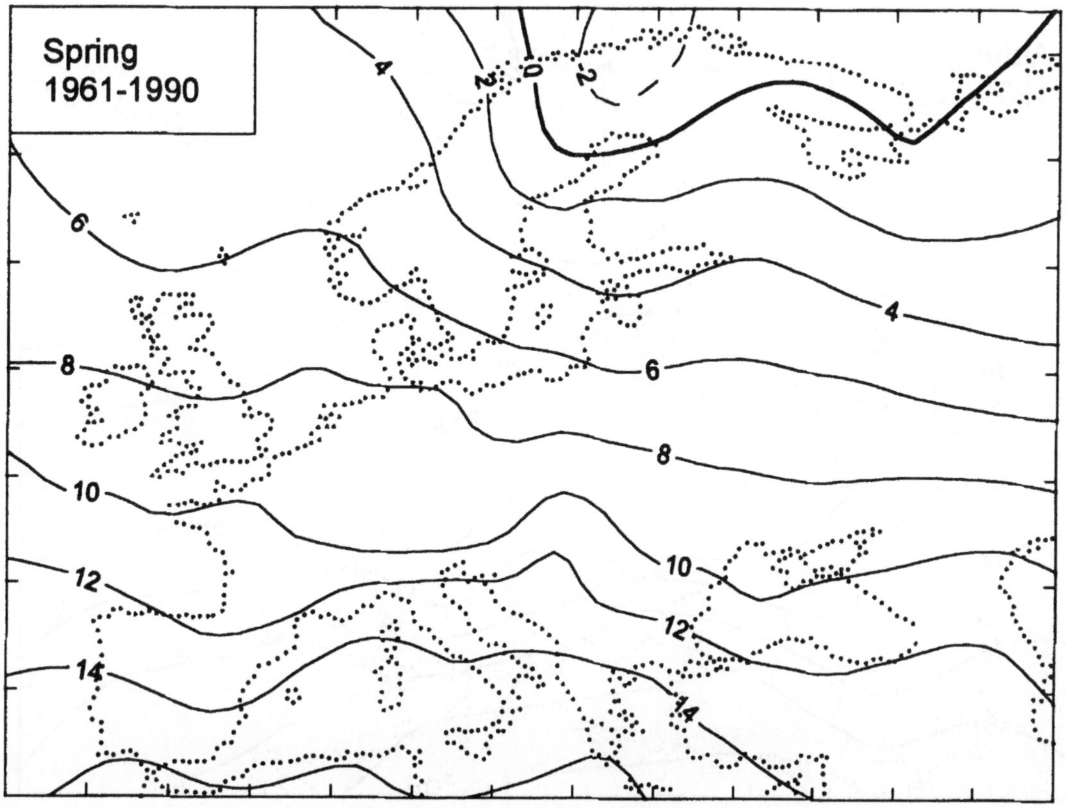

Temperature, average [°C]

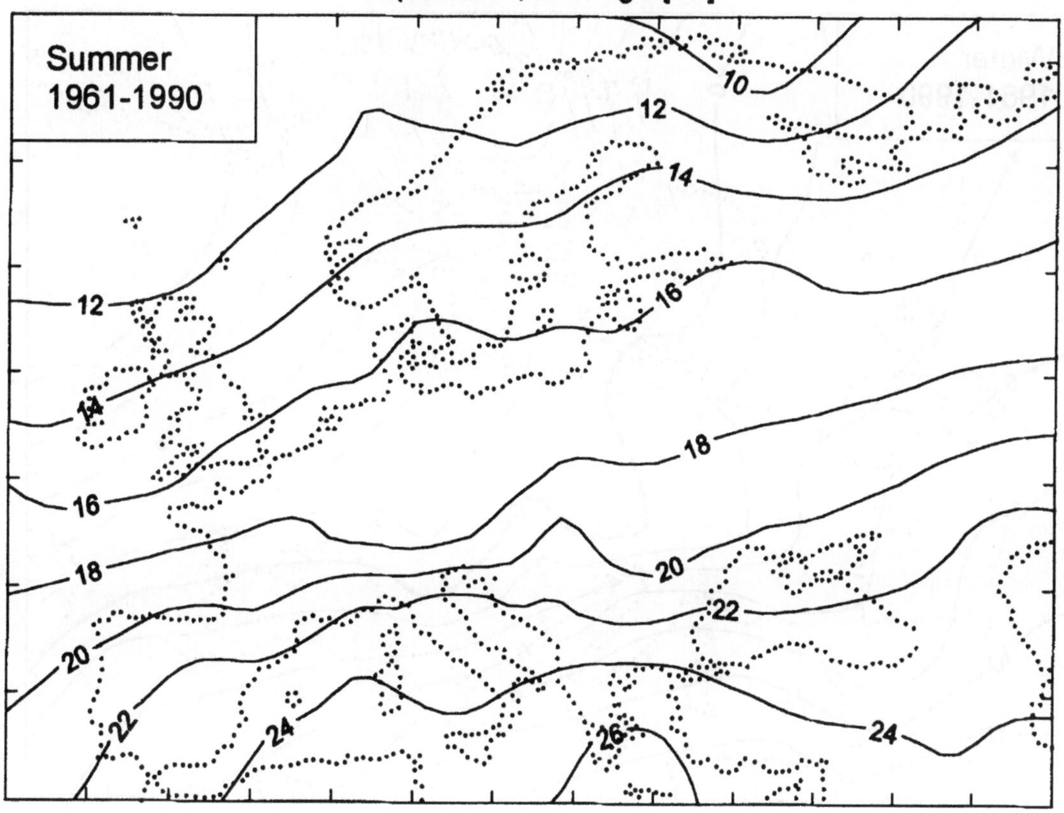

Temperature, average [°C]

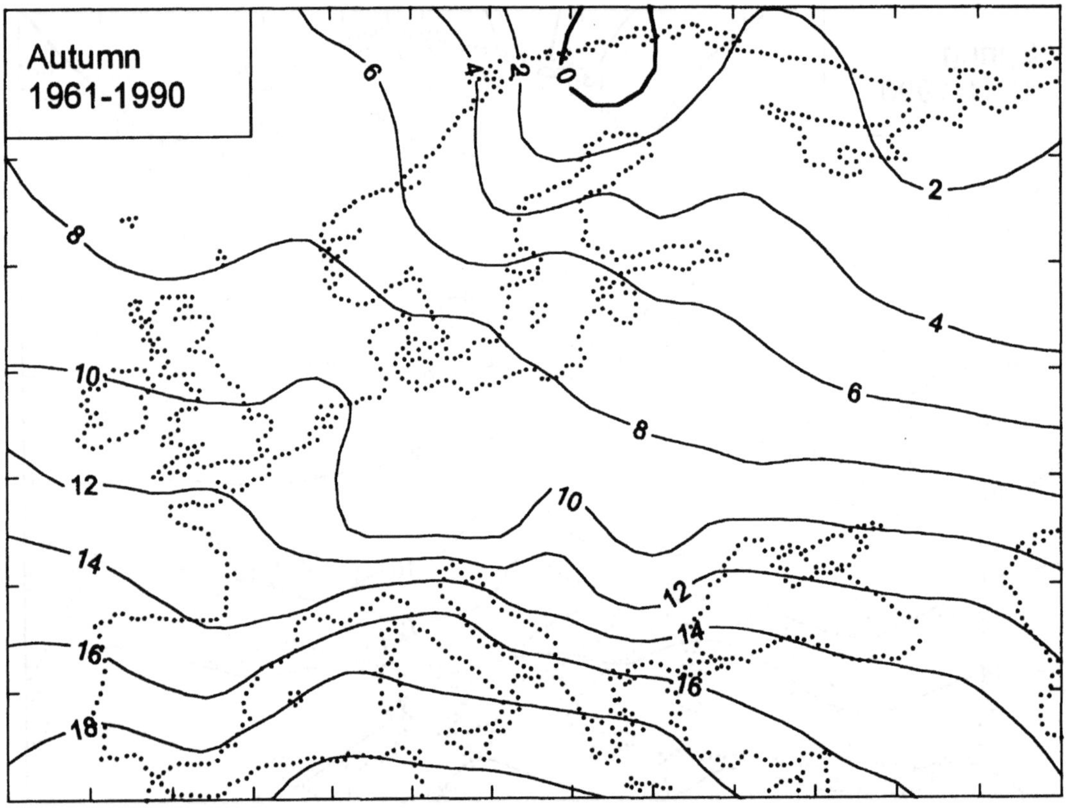

Temperature, average [°C]

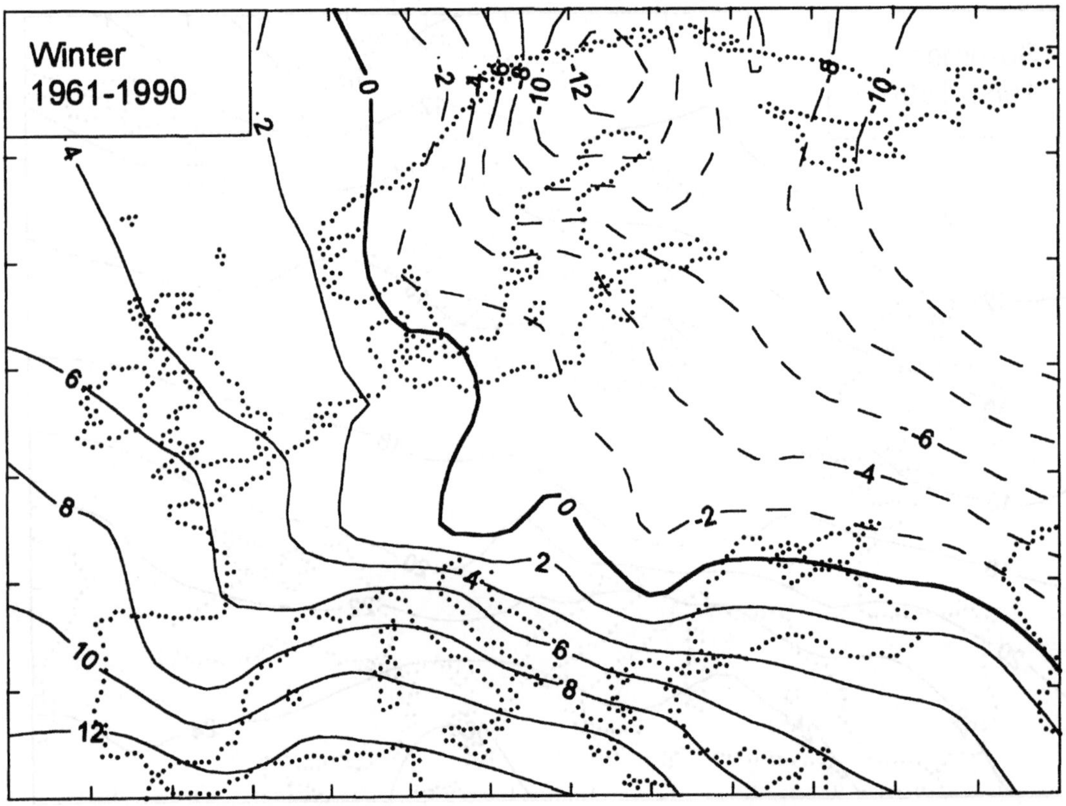

Temperature, average [°C]

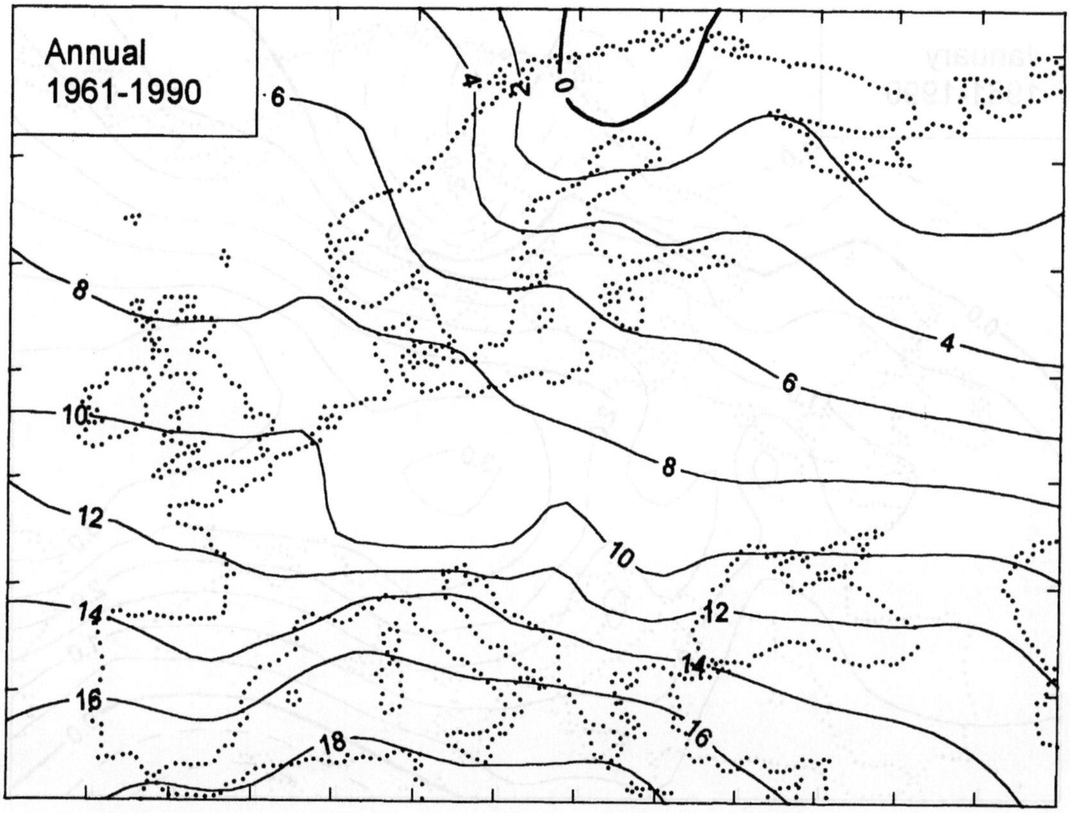

Temperature, linear trend [K]

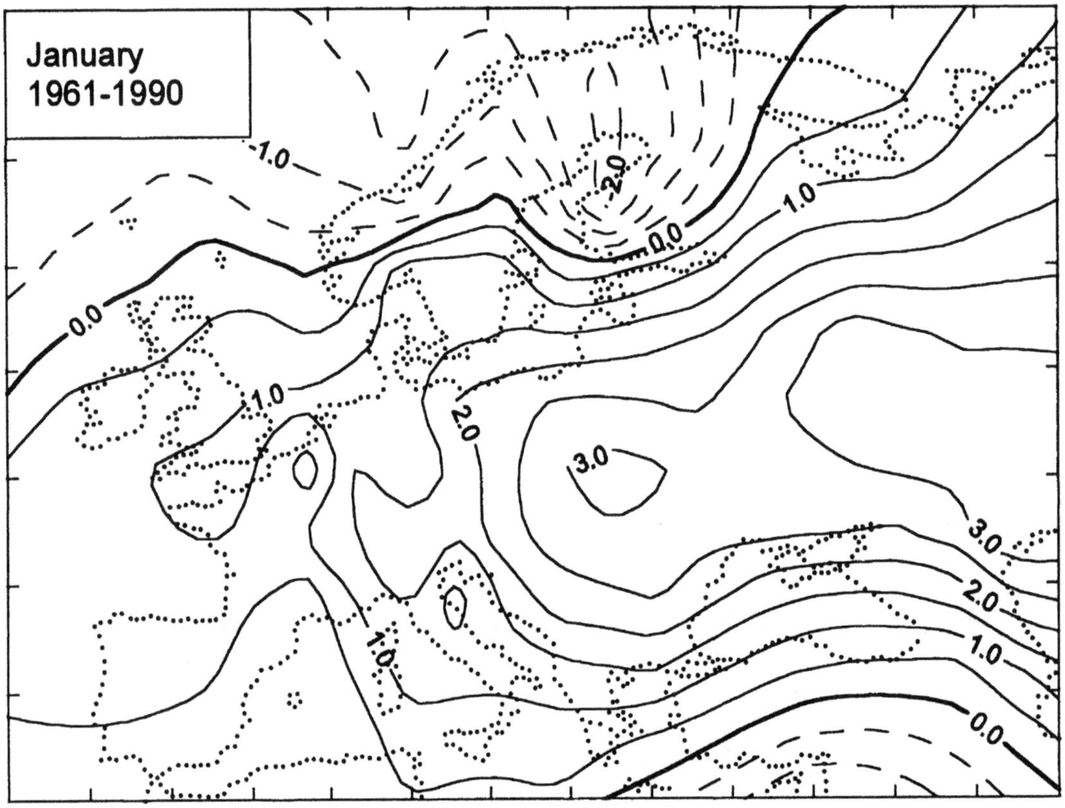

Temperature, trend test

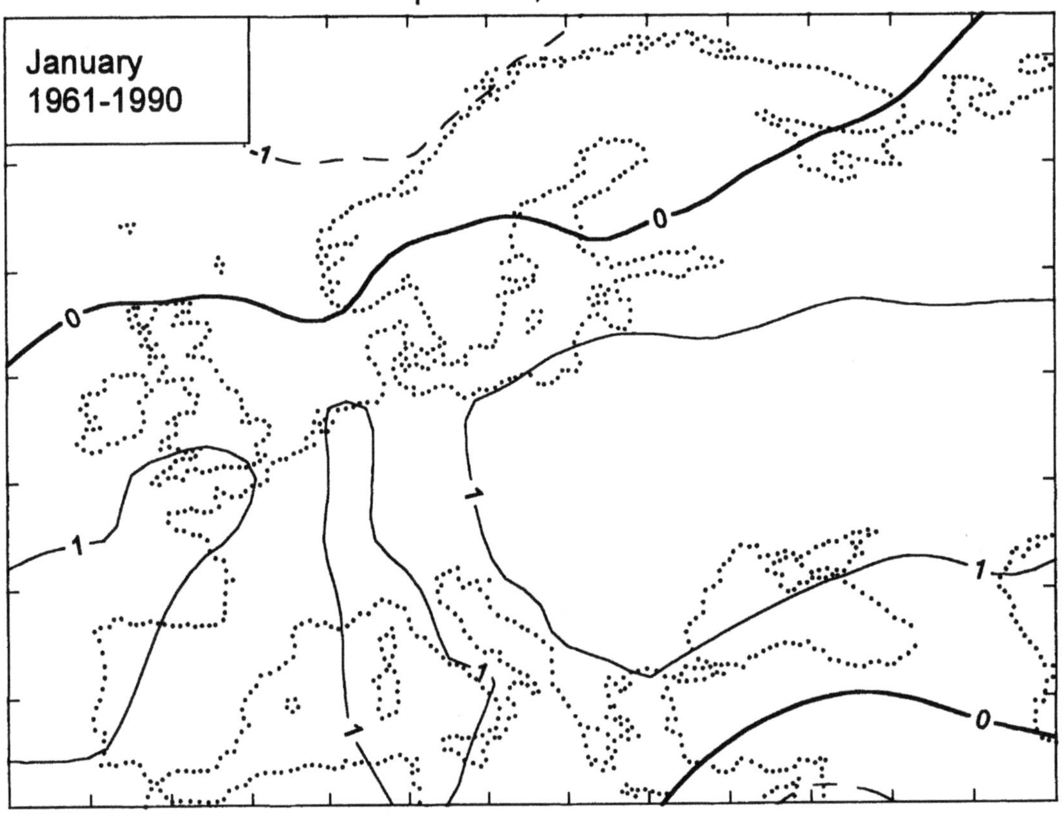

Temperature, linear trend [K]

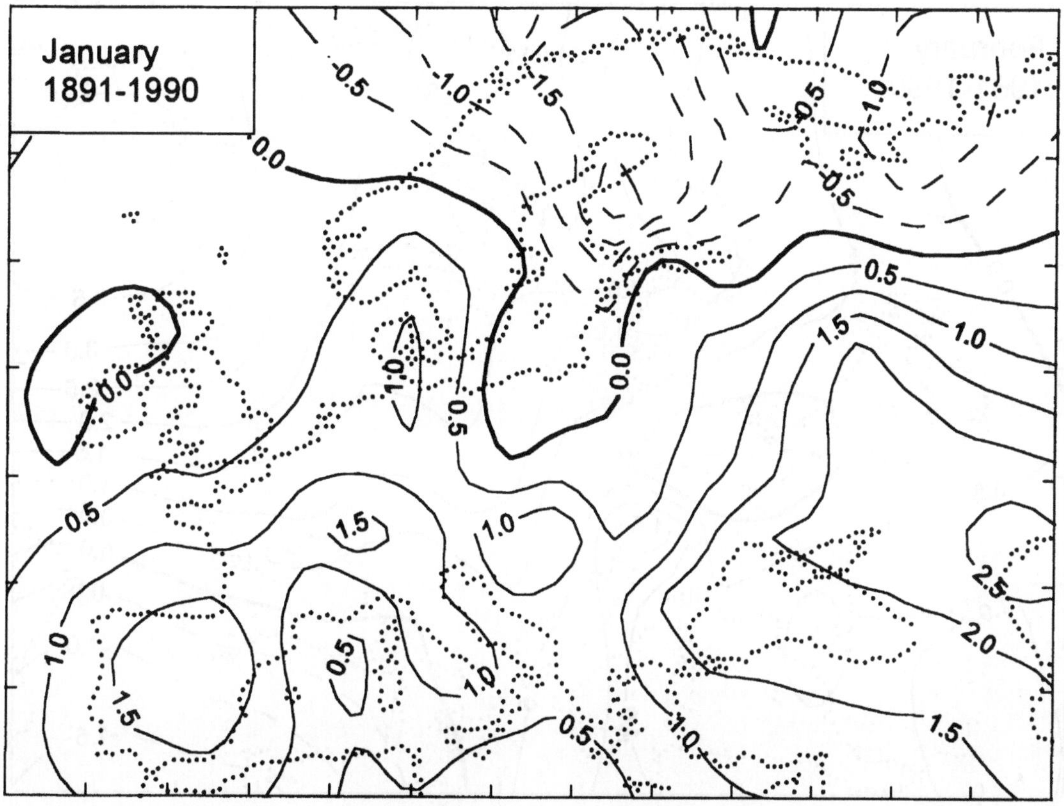

Temperature, trend test

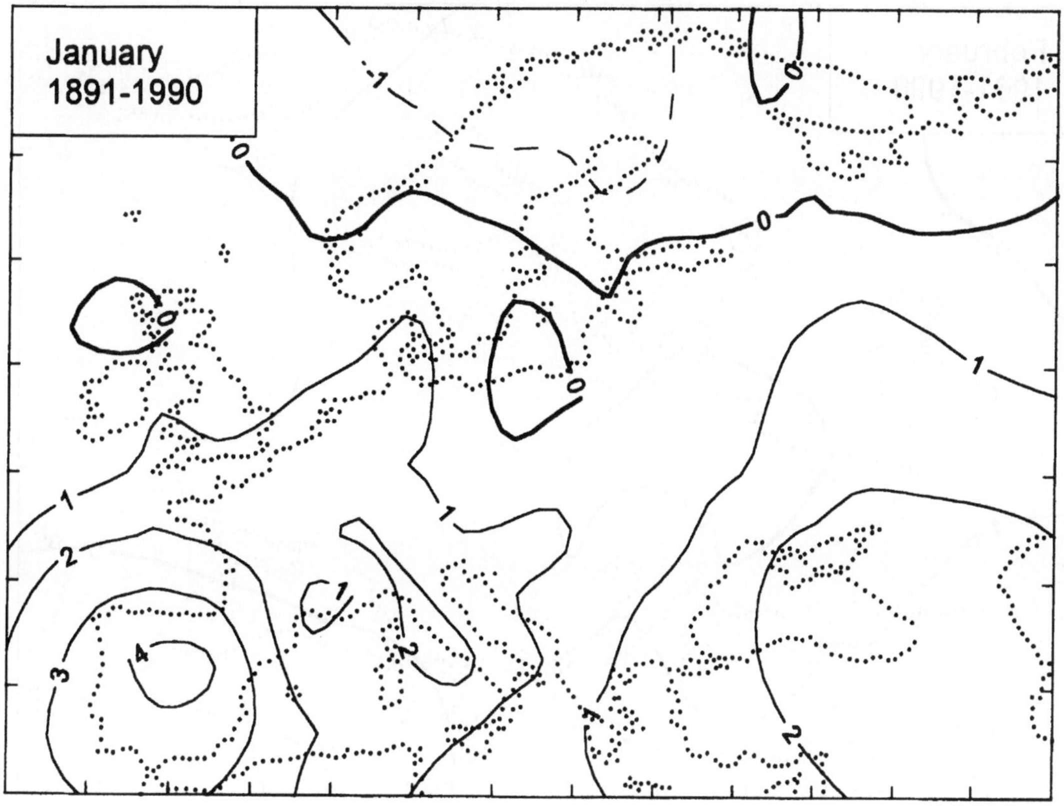

Temperature, linear trend [K]

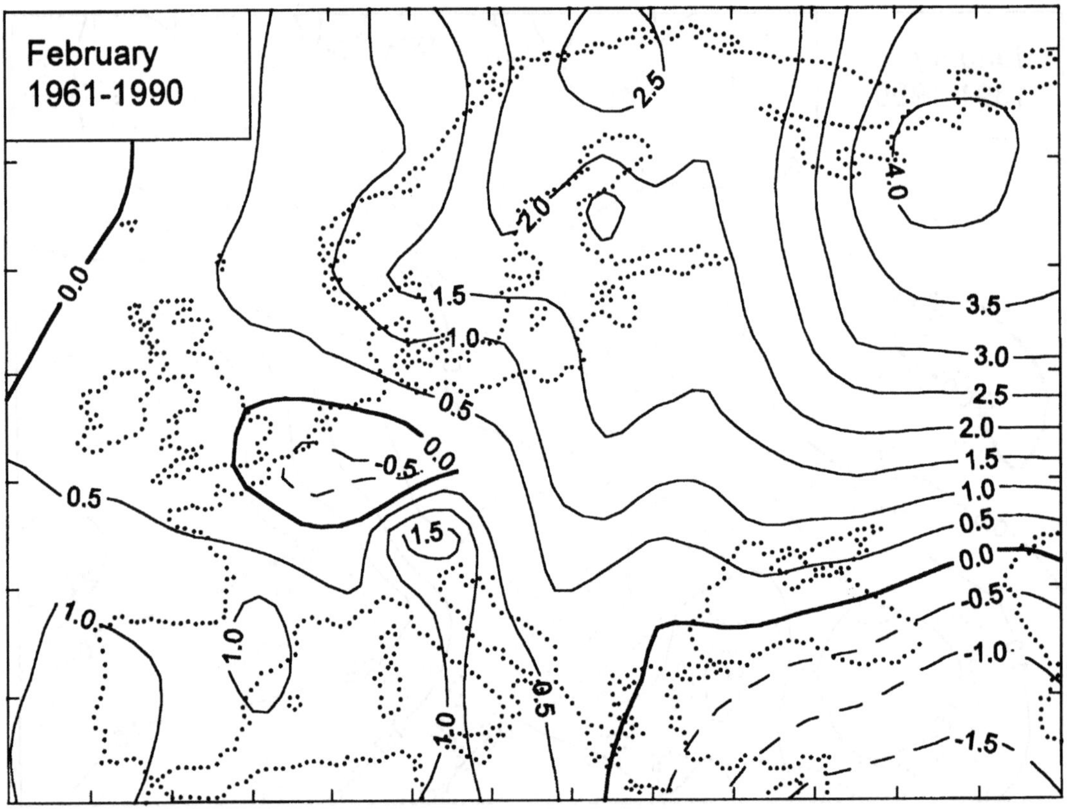

Temperature, trend test

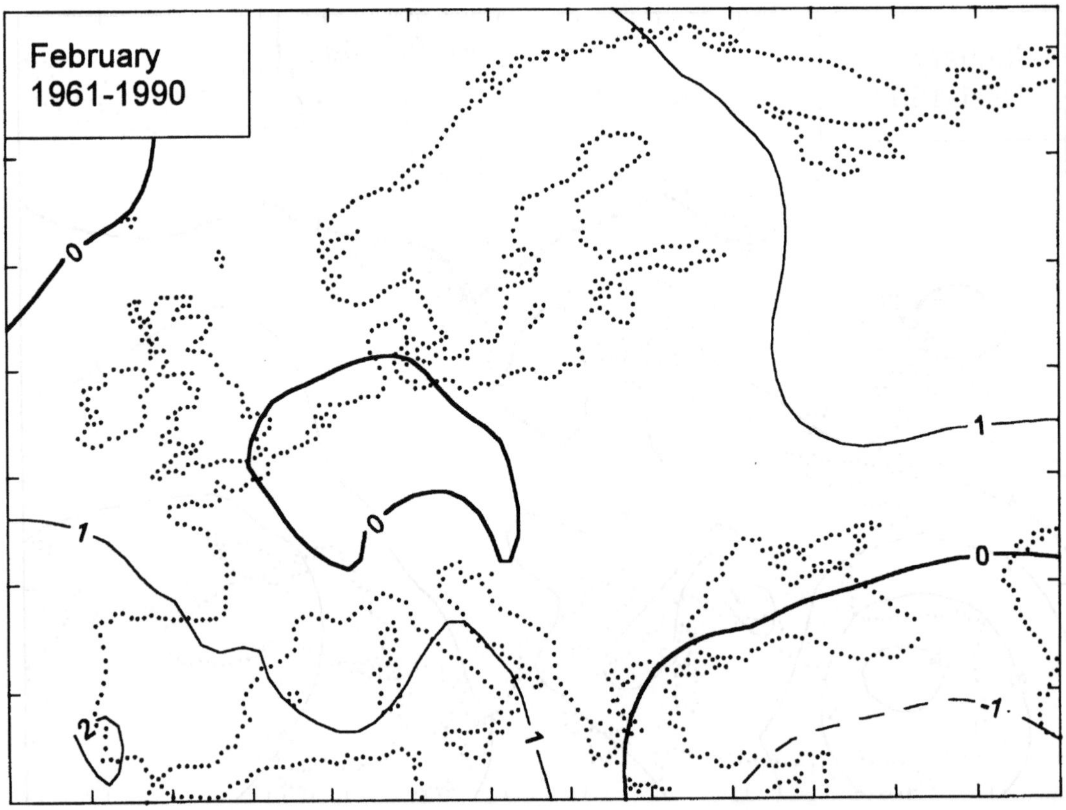

Temperature, linear trend [K]

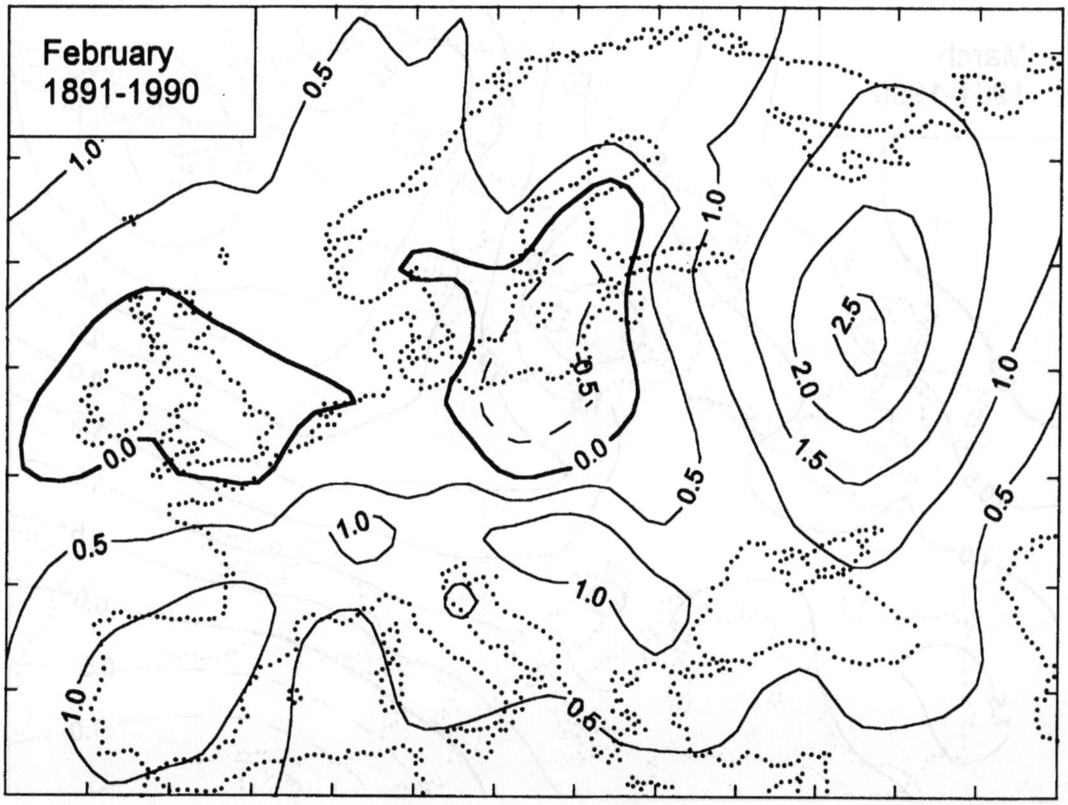

Temperature, trend test

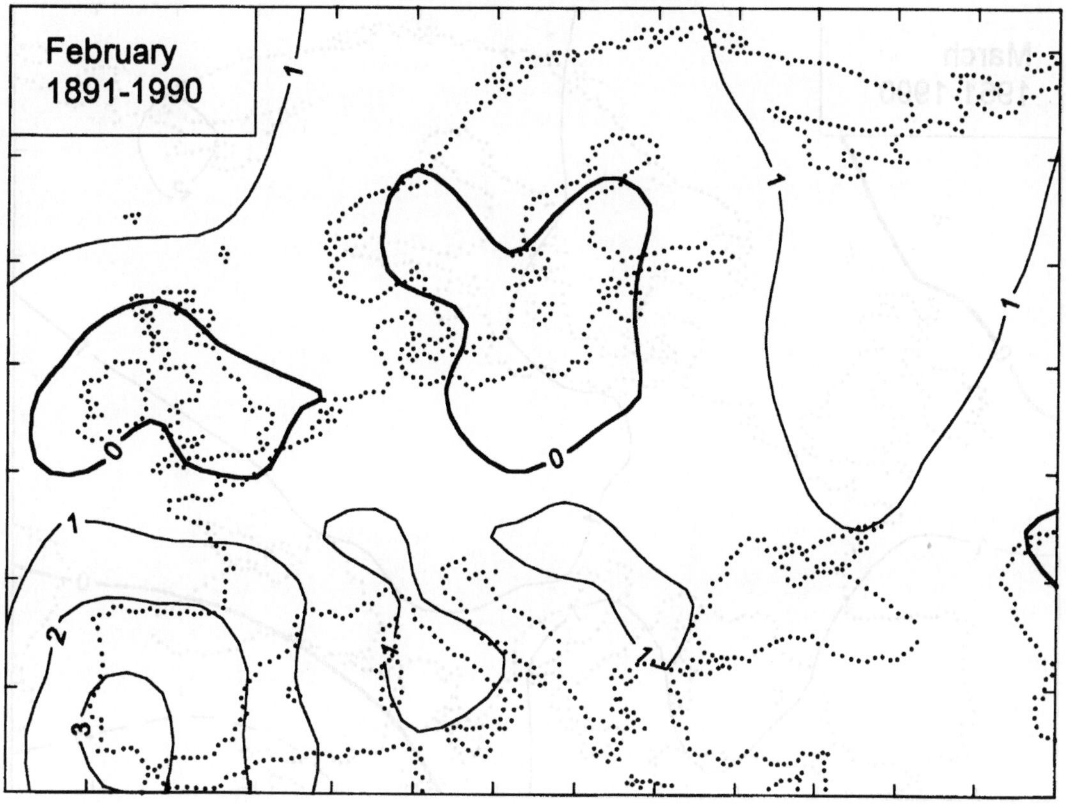

Temperature, linear trend [K]

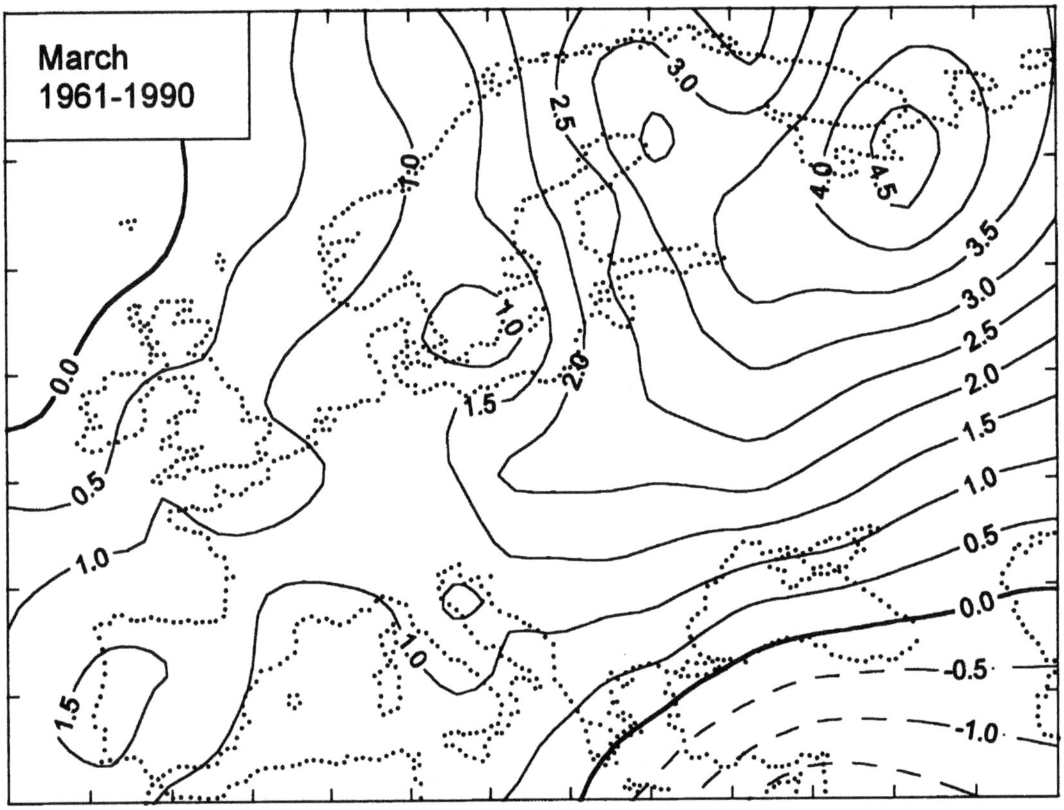

Temperature, trend test

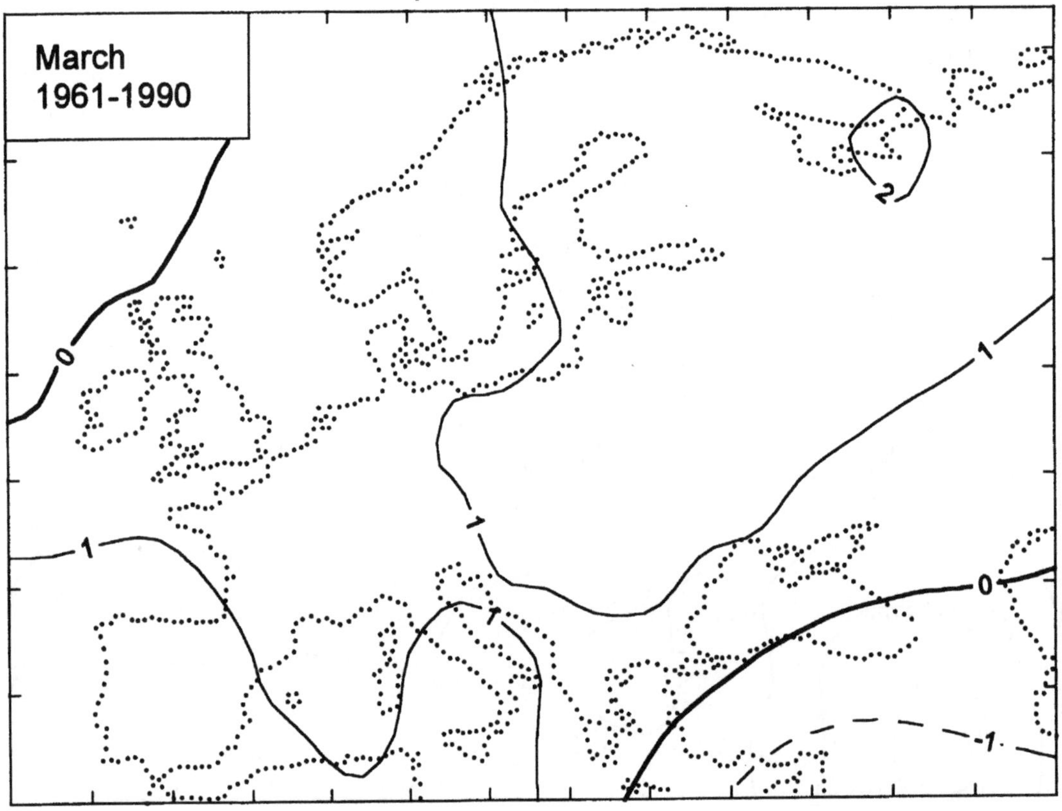

Temperature, linear trend [K]

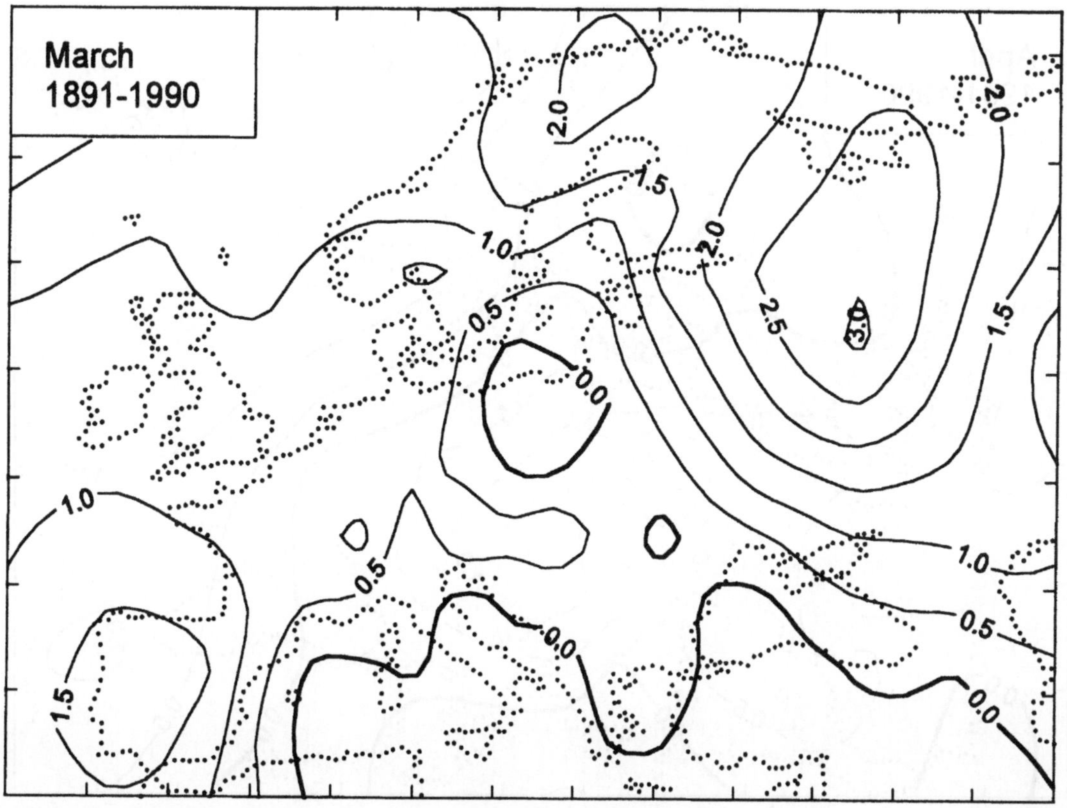

Temperature, trend test

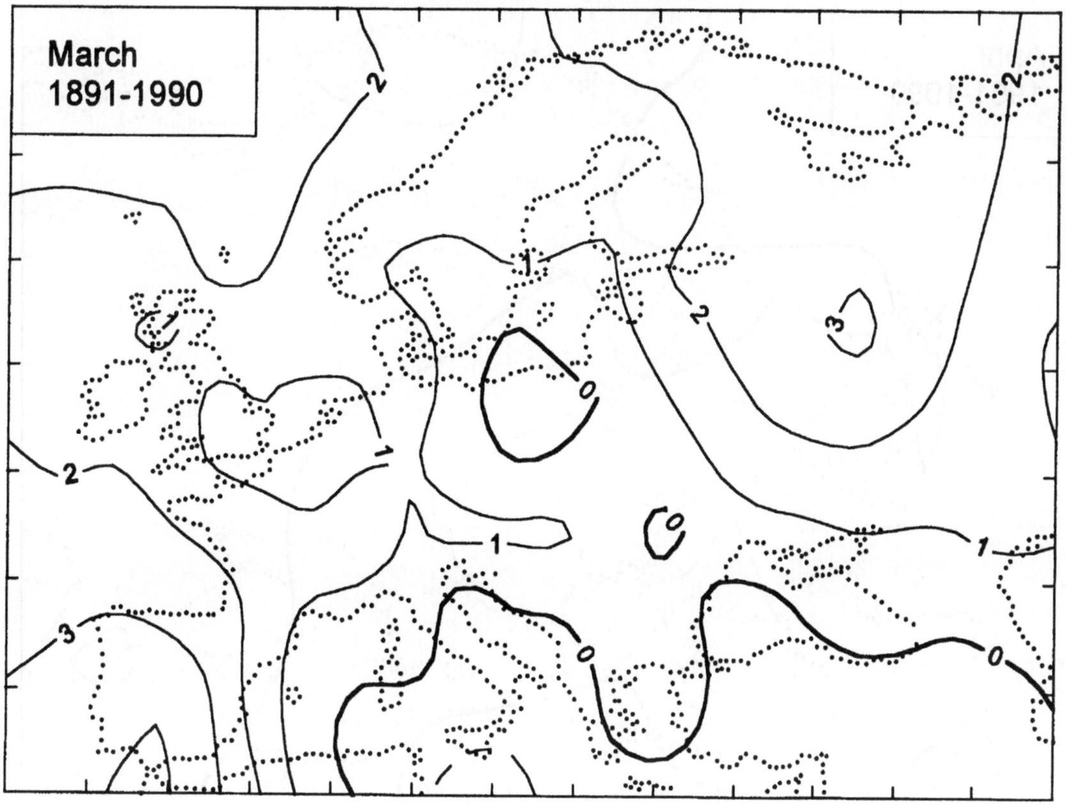

Temperature, linear trend [K]

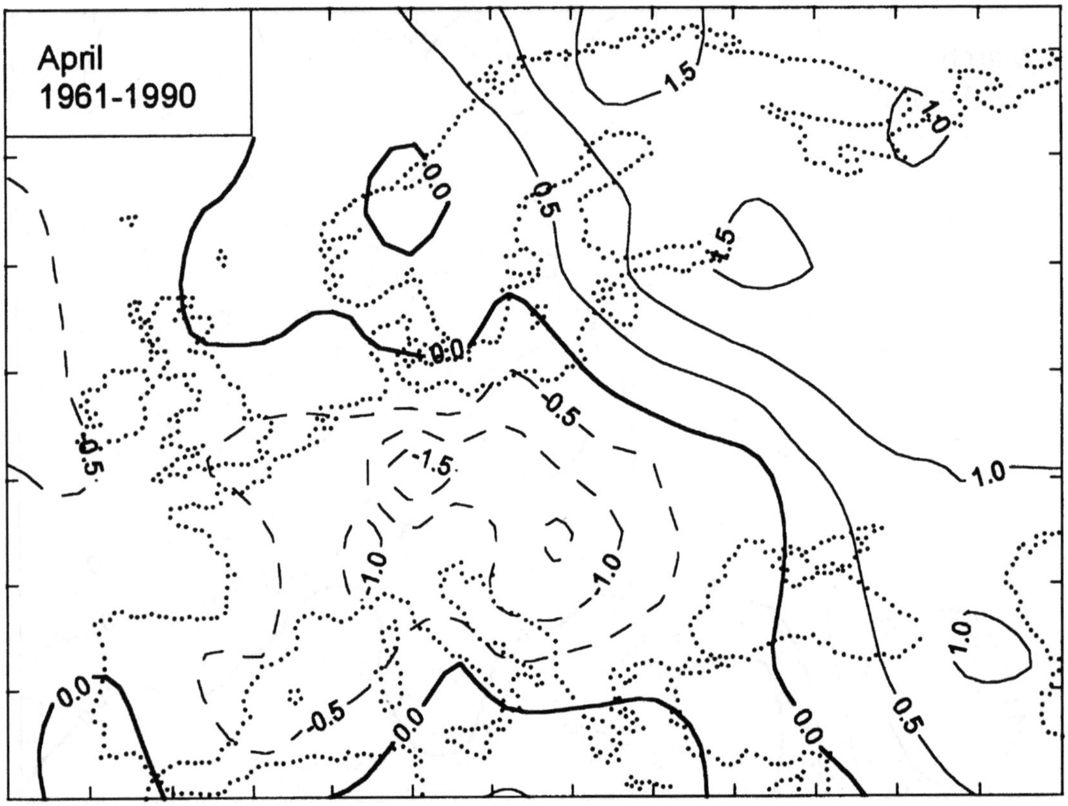

Temperature, trend test

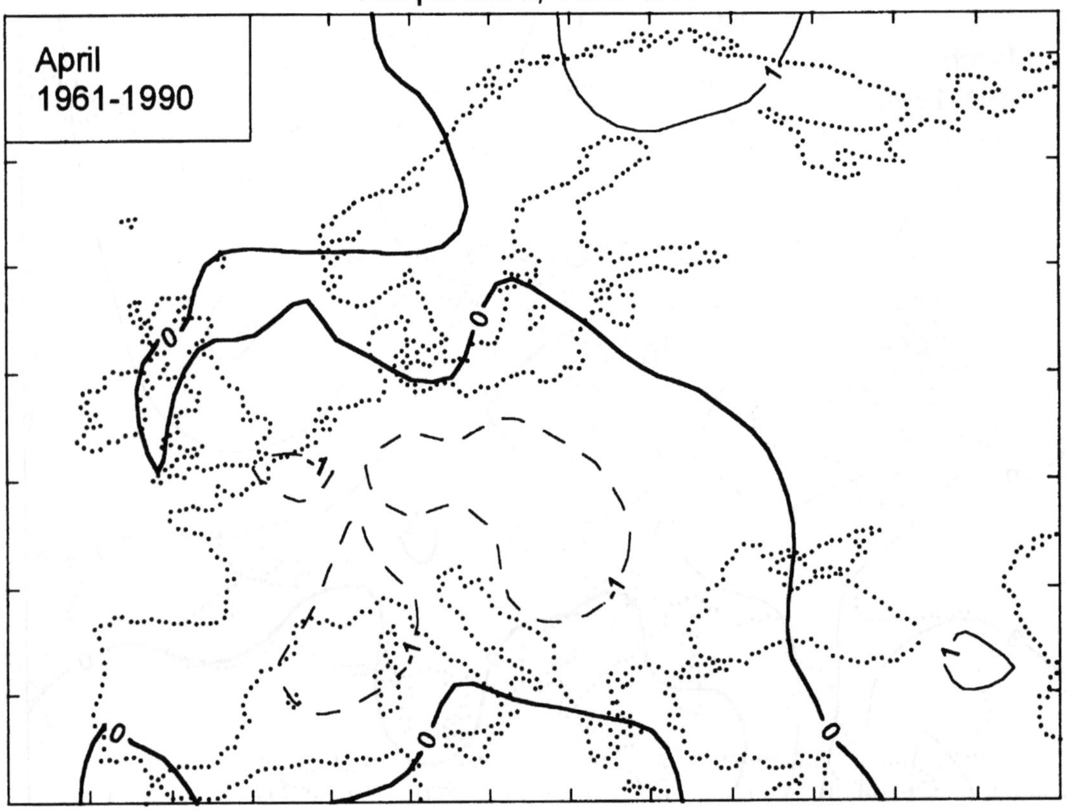

Temperature, linear trend [K]

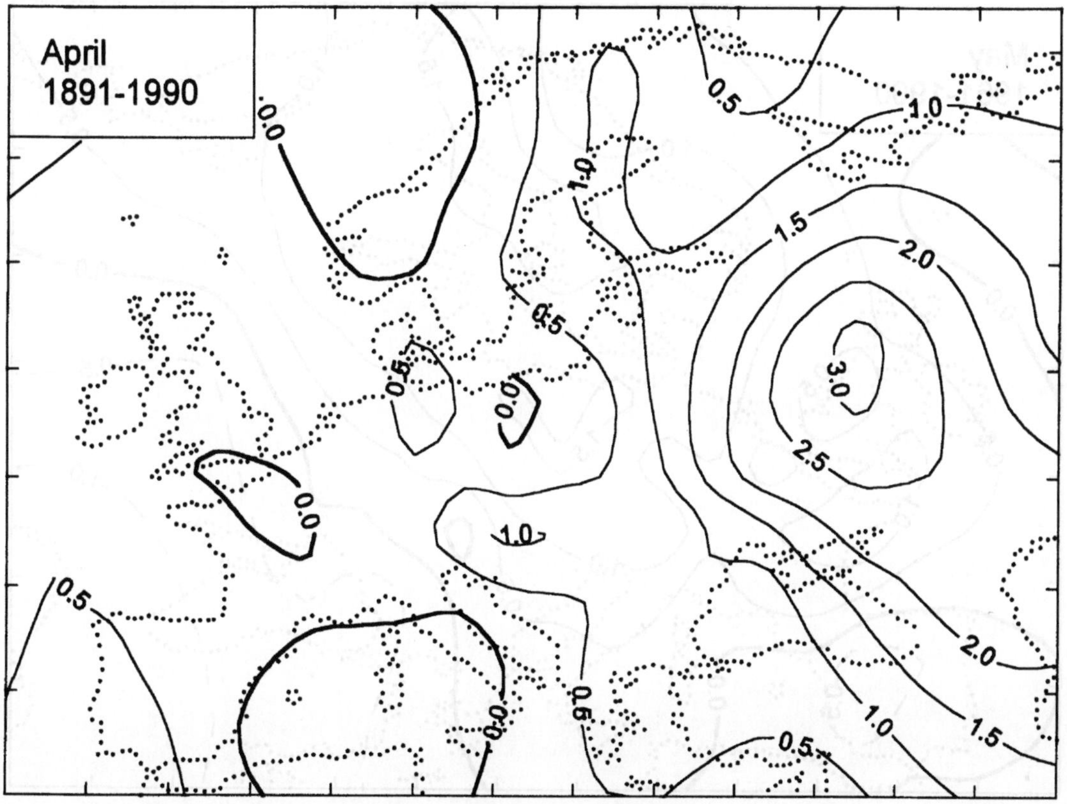

Temperature, trend test

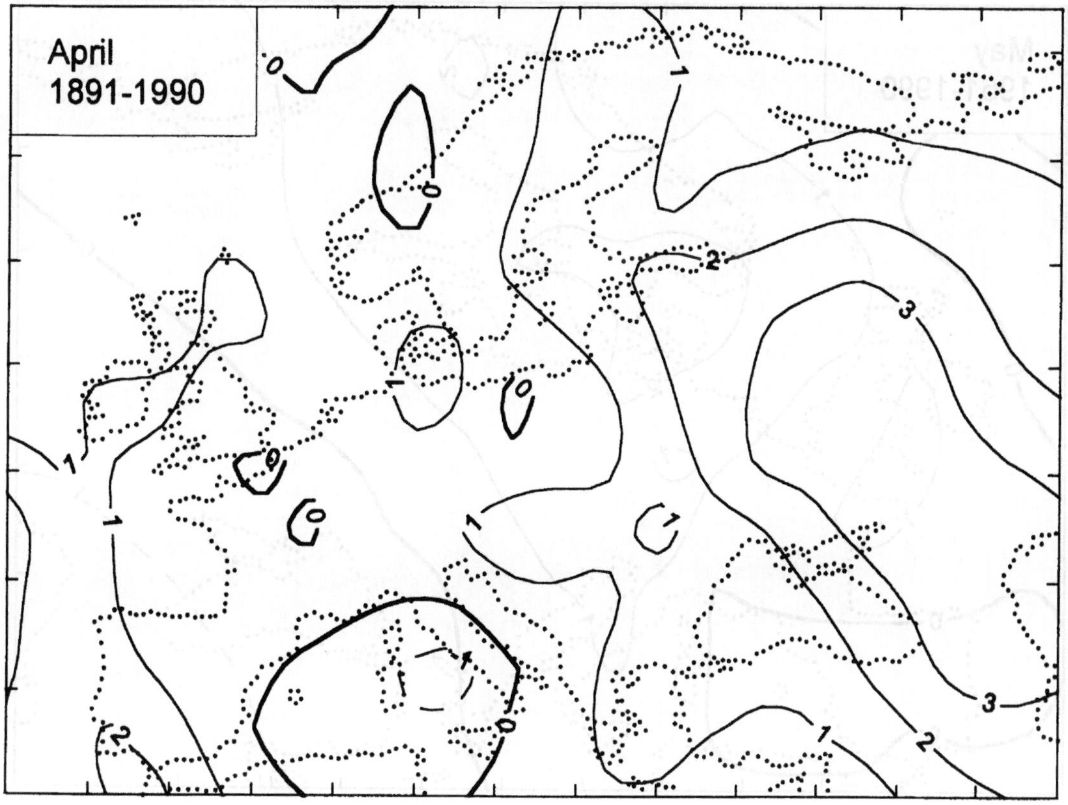

Temperature, linear trend [K]

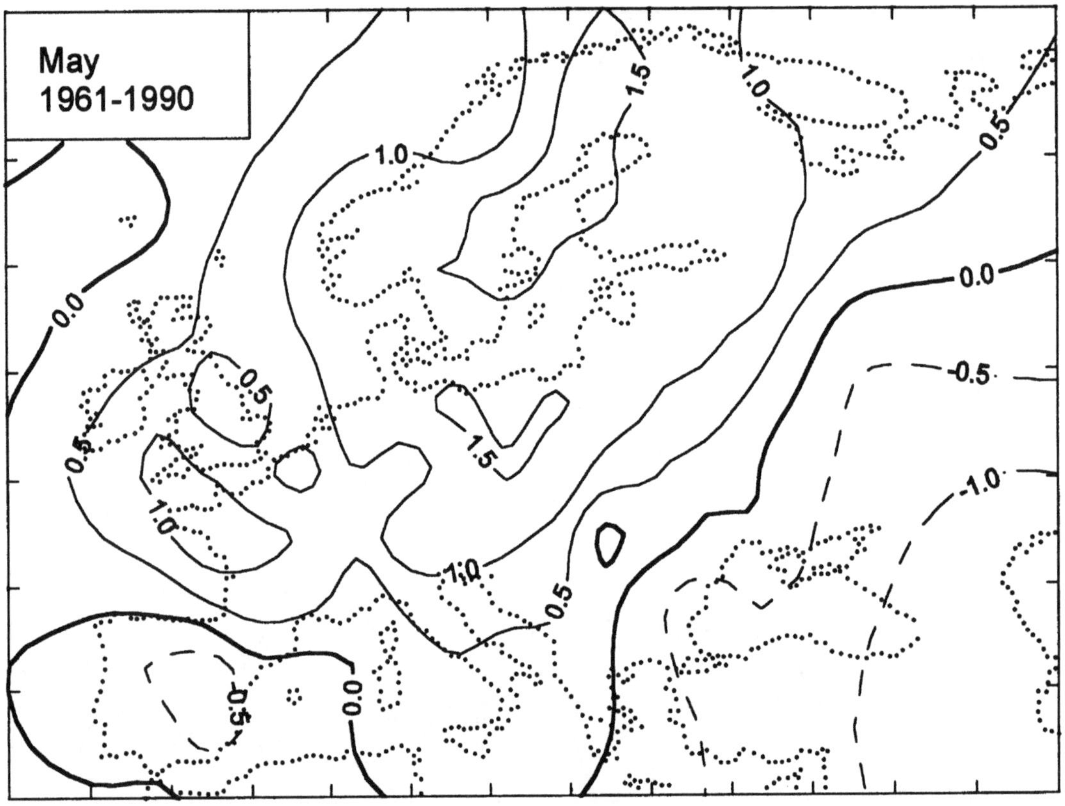

Temperature, trend test

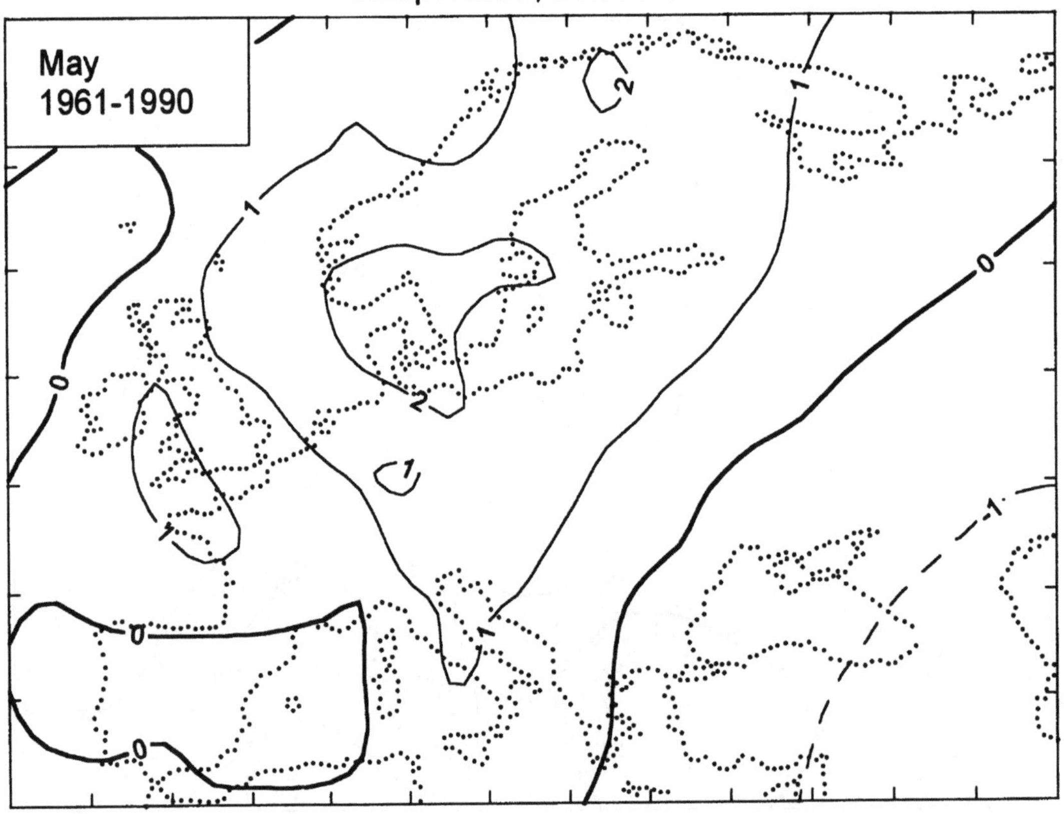

Temperature, linear trend [K]

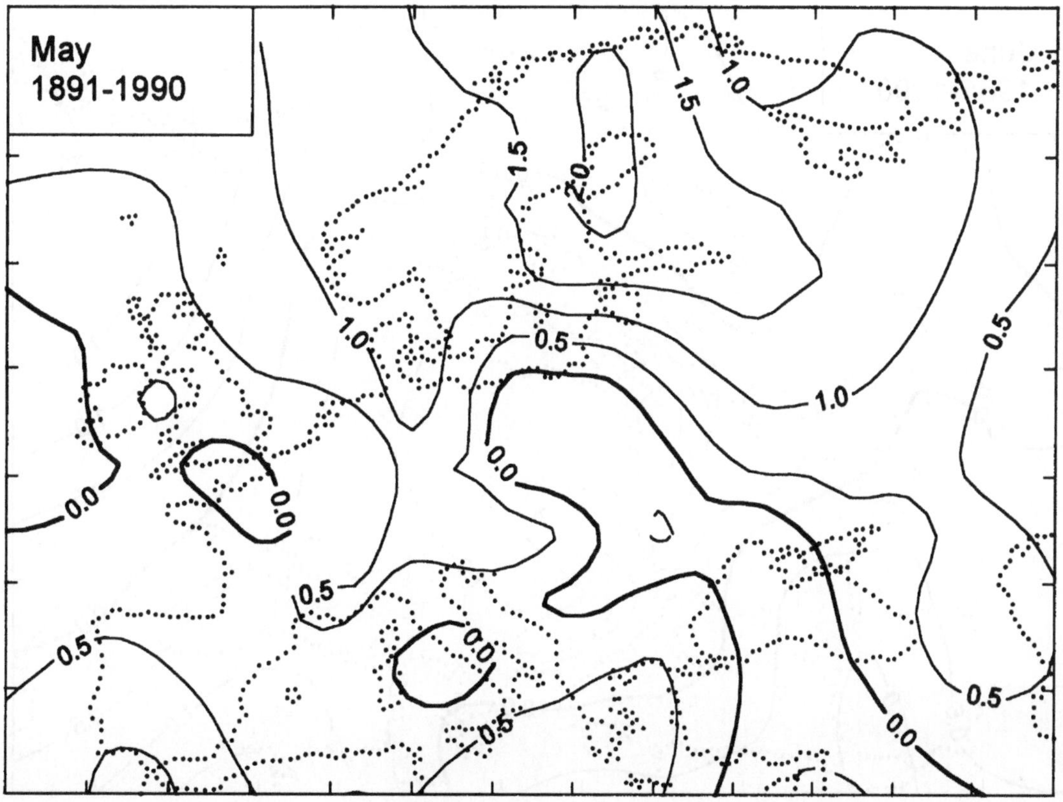

Temperature, trend test

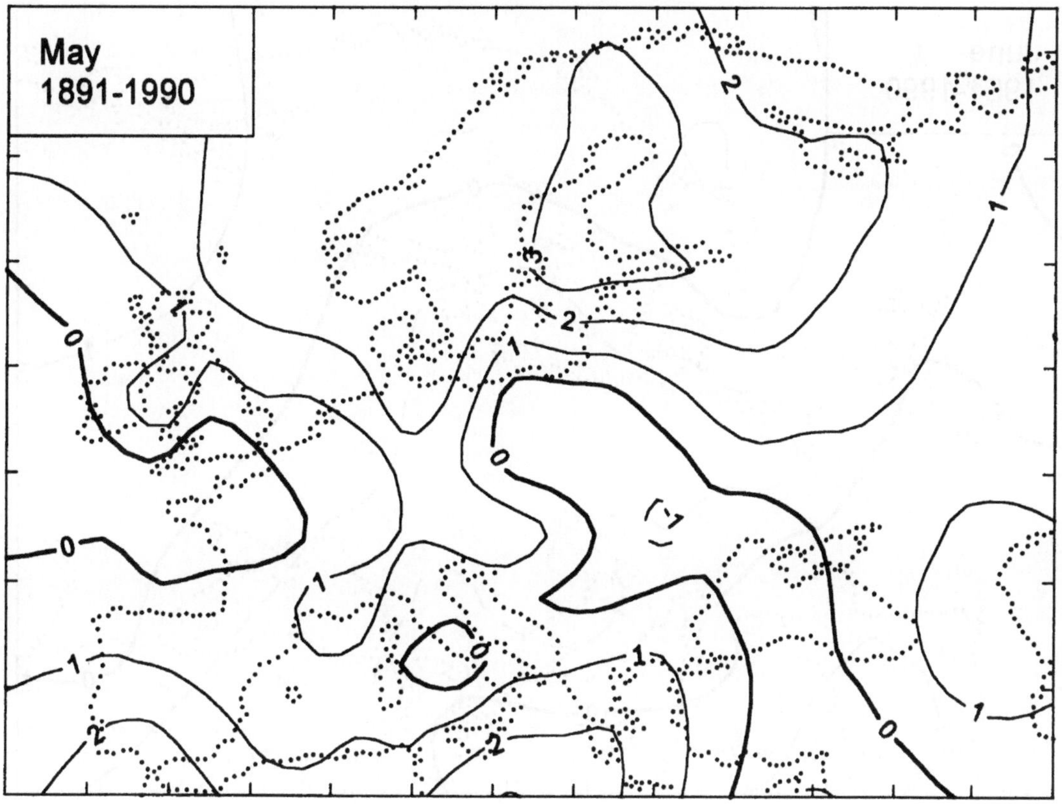

Temperature, linear trend [K]

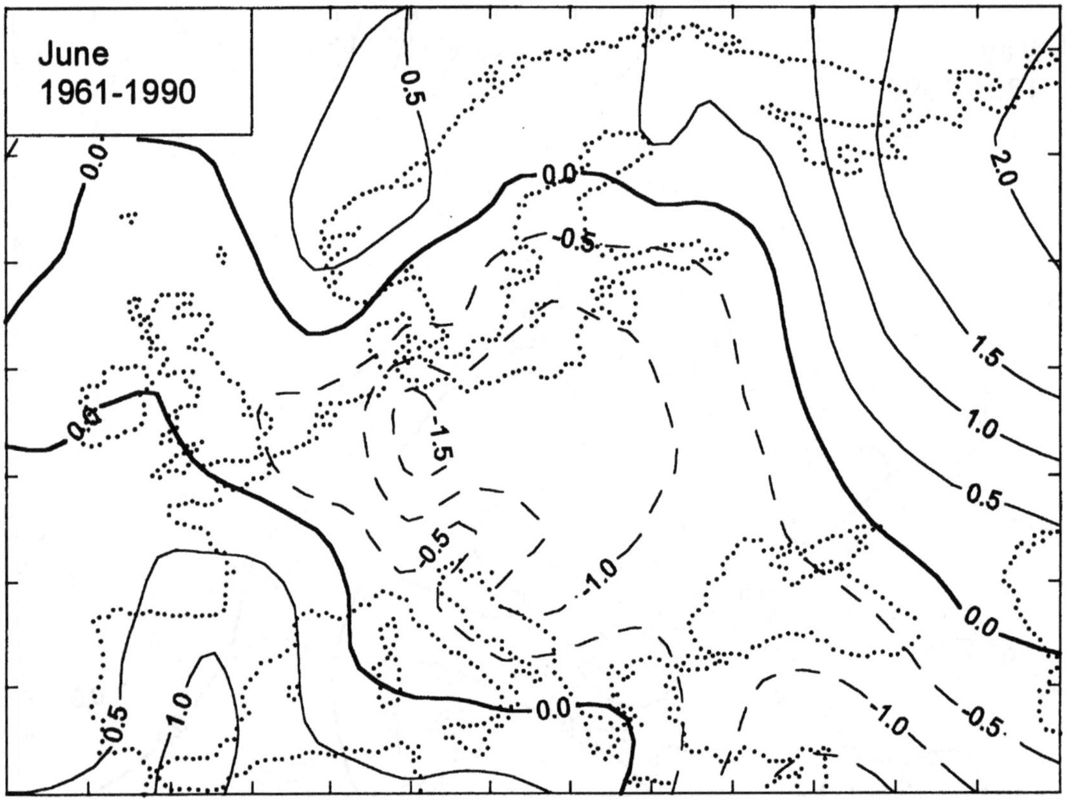

Temperature, trend test

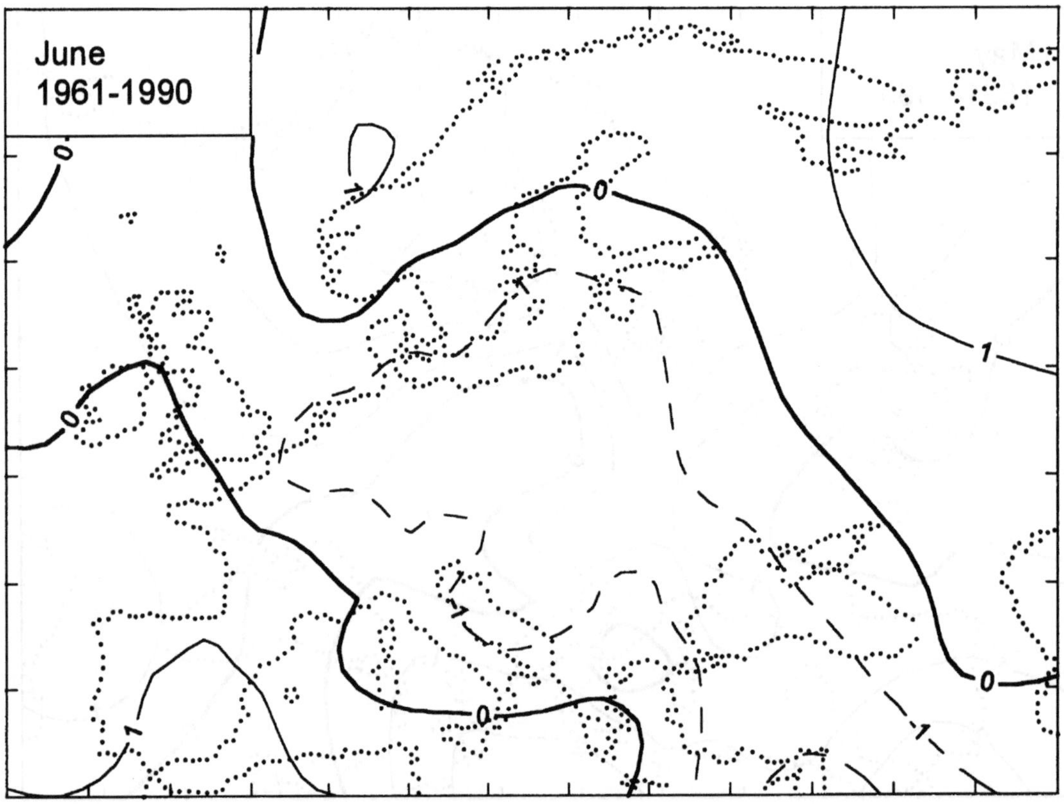

Temperature, linear trend [K]

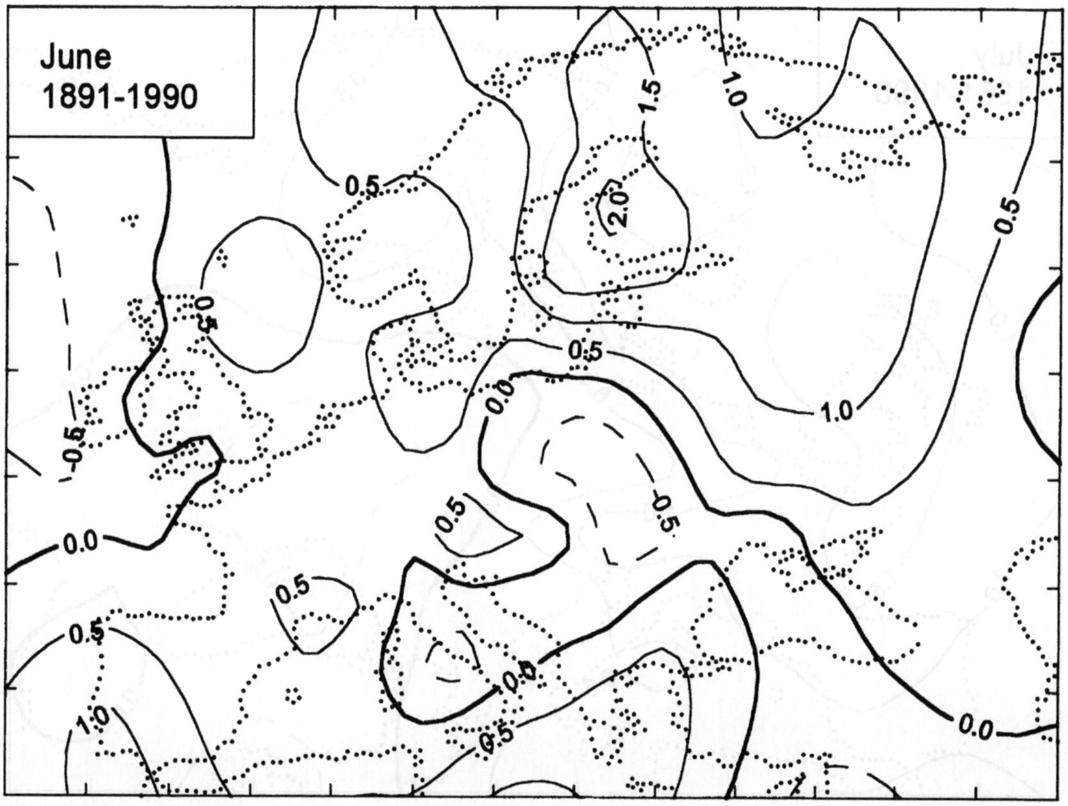

Temperature, trend test

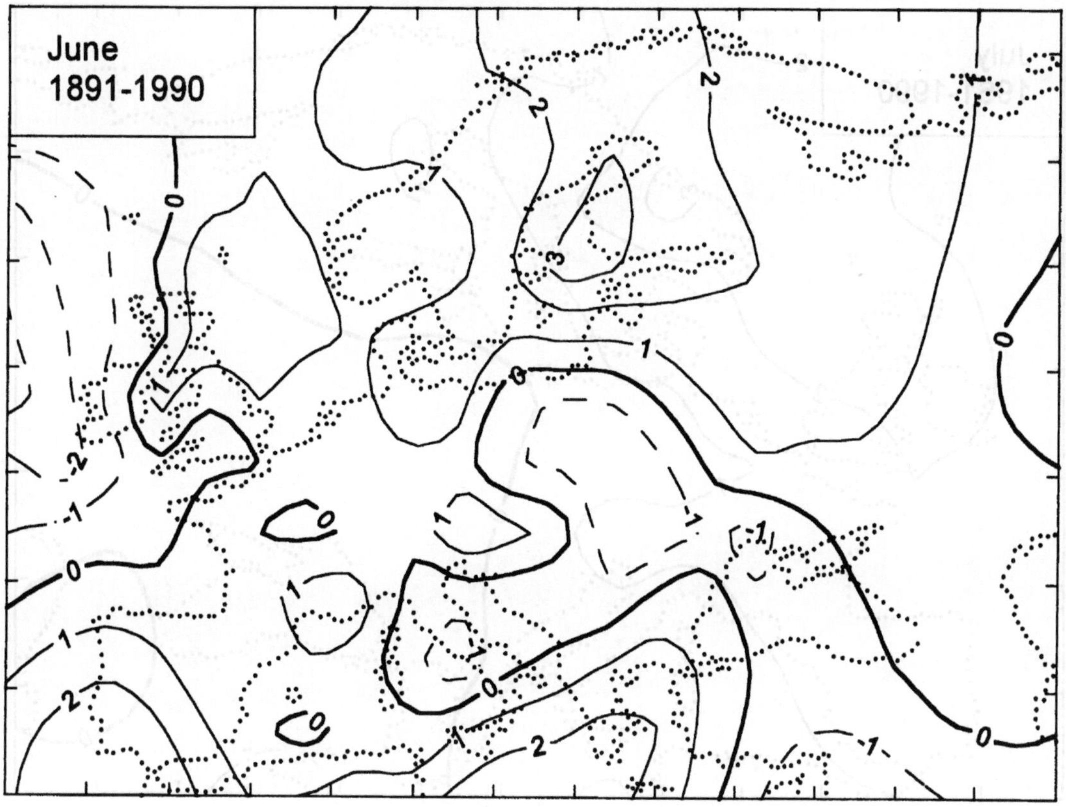

Temperature, linear trend [K]

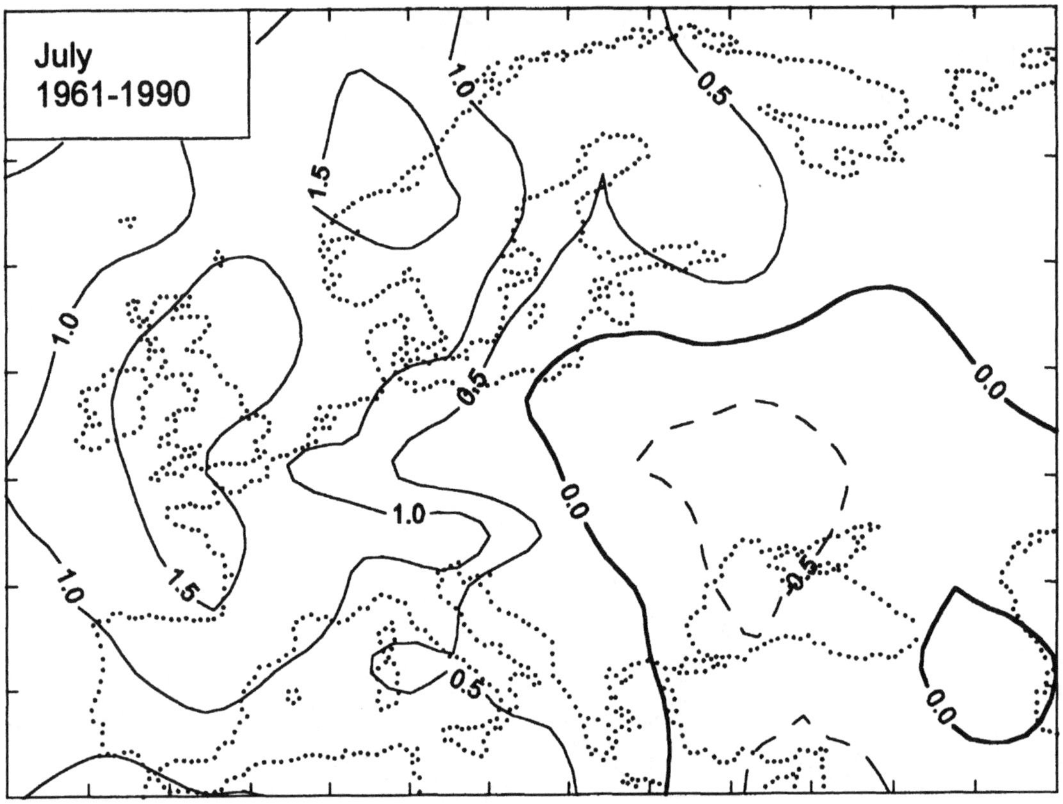

Temperature, trend test

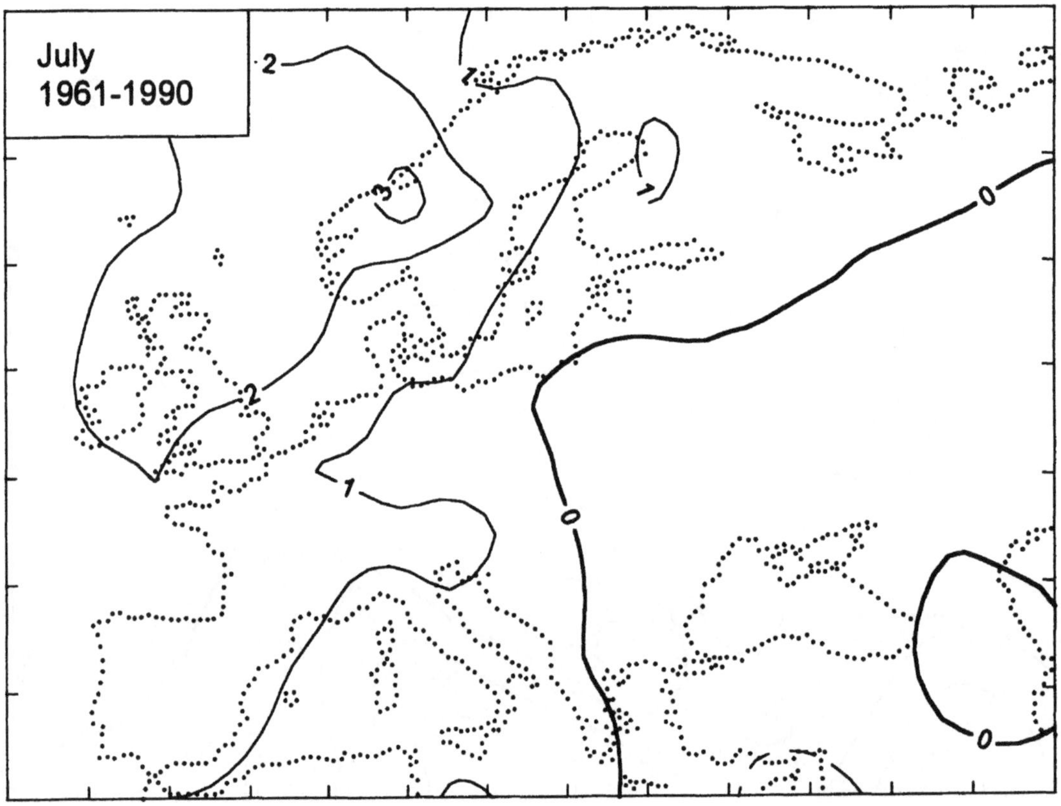

Temperature, linear trend [K]

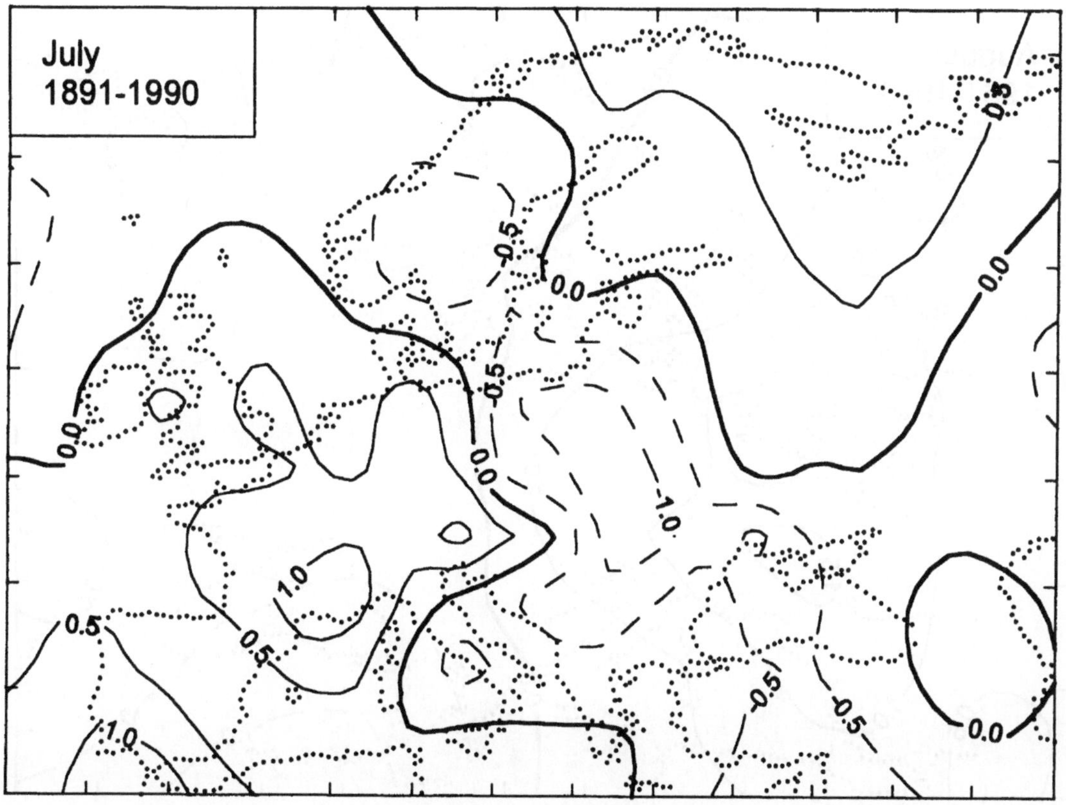

Temperature, trend test

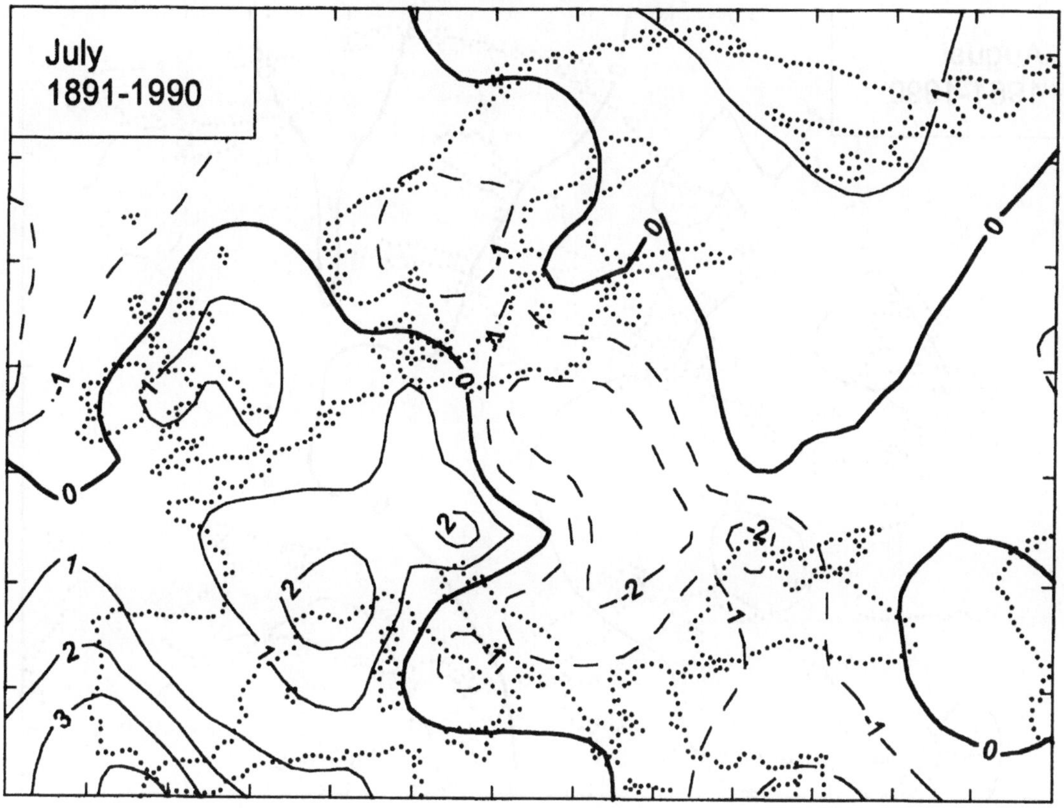

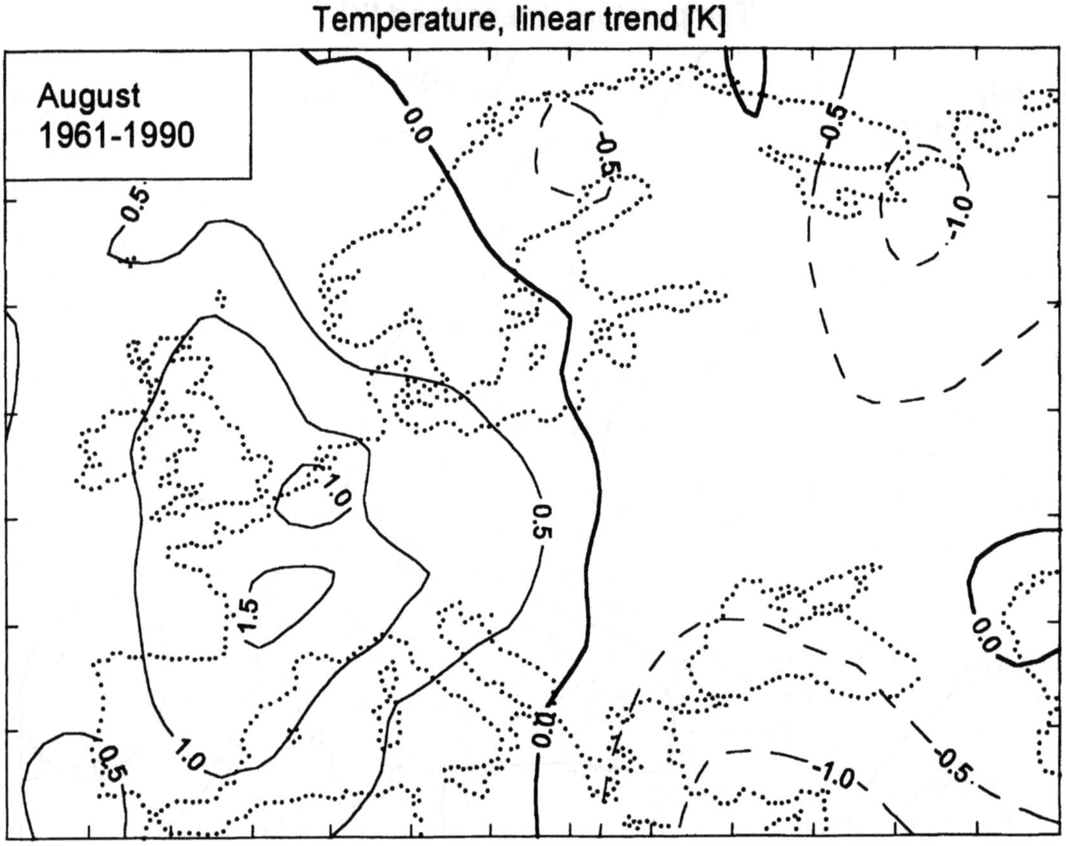
Temperature, linear trend [K]
August
1961-1990
0.0
-0.5
-0.5
-1.0
0.5
1.0
1.5
0.5
0.0
1.0
0.5
0.0
-1.0
-0.5

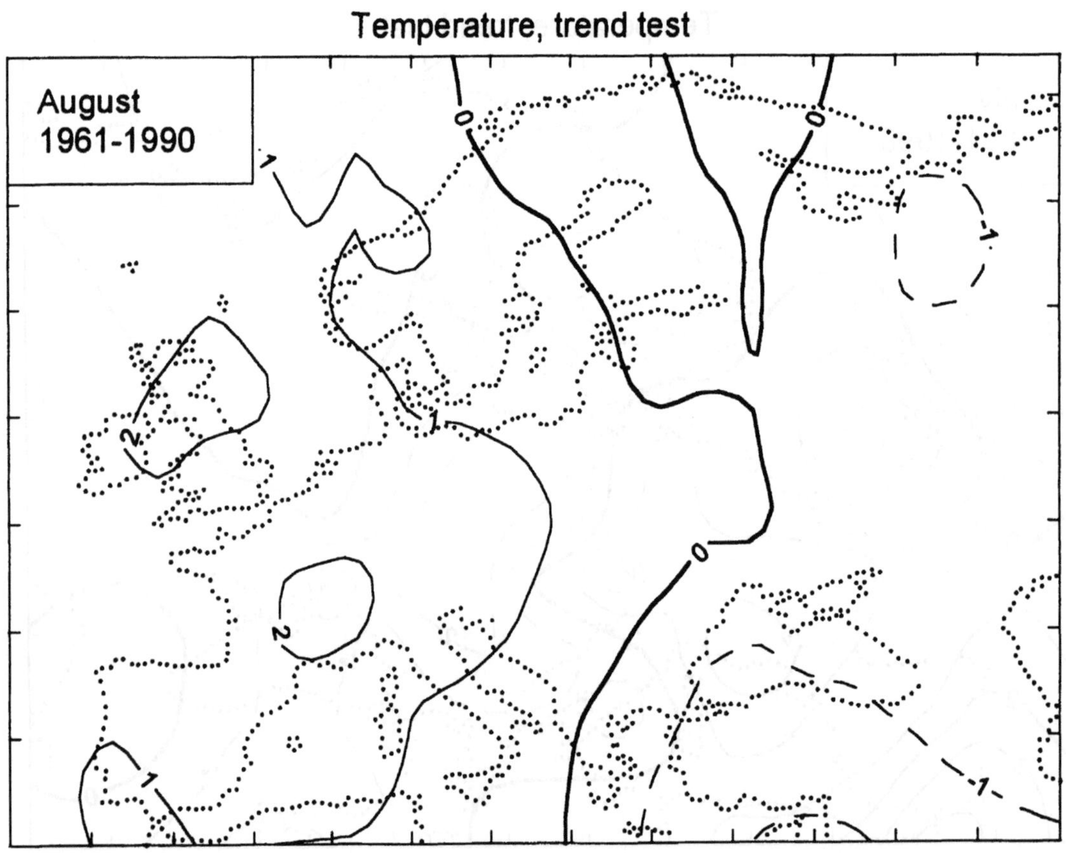
Temperature, trend test
August
1961-1990
0
0
1
-1
2
1
0
2
1
-1

Temperature, linear trend [K]

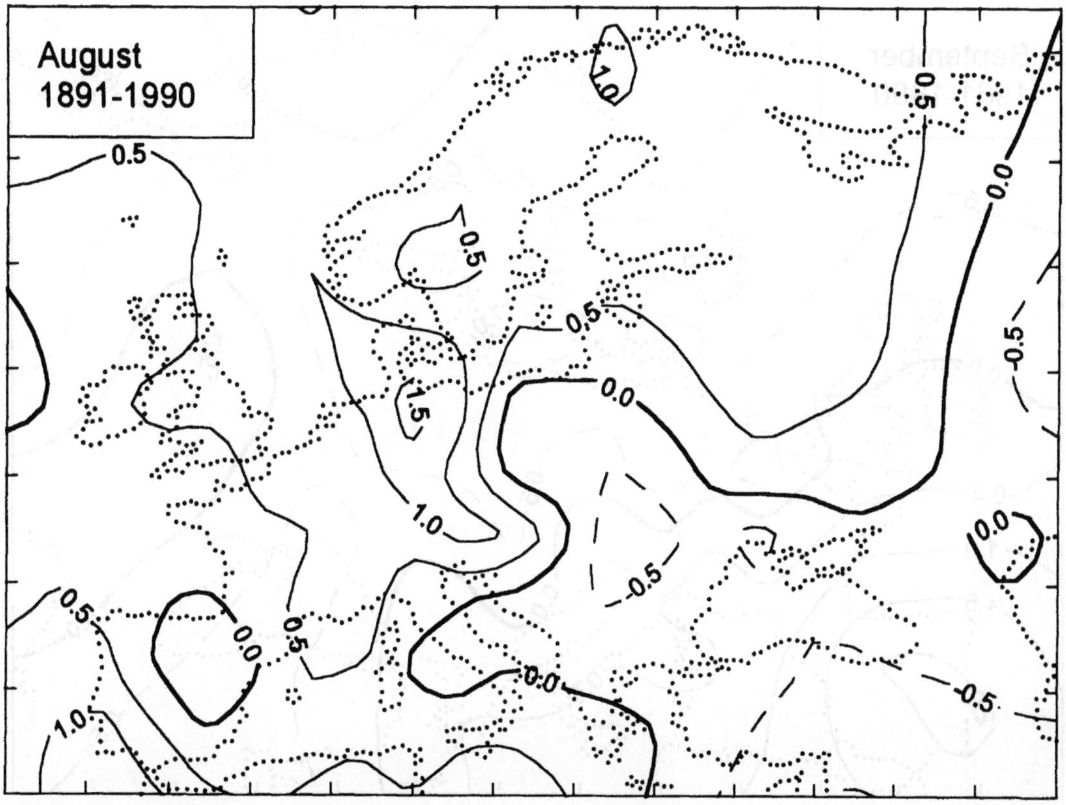

Temperature, trend test

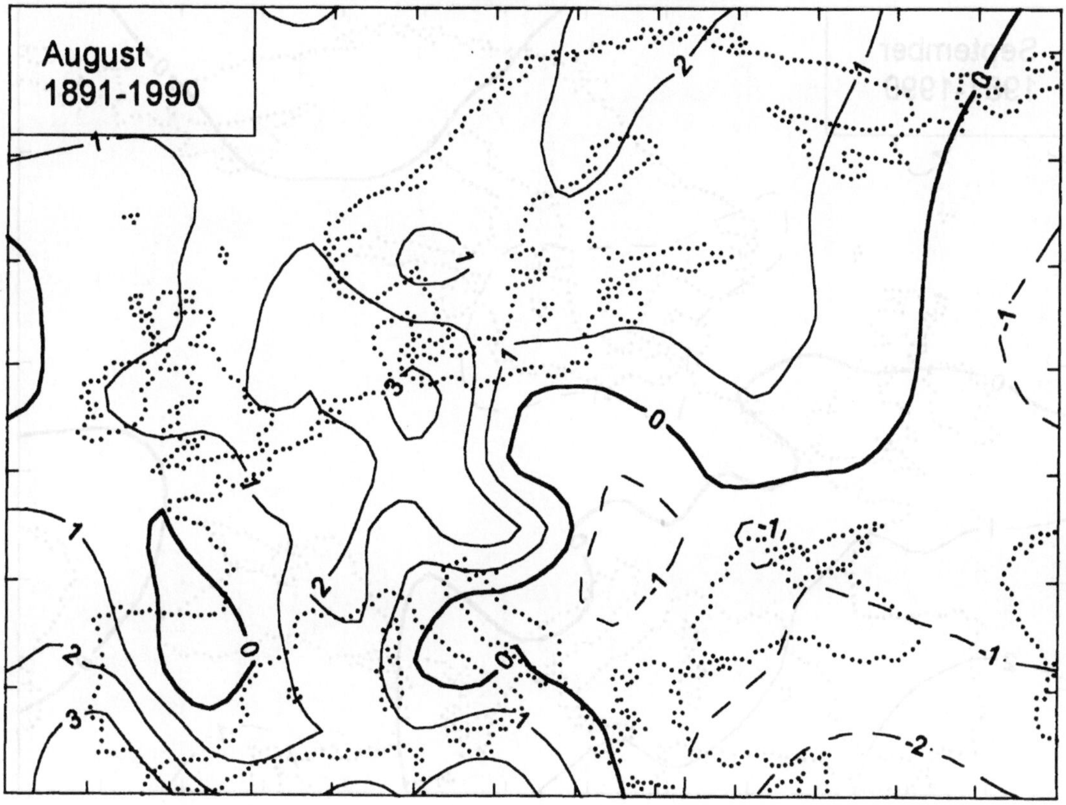

Temperature, linear trend [K]

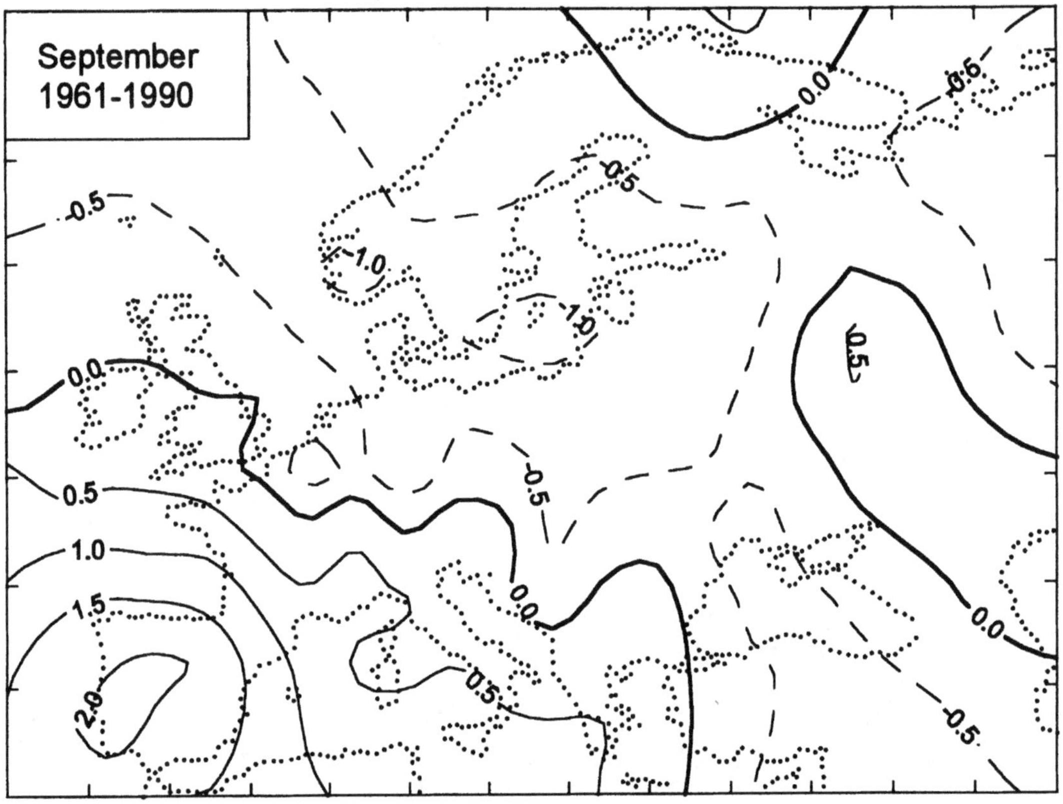

Temperature, trend test

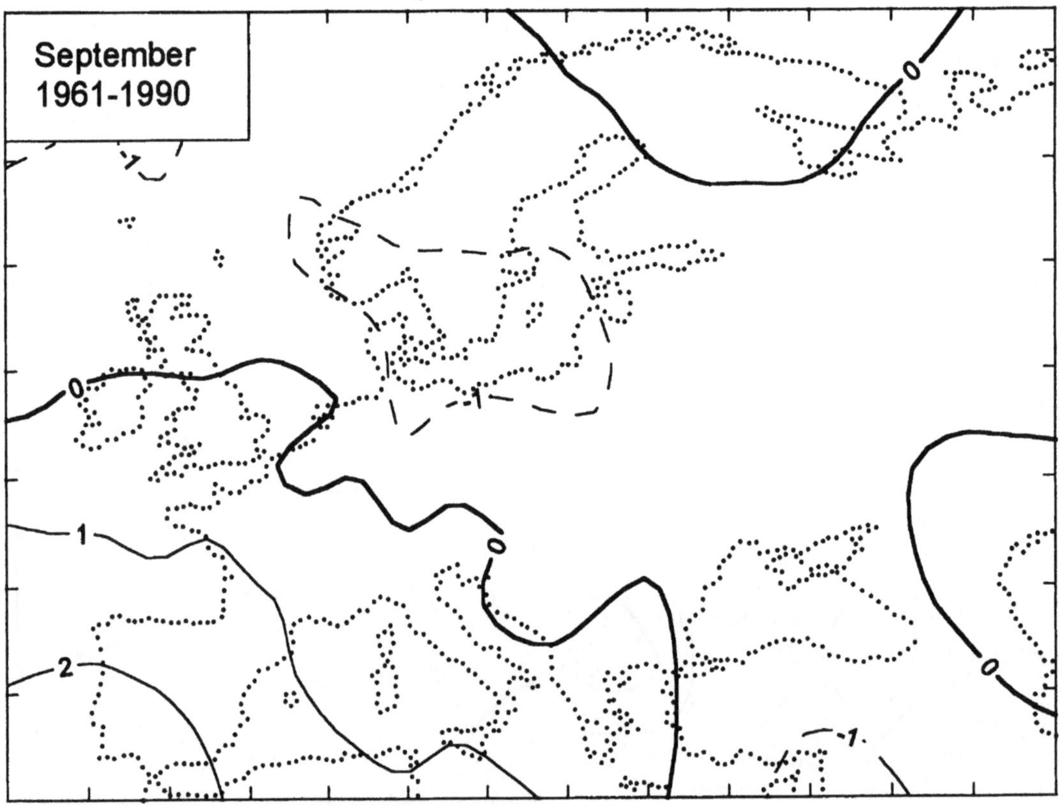

Temperature, linear trend [K]

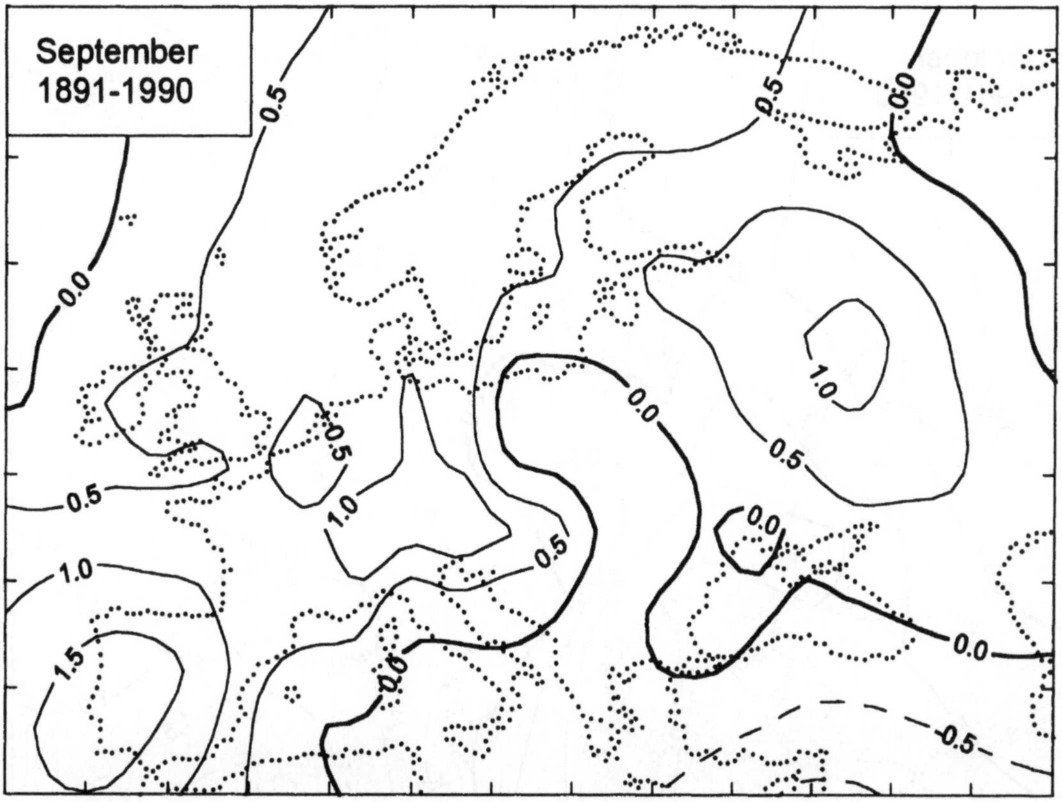

Temperature, trend test

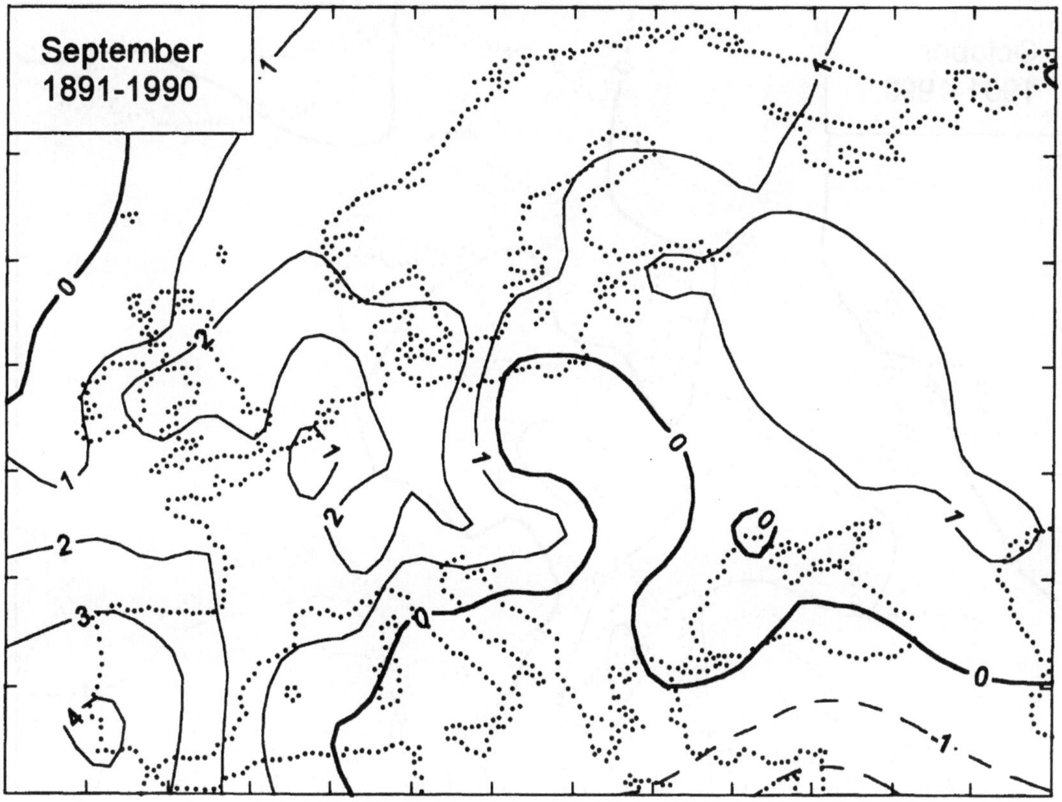

Temperature, linear trend [K]

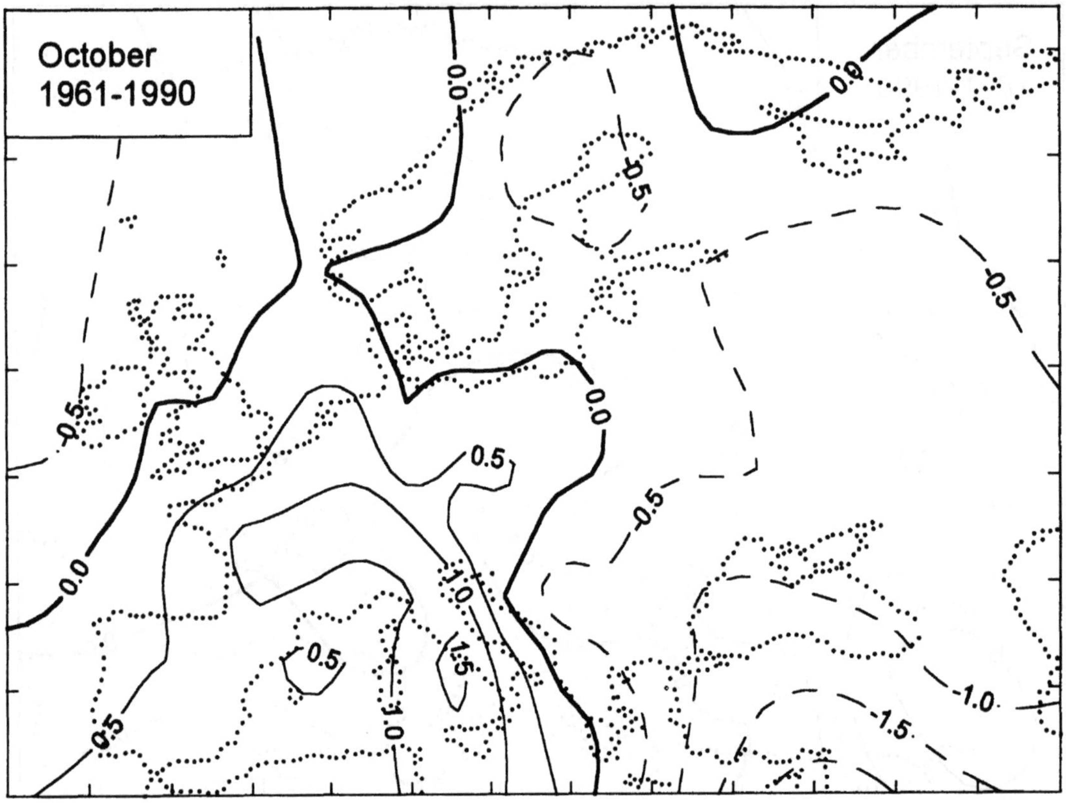

Temperature, trend test

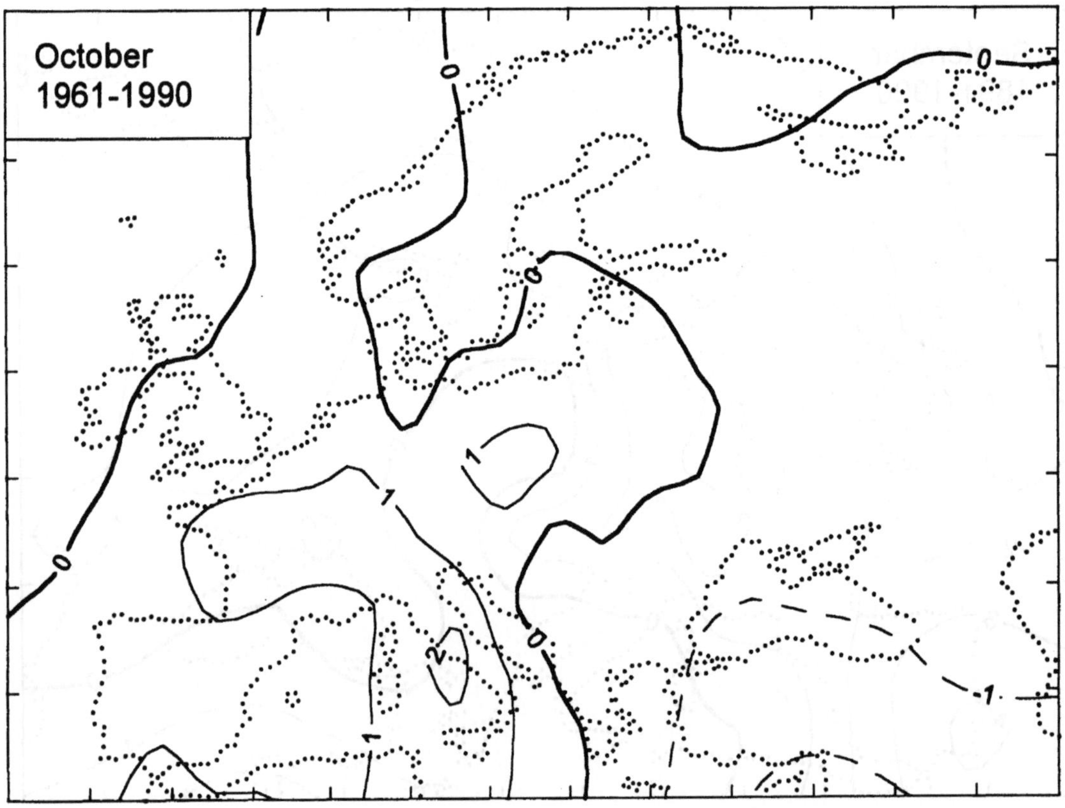

Temperature, linear trend [K]

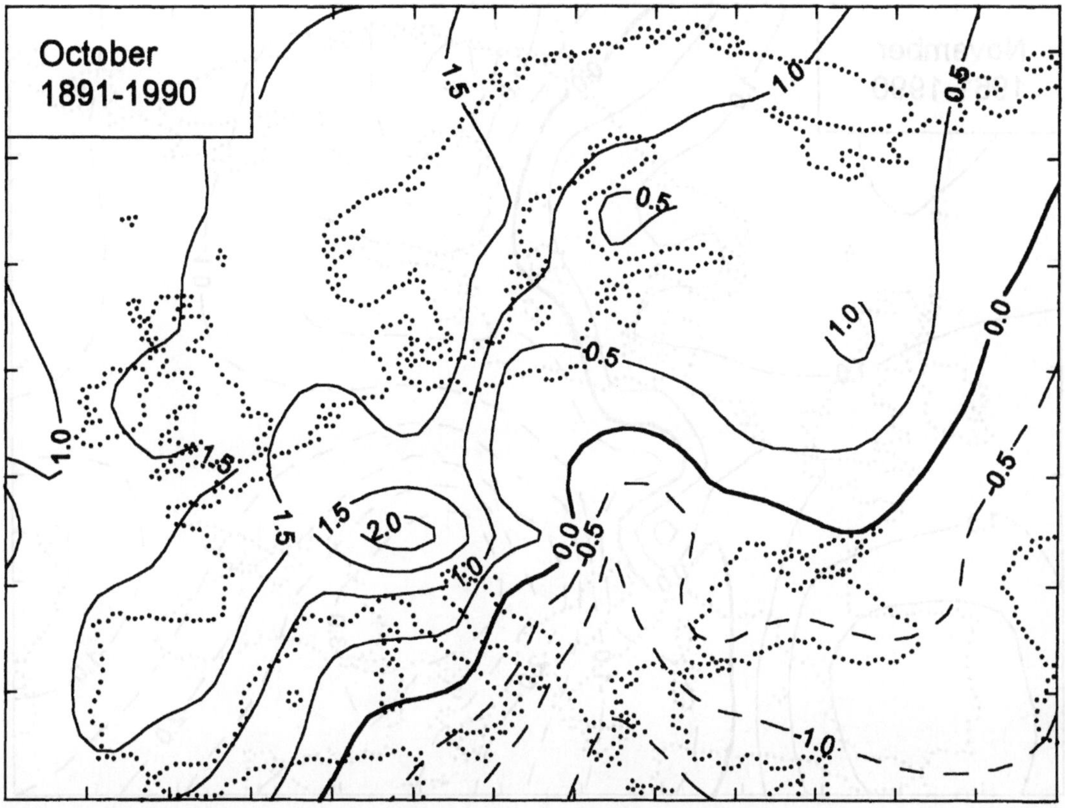

Temperature, trend test

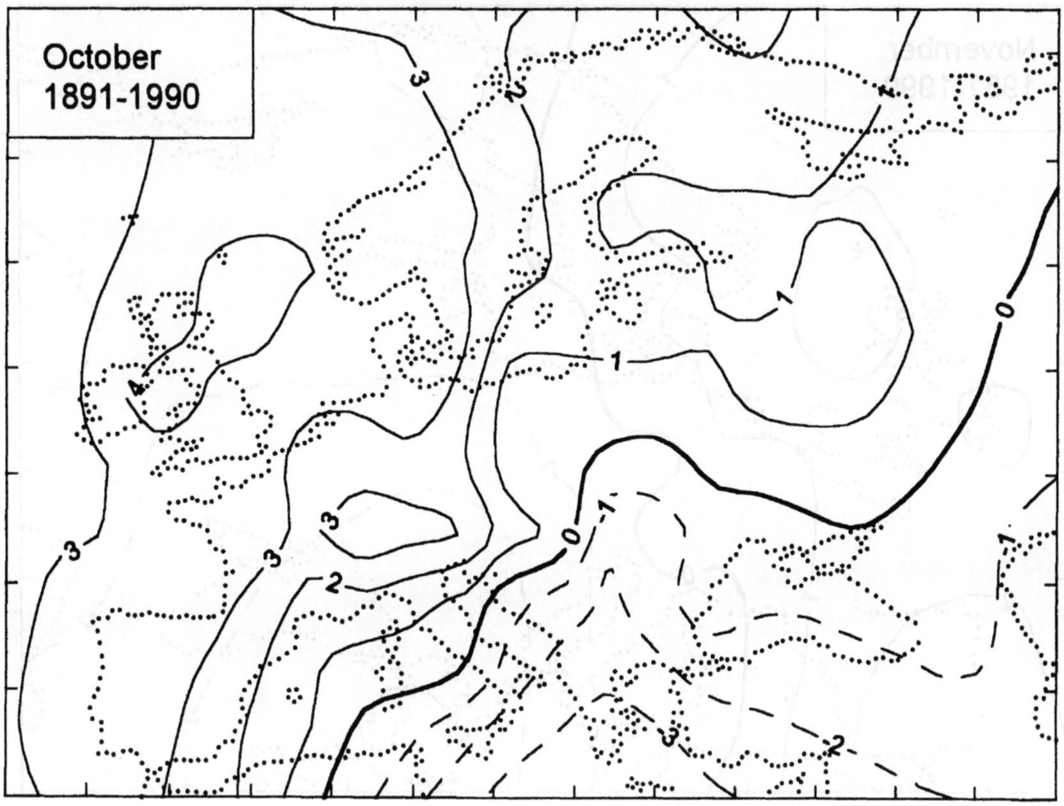

Temperature, linear trend [K]

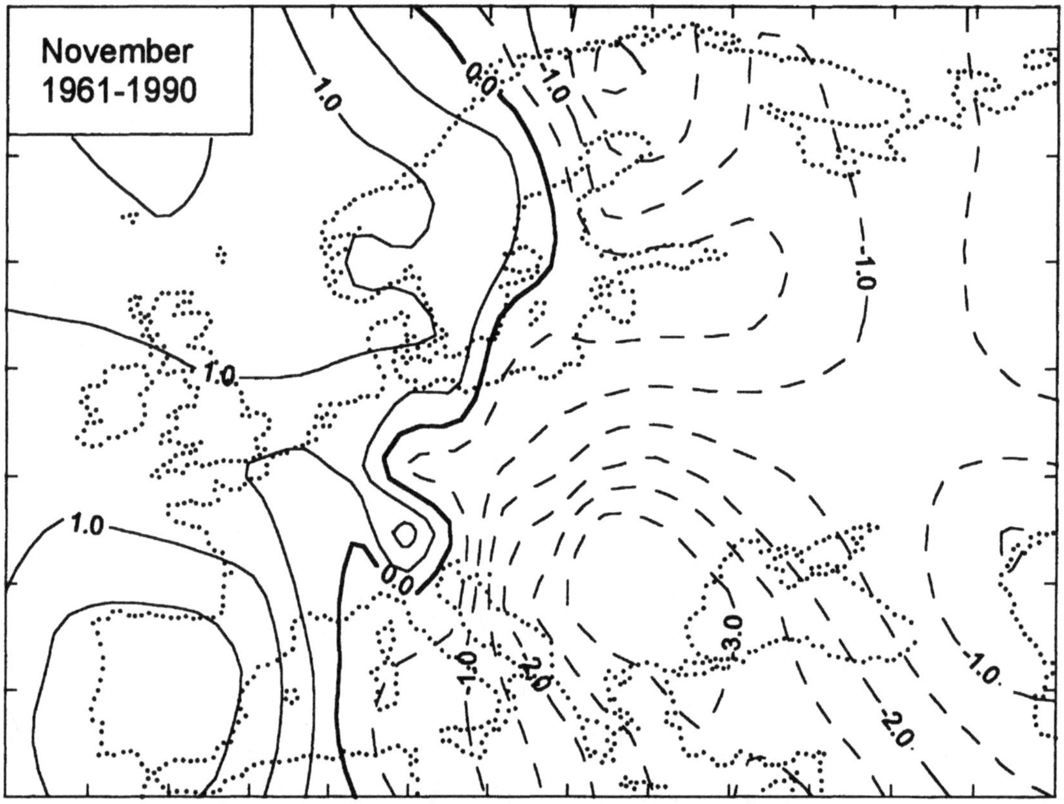

Temperature, trend test

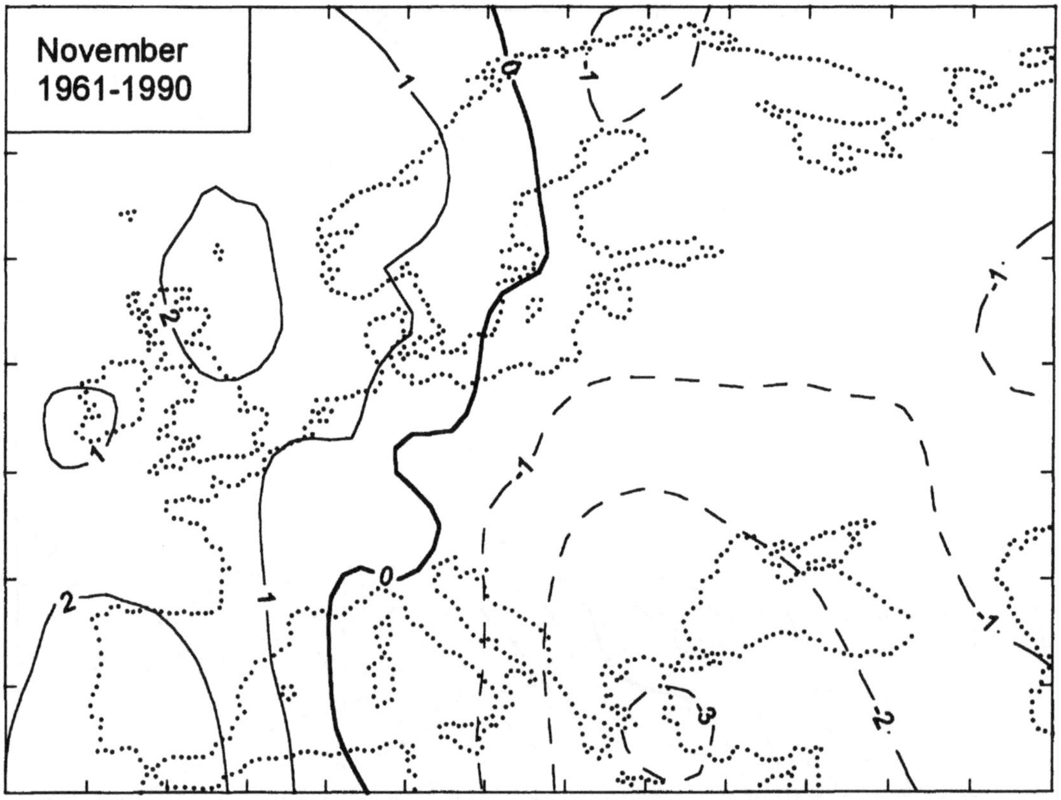

Temperature, linear trend [K]

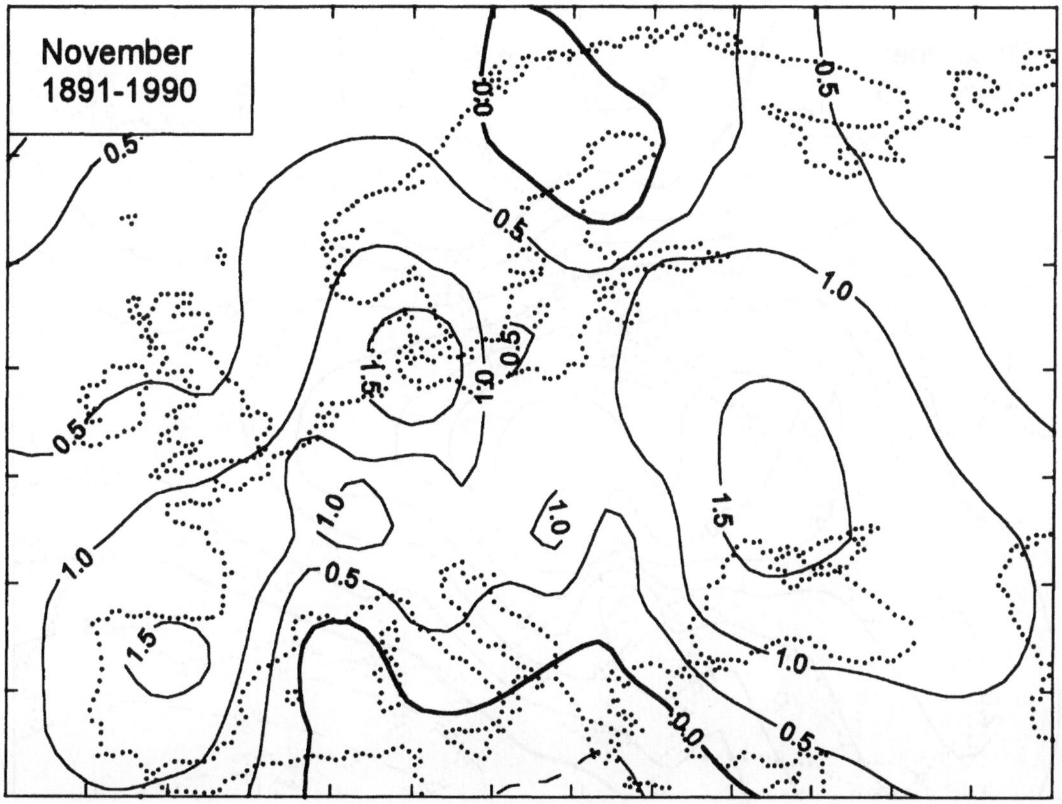

Temperature, trend test

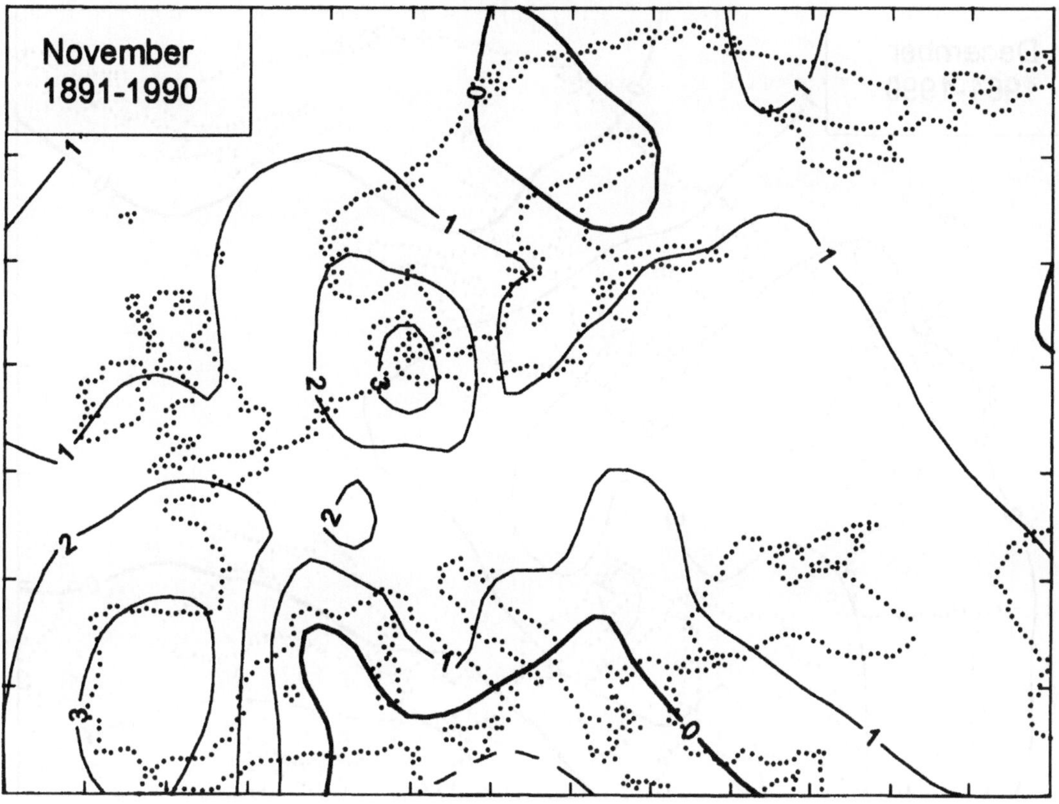

Temperature, linear trend [K]

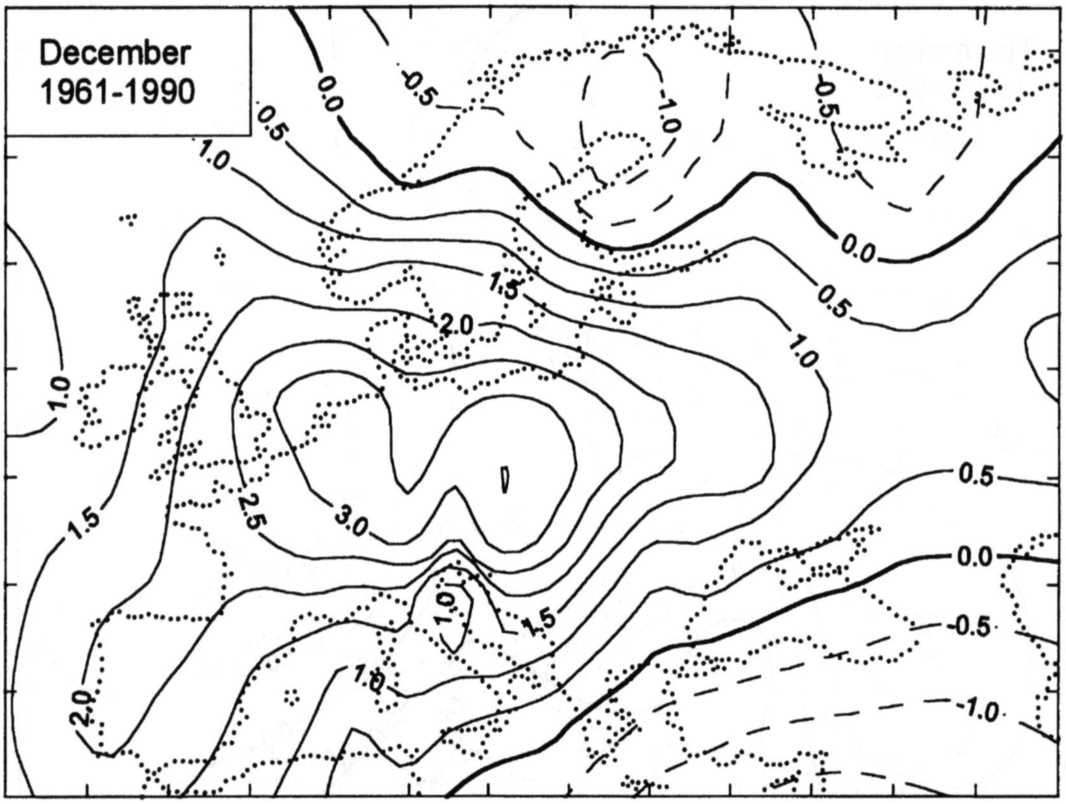

Temperature, trend test

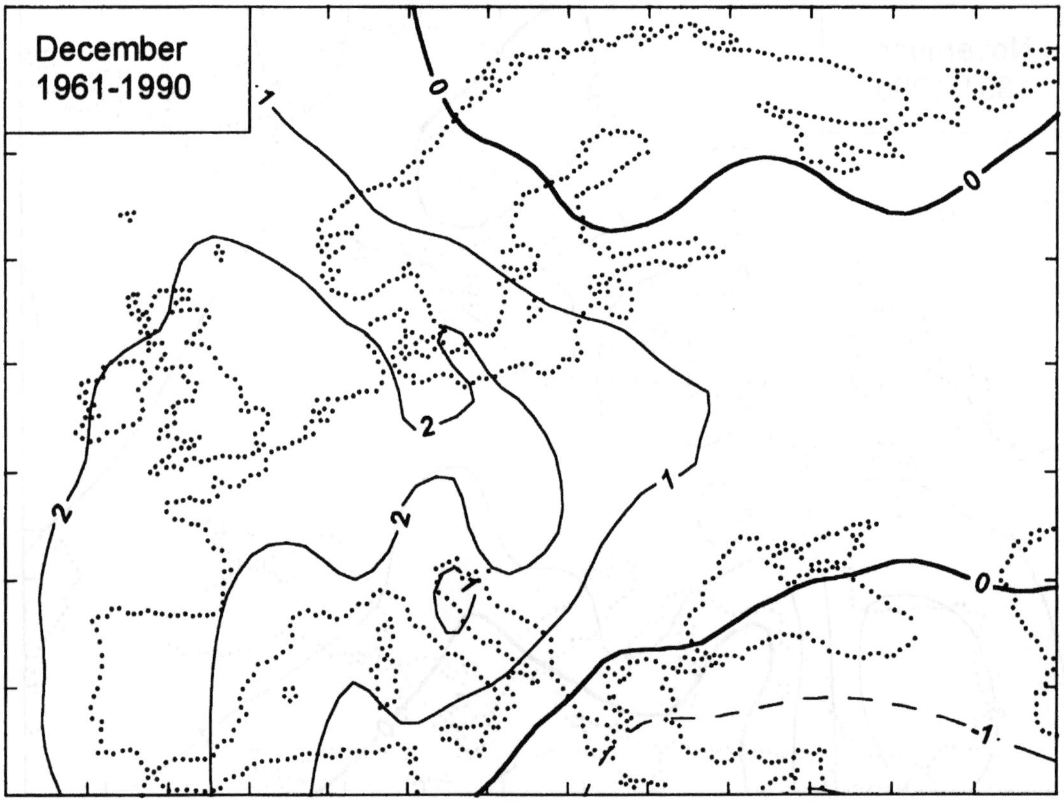

Temperature, linear trend [K]

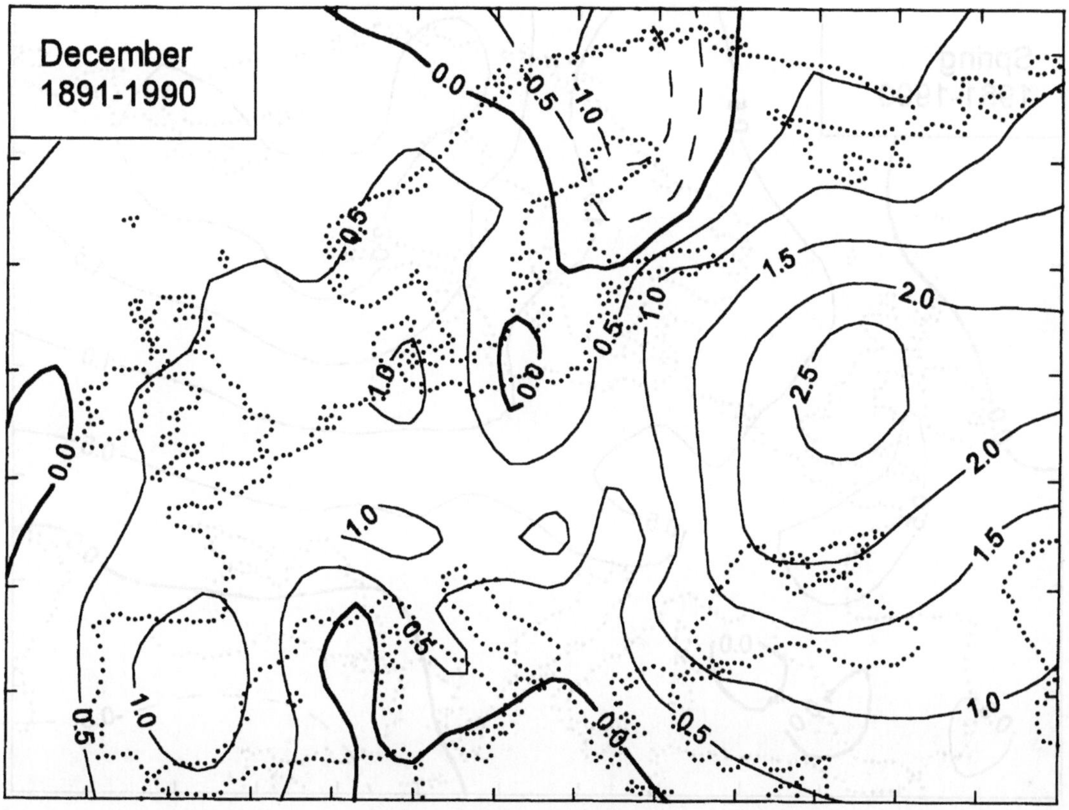

Temperature, trend test

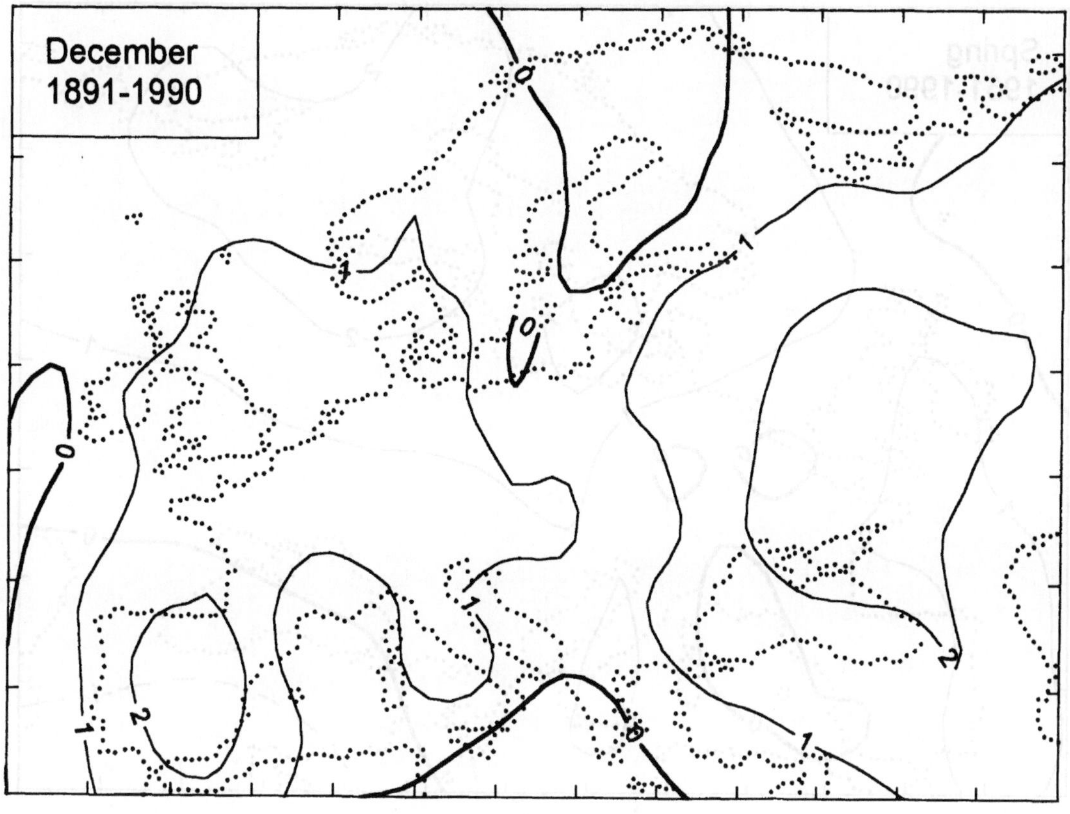

Temperature, linear trend [K]

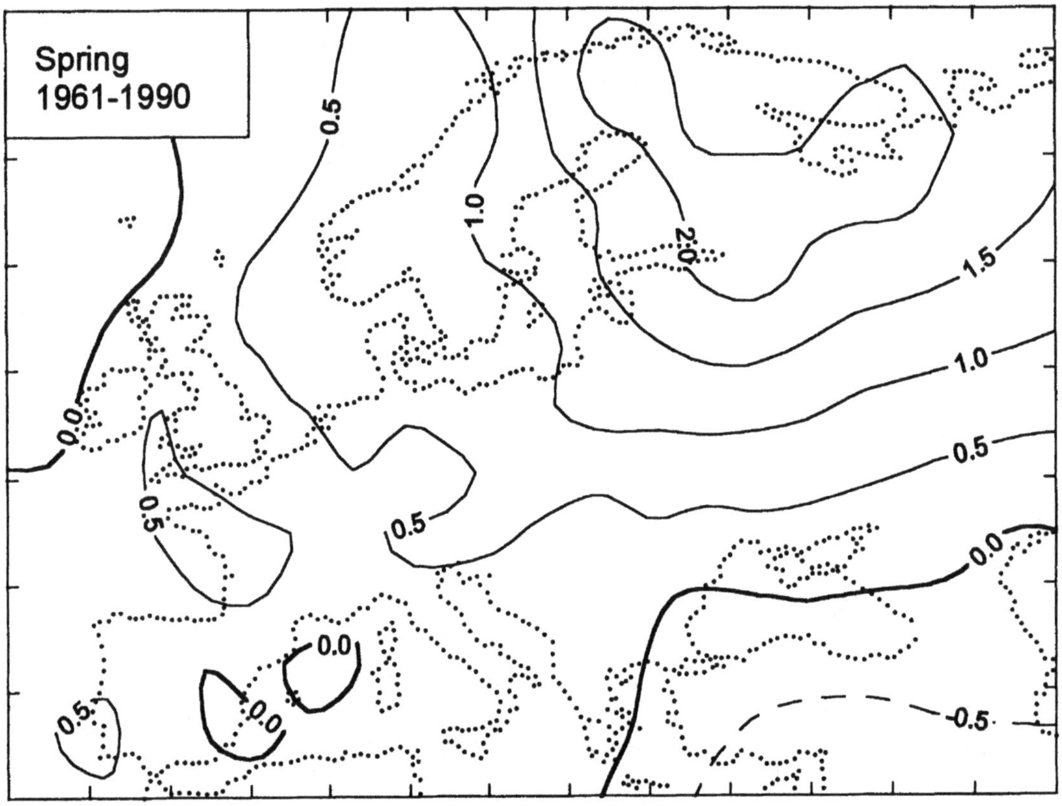

Temperature, trend test

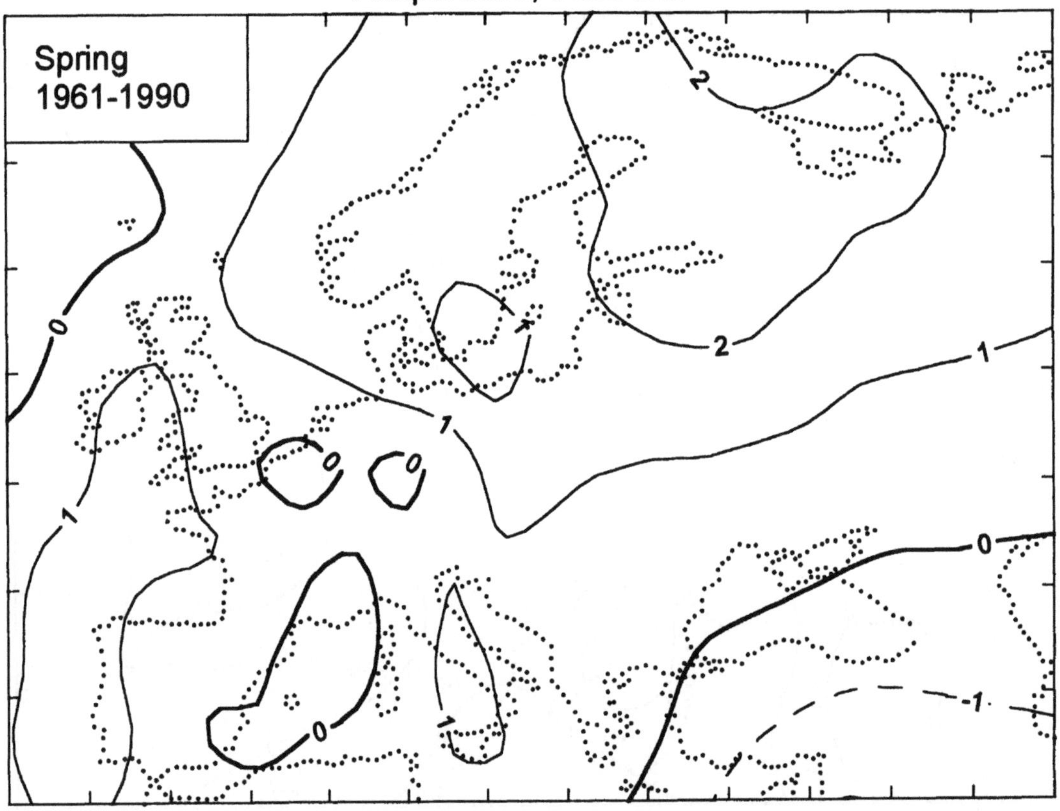

Temperature, linear trend [K]

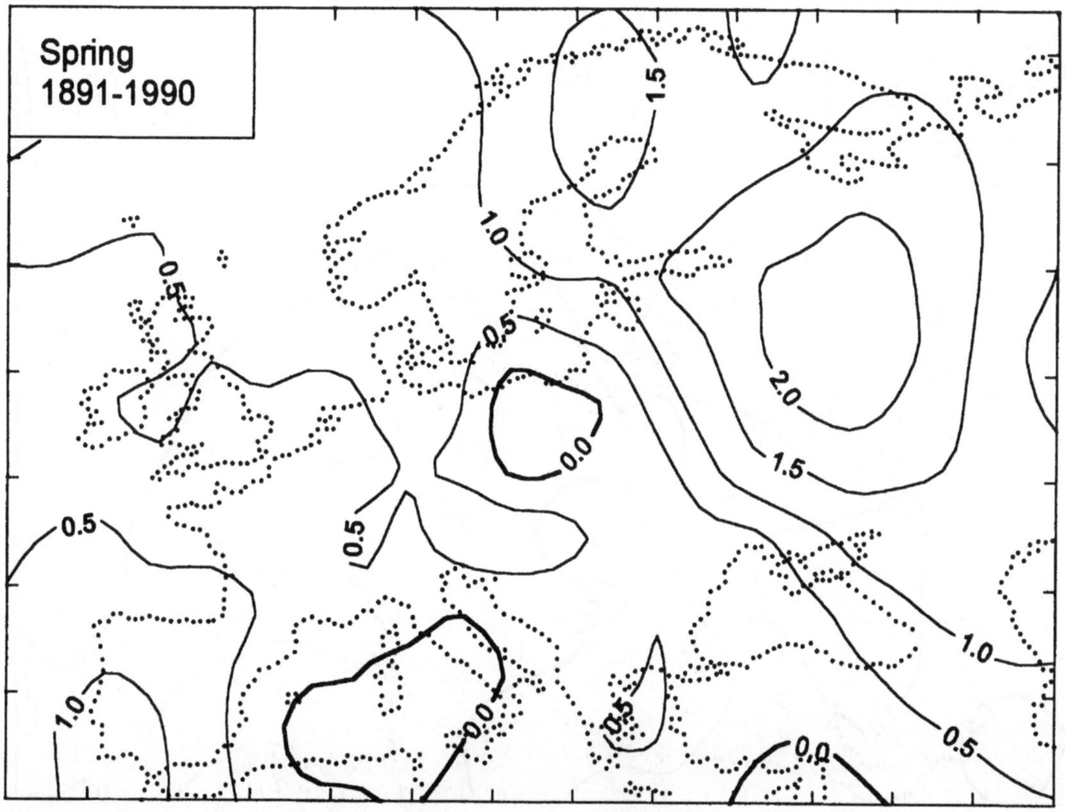

Temperature, trend test

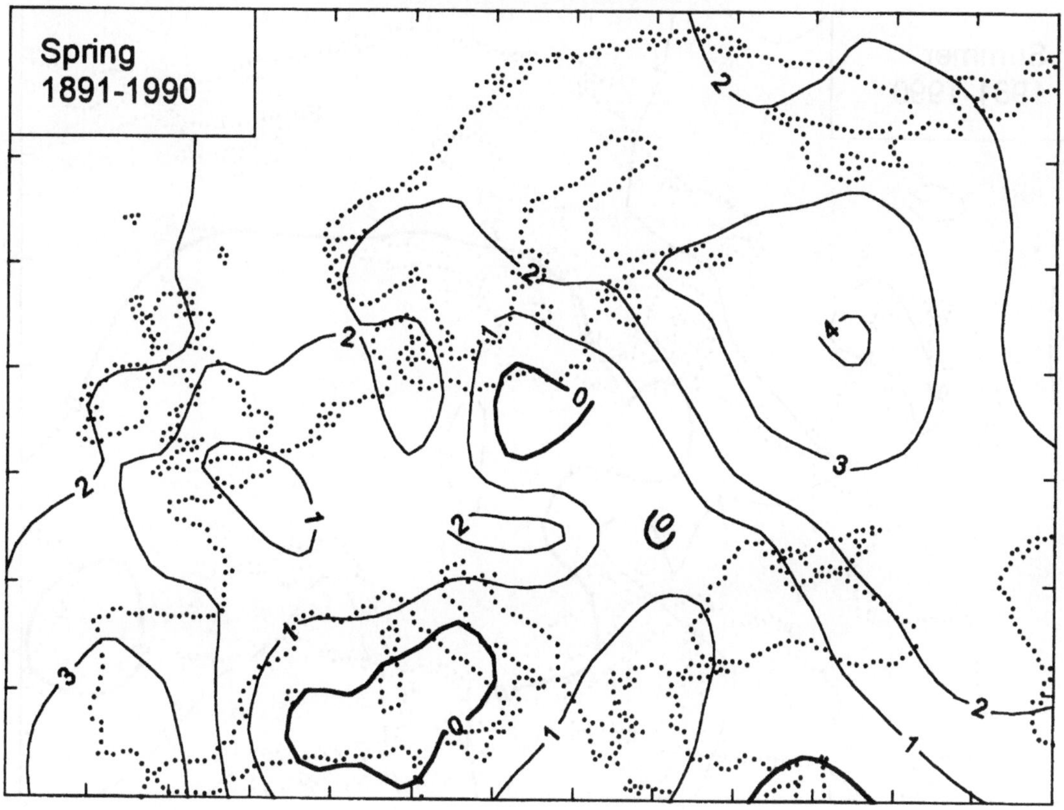

Temperature, linear trend [K]

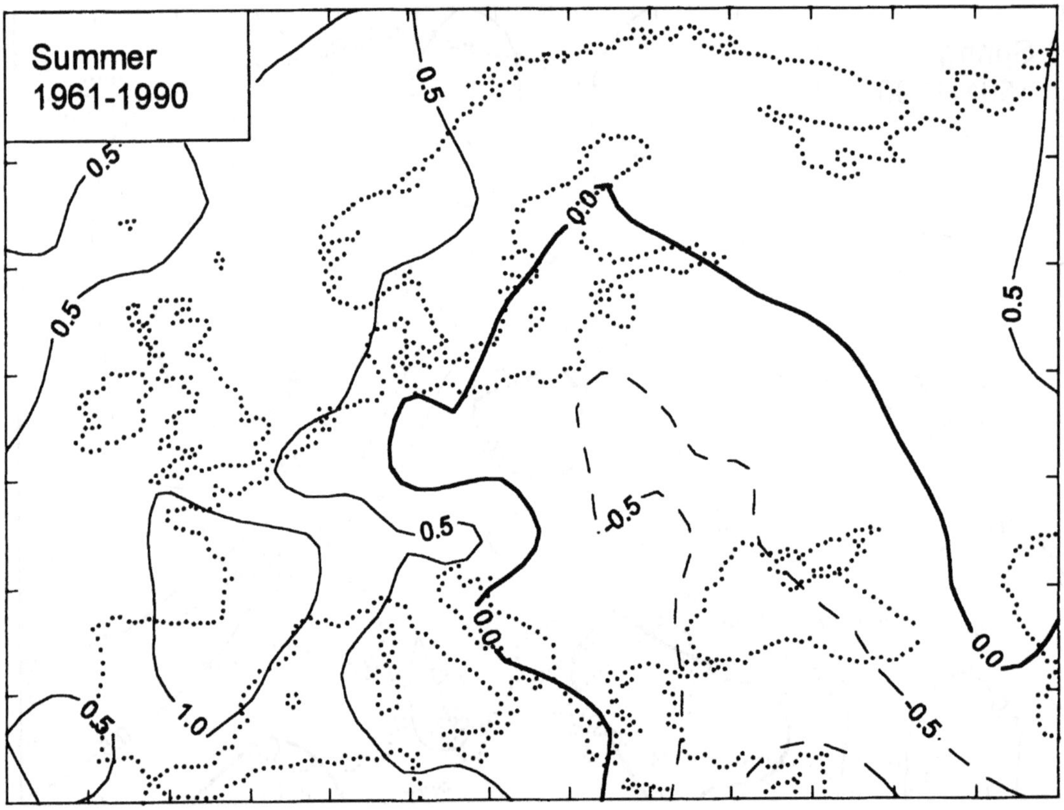

Temperature, trend test

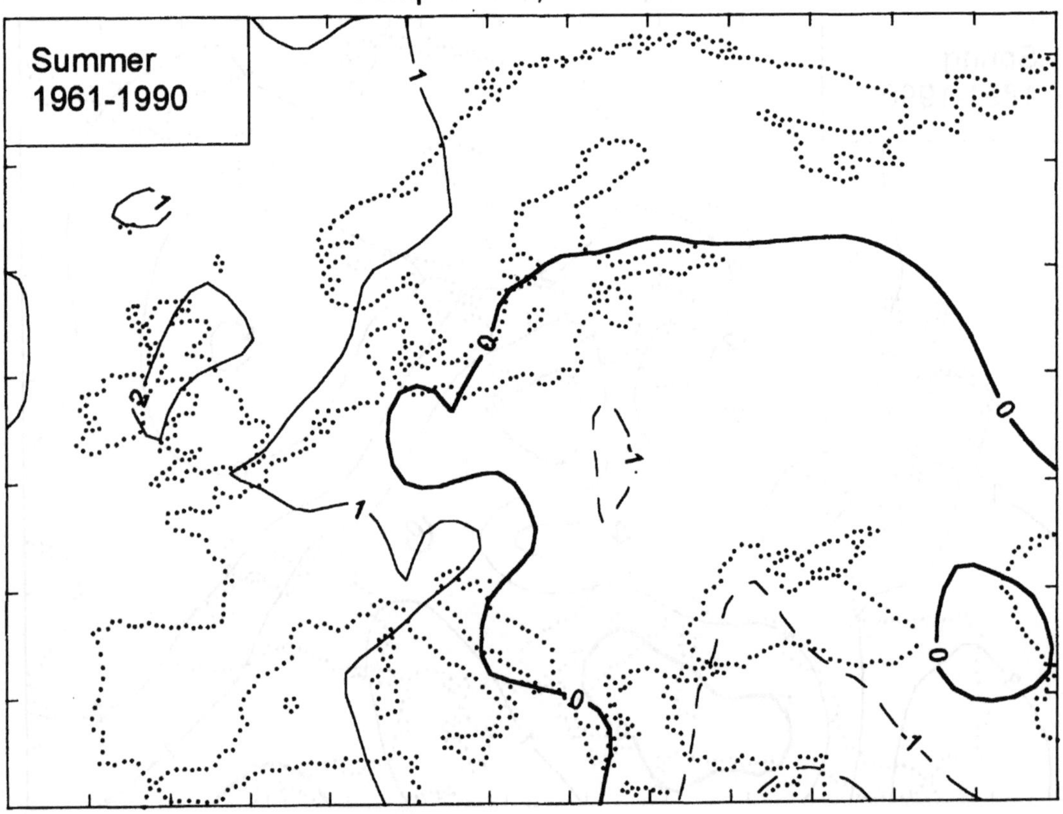

Temperature, linear trend [K]

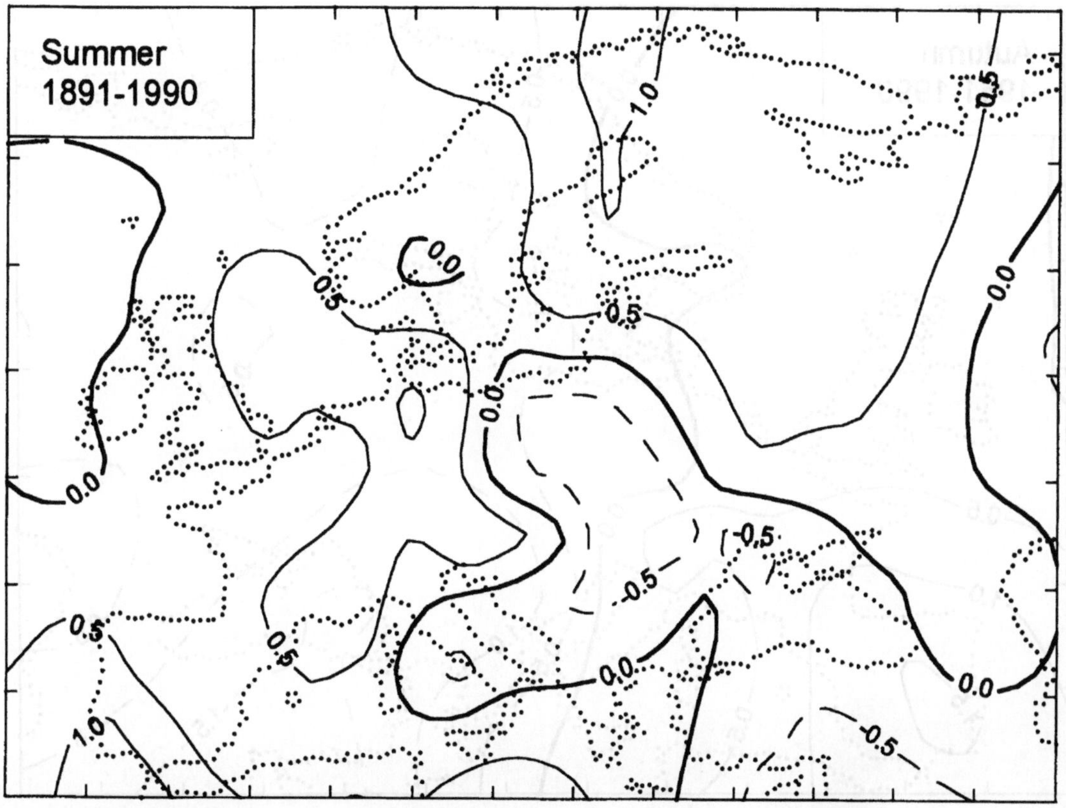

Temperature, trend test

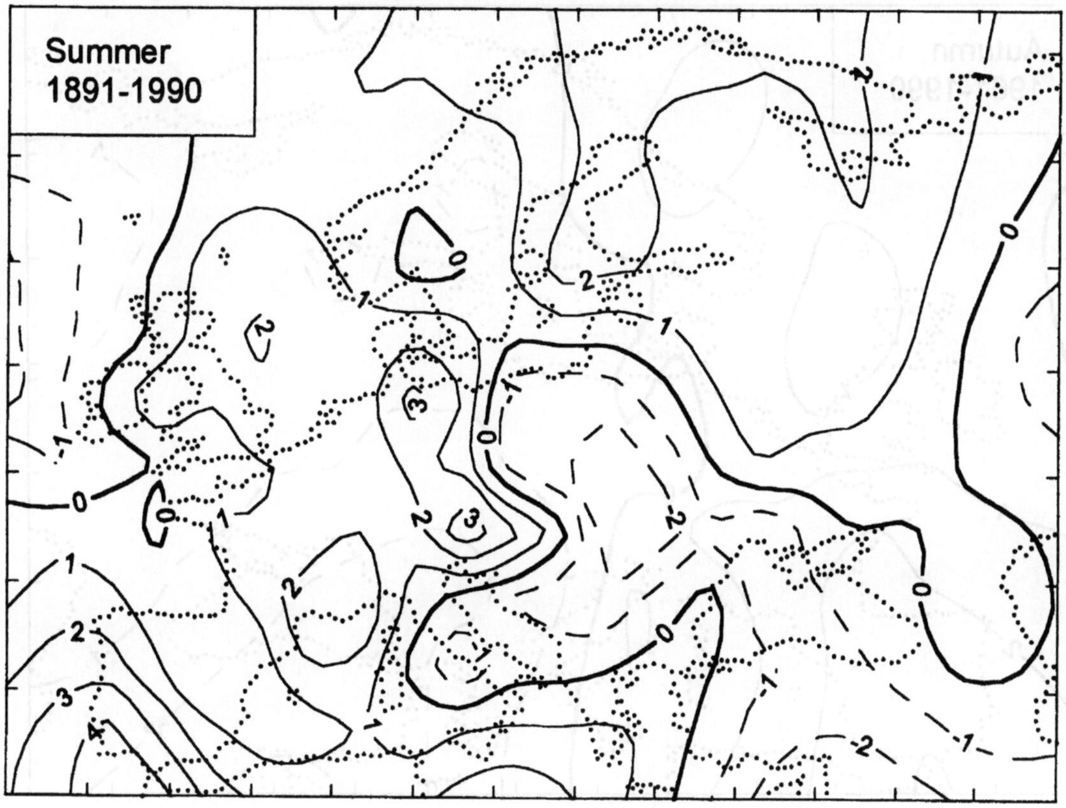

Temperature, linear trend [K]

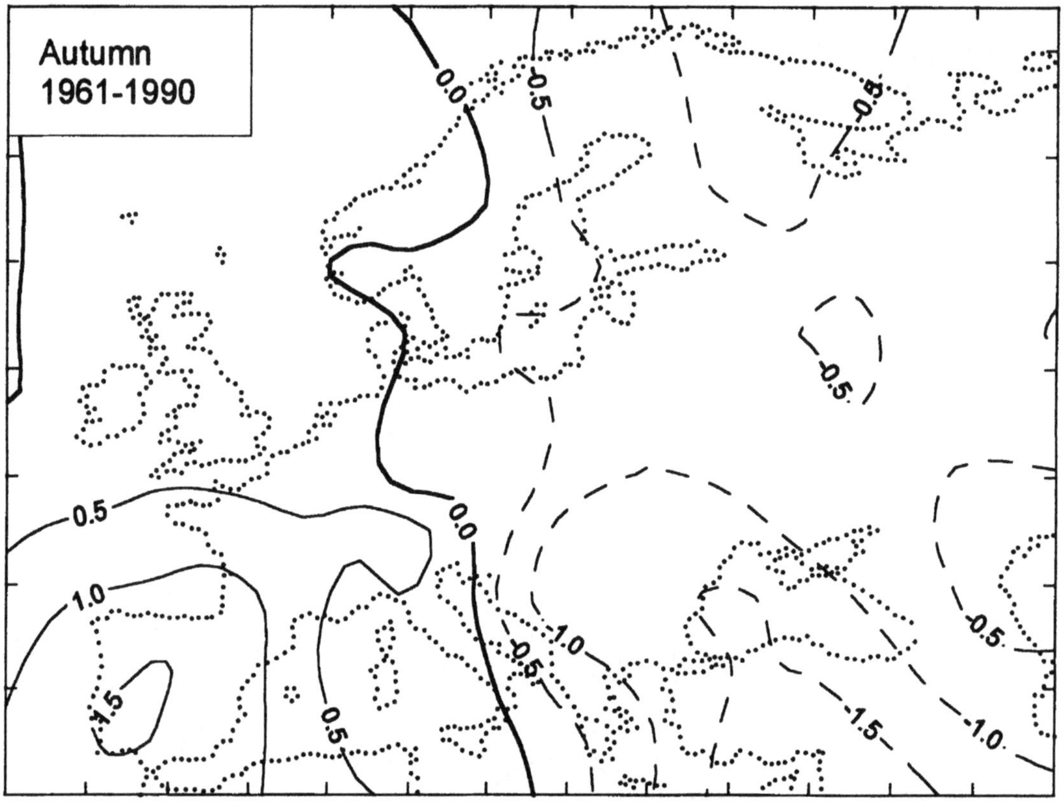

Temperature, trend test

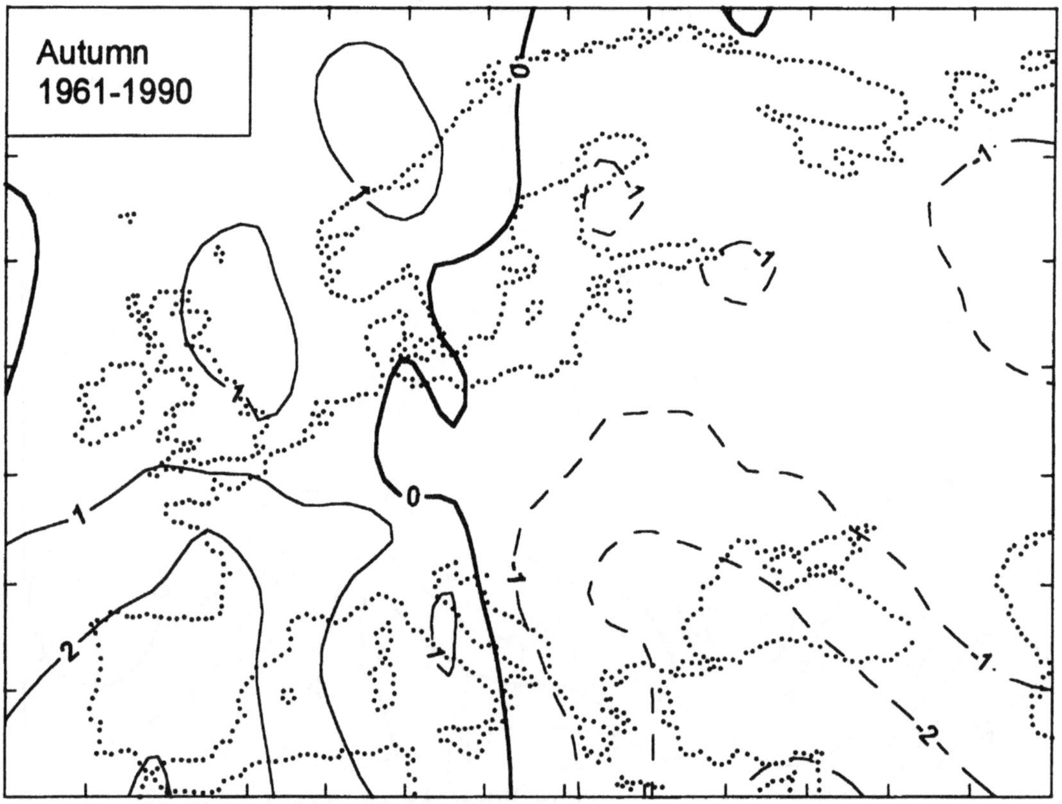

Temperature, linear trend [K]

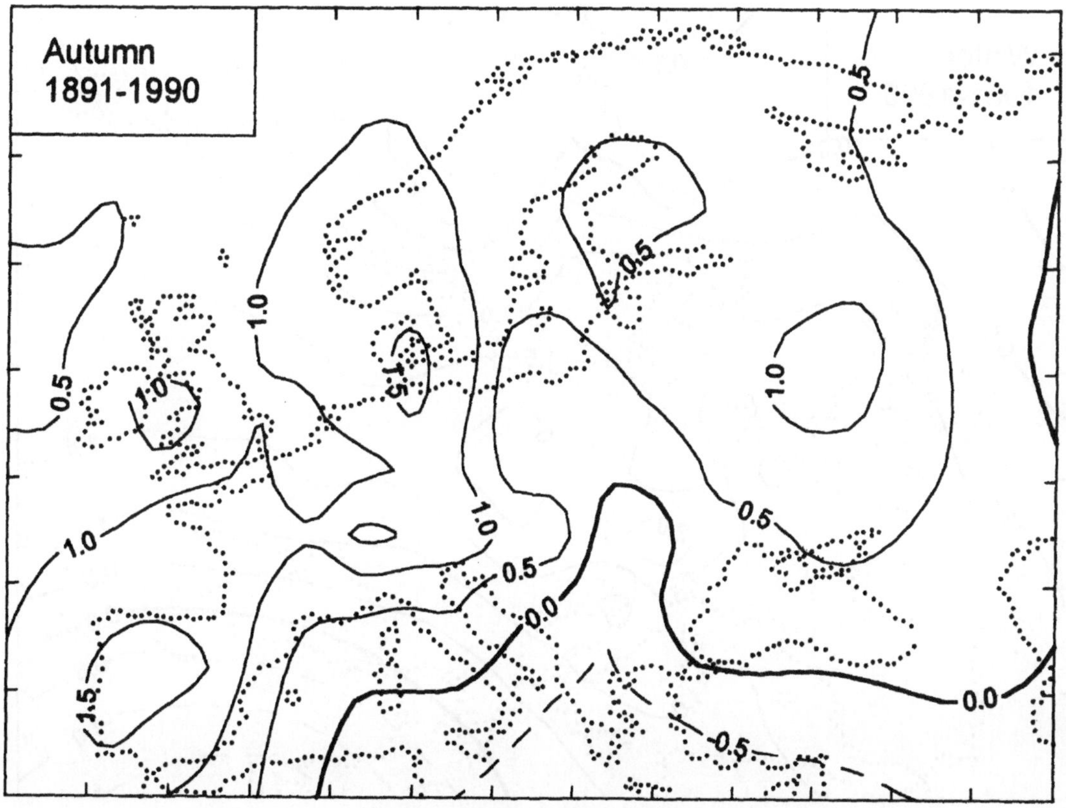

Temperature, trend test

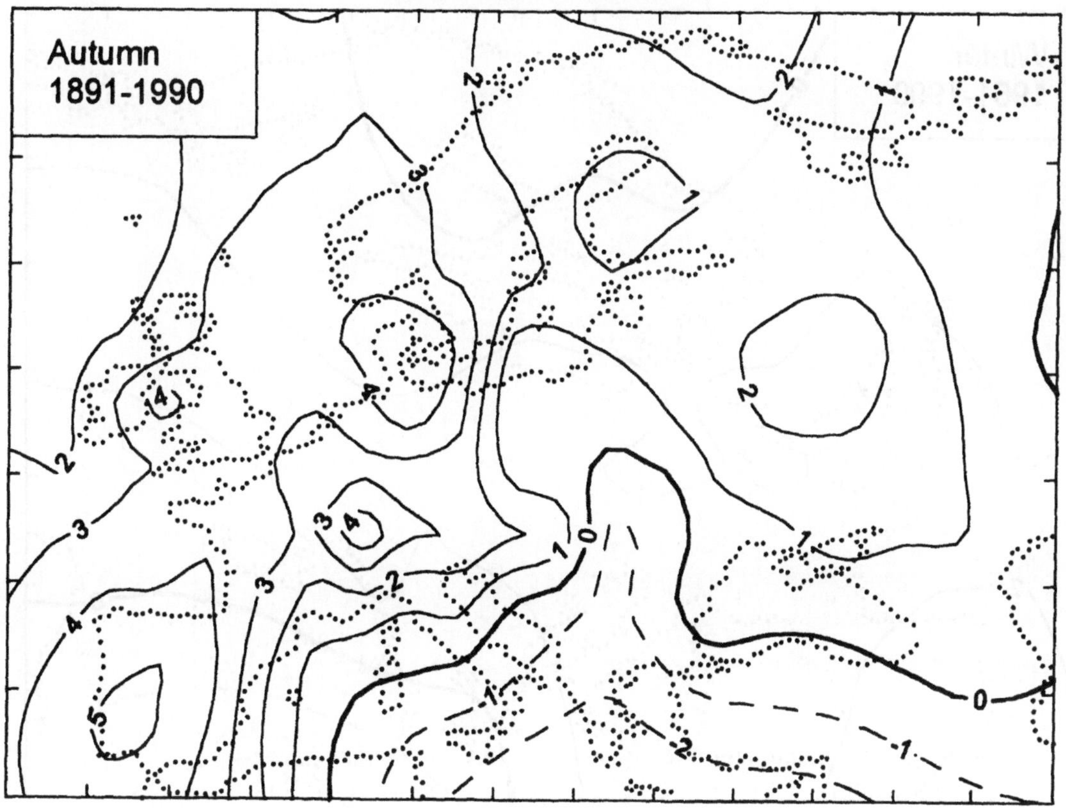

Temperature, linear trend [K]

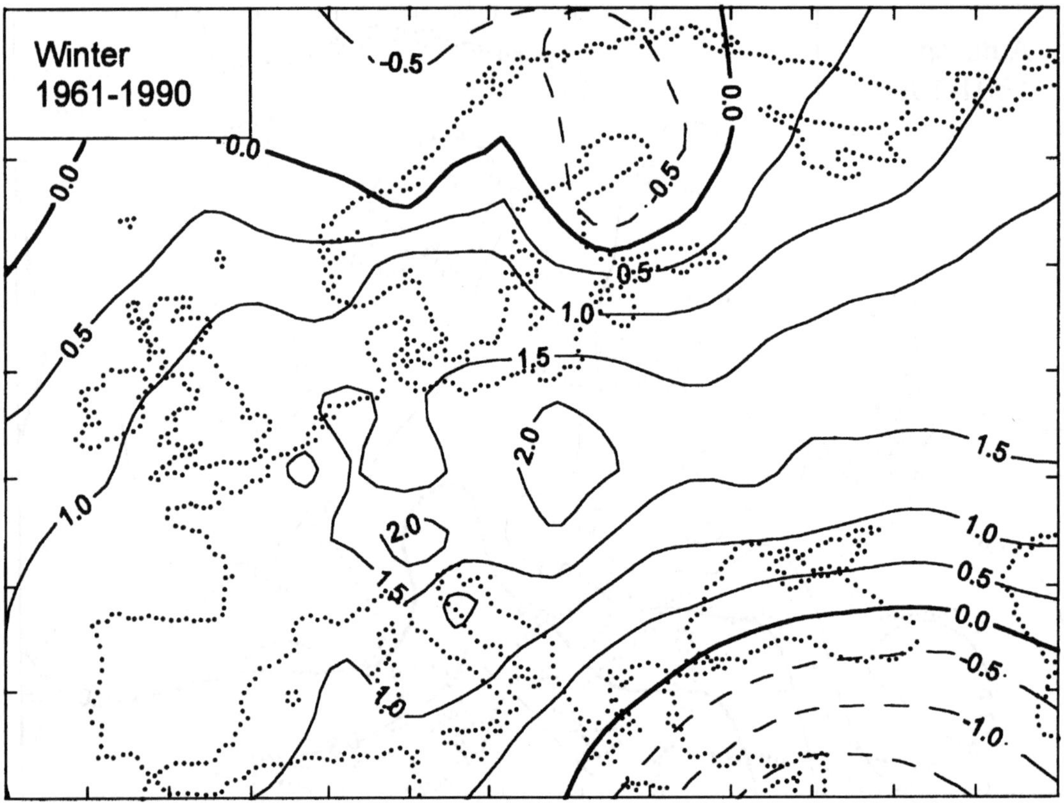

Temperature, trend test

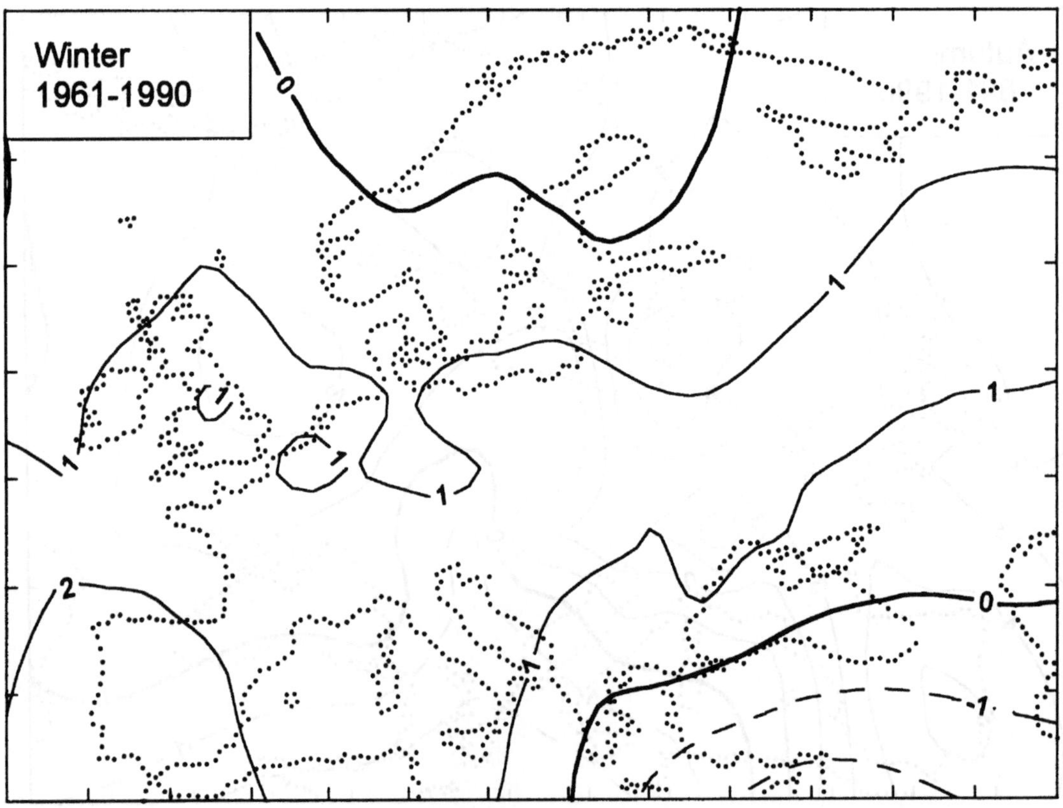

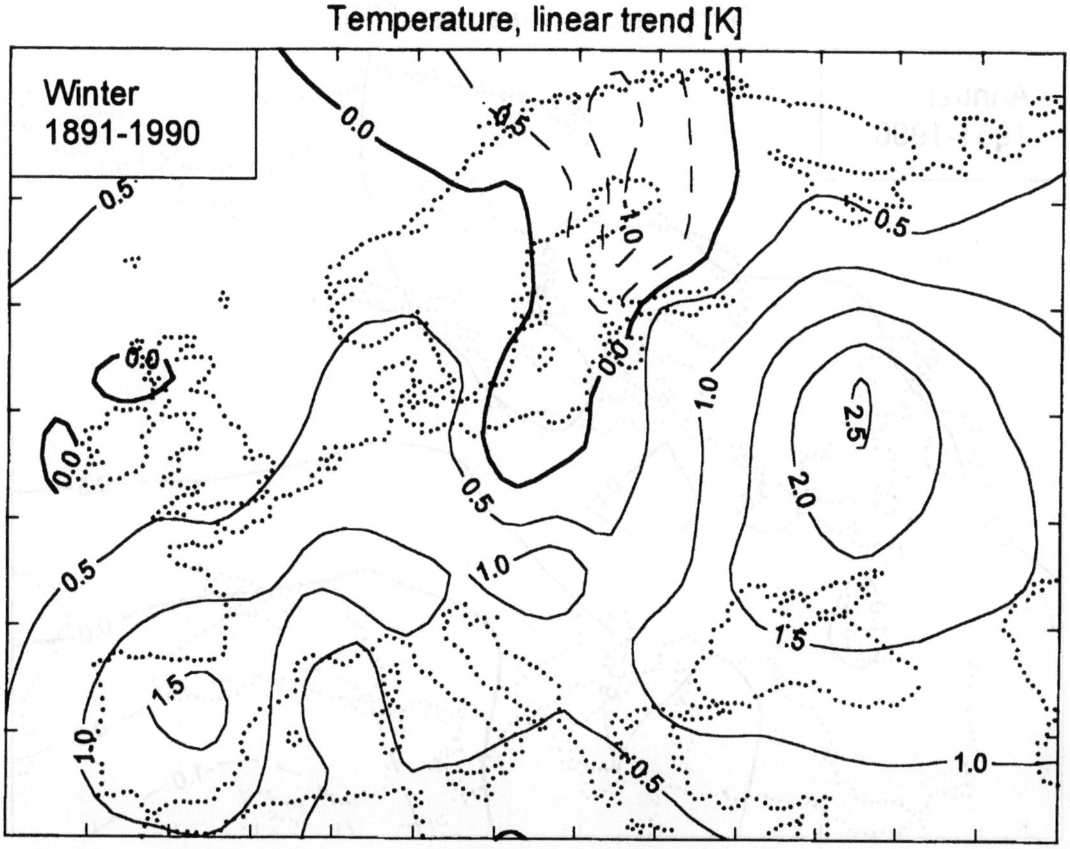
Temperature, linear trend [K]
Winter
1891-1990
0.0
0.5
-0.5
-1.0
1.0
2.5
2.0
1.5

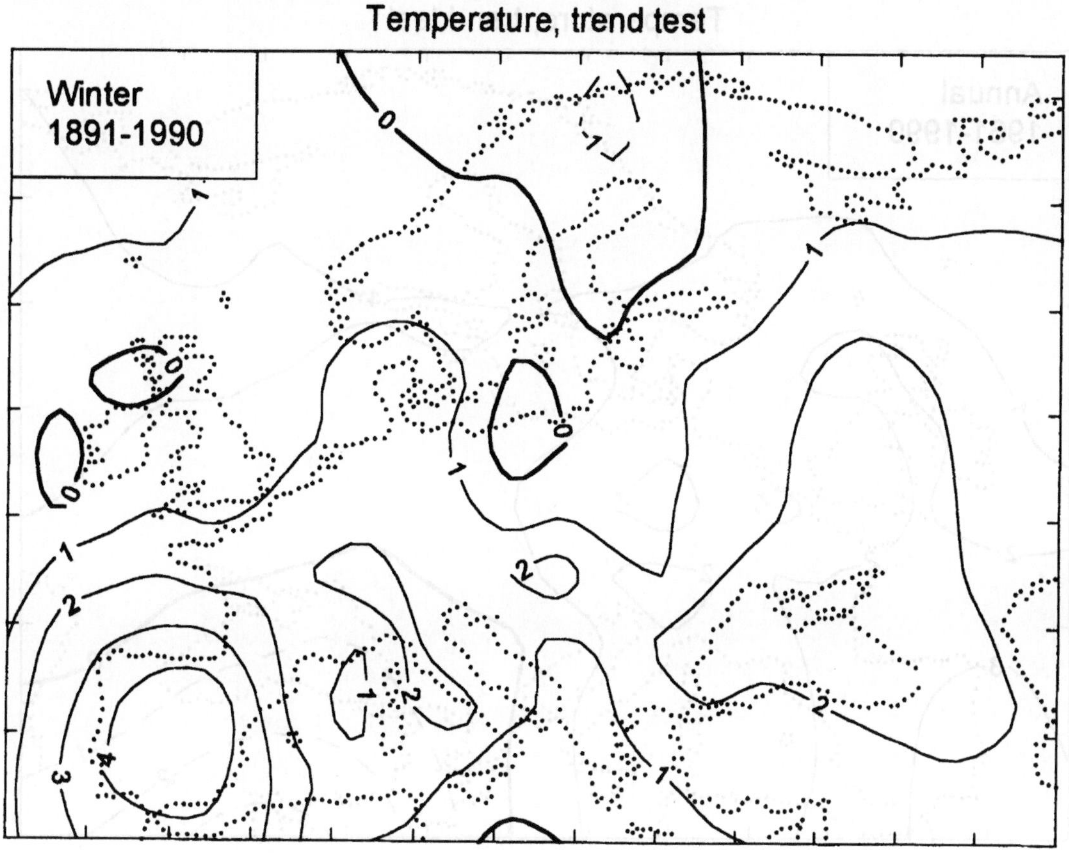
Temperature, trend test
Winter
1891-1990
0
1
-1
2
3
4

Temperature, linear trend [K]

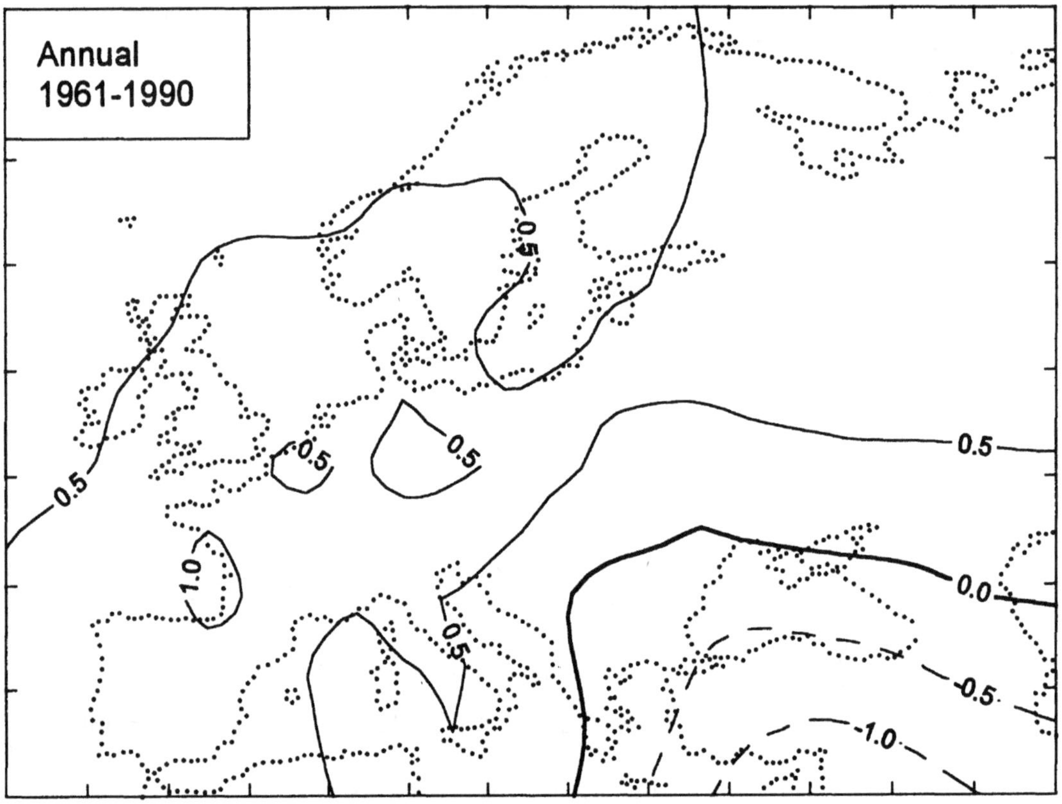

Temperature, trend test

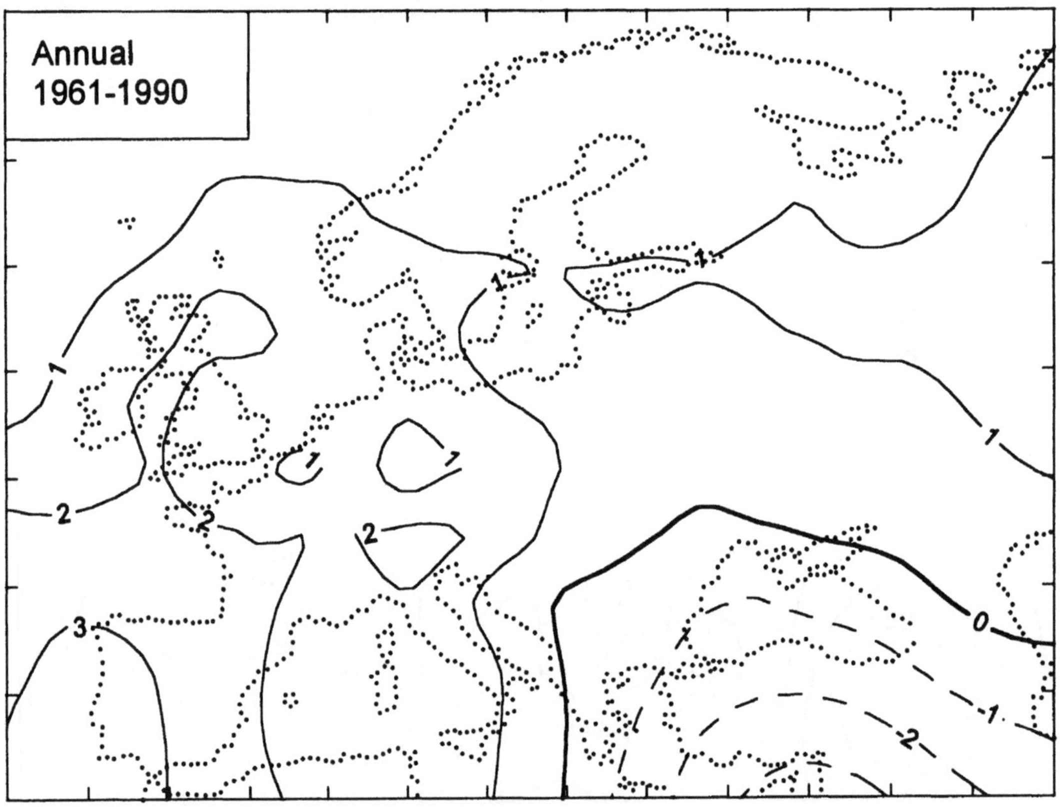

Temperature, linear trend [K]

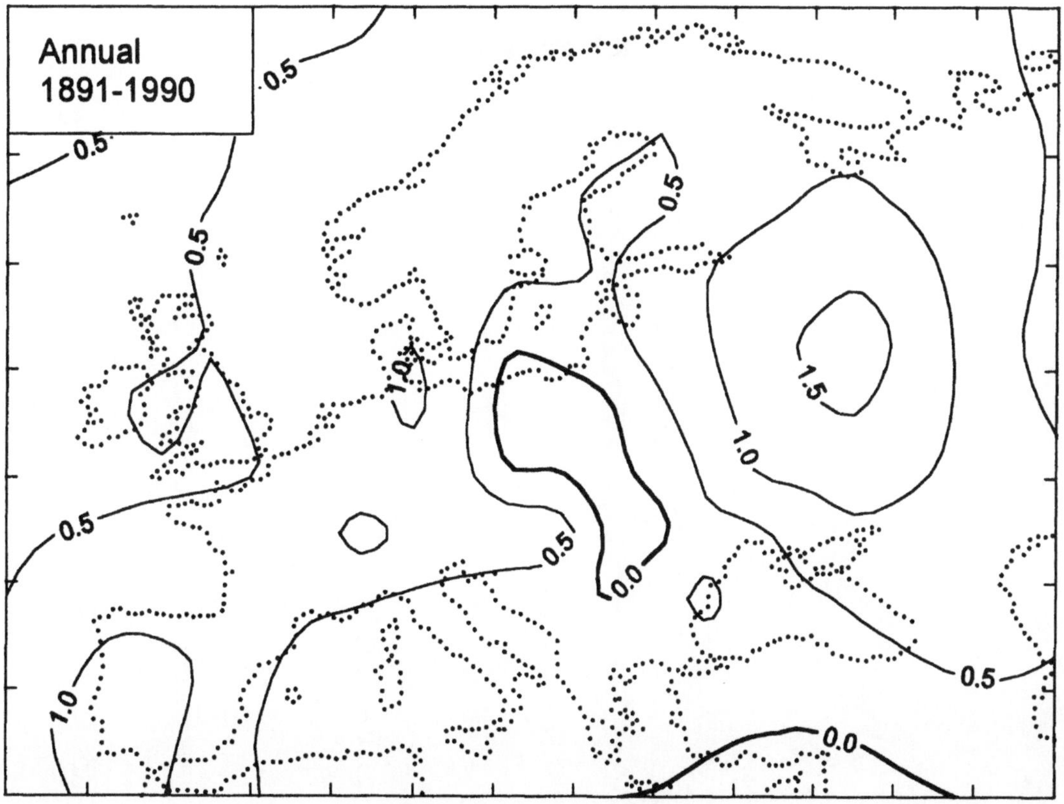

Temperature, trend test

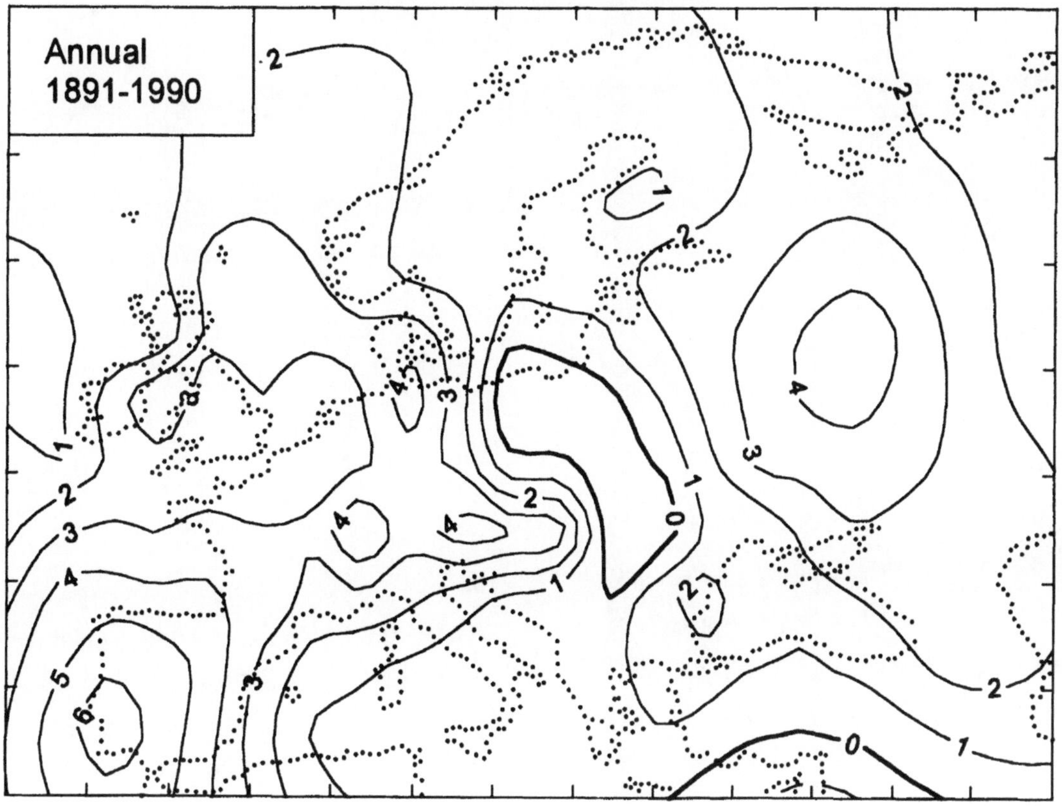

8.2. Pressure

Sea level pressure, average [hpa]

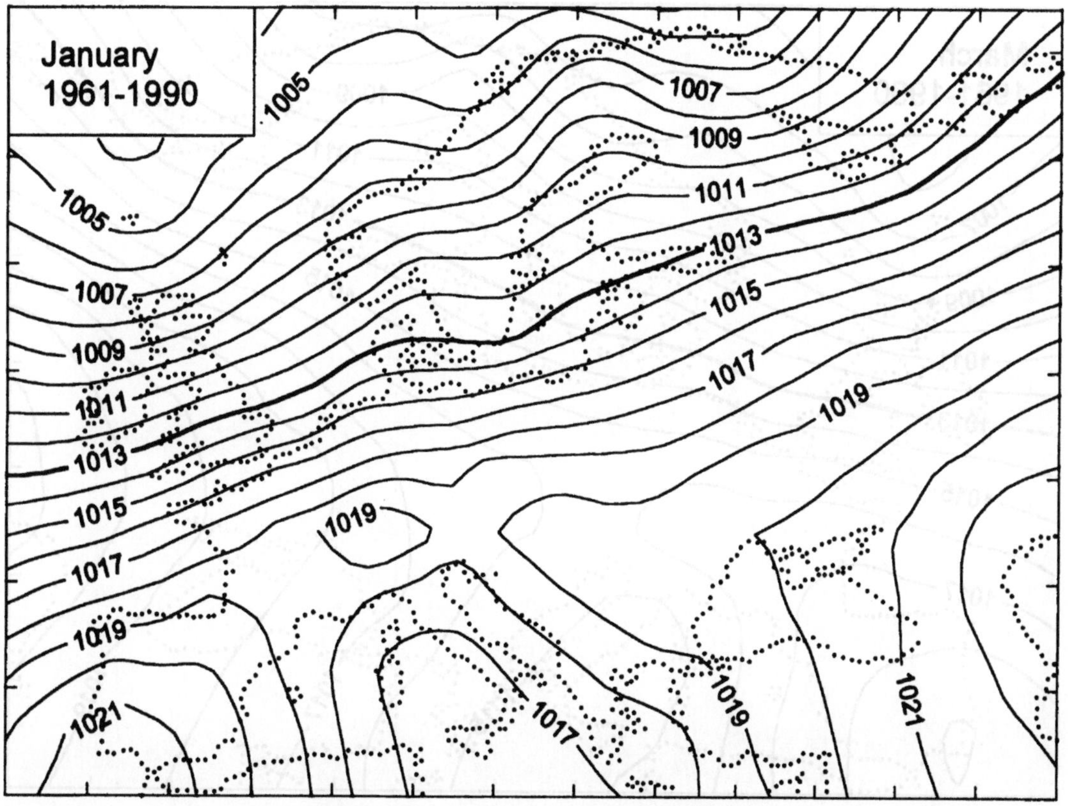

Sea level pressure, average [hpa]

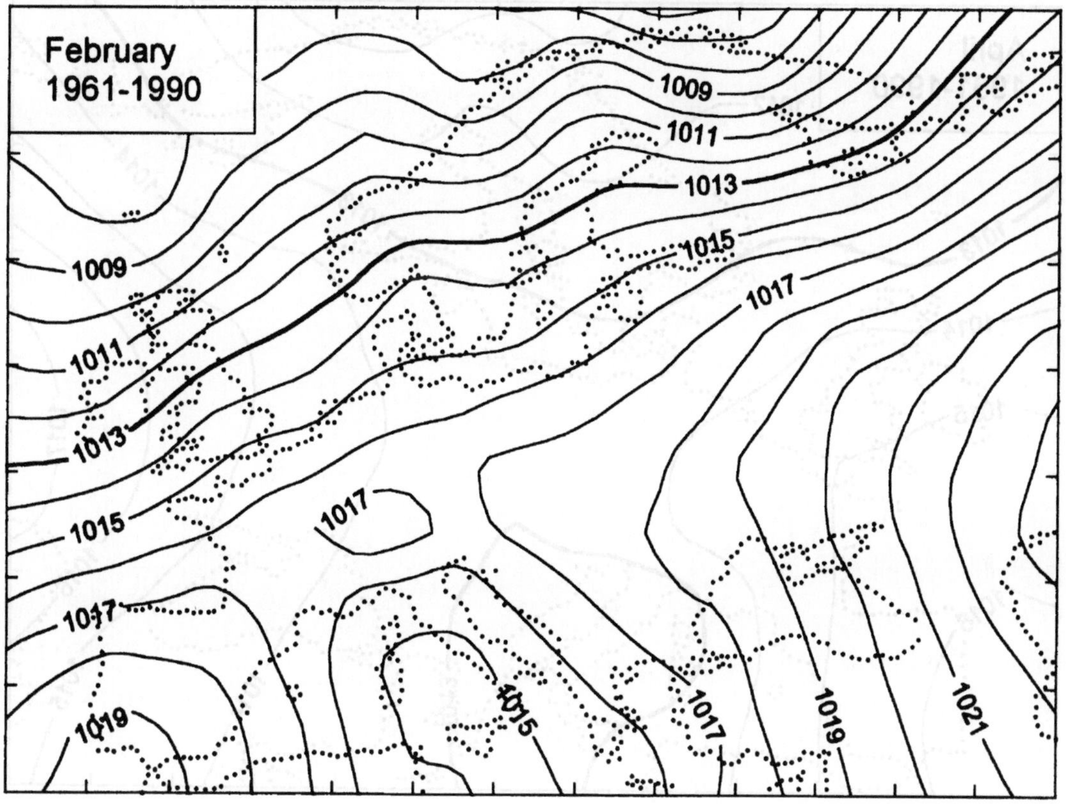

Sea level pressure, average [hpa]

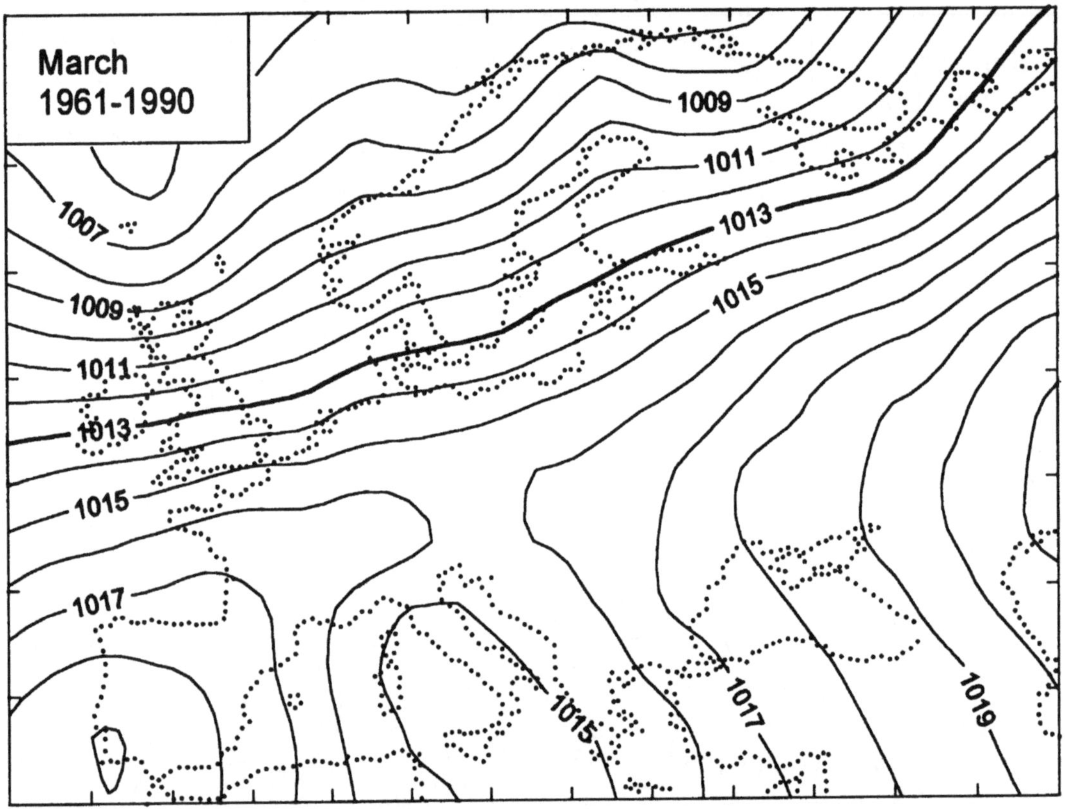

Sea level pressure, average [hpa]

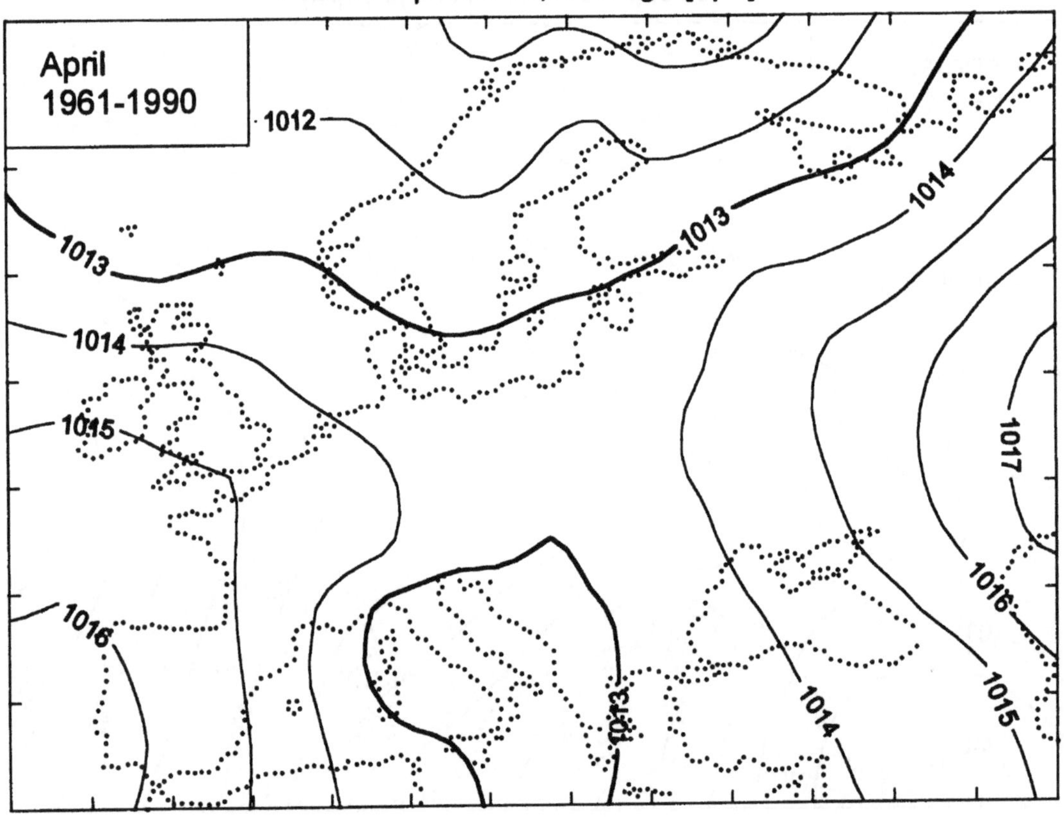

Sea level pressure, average [hpa]

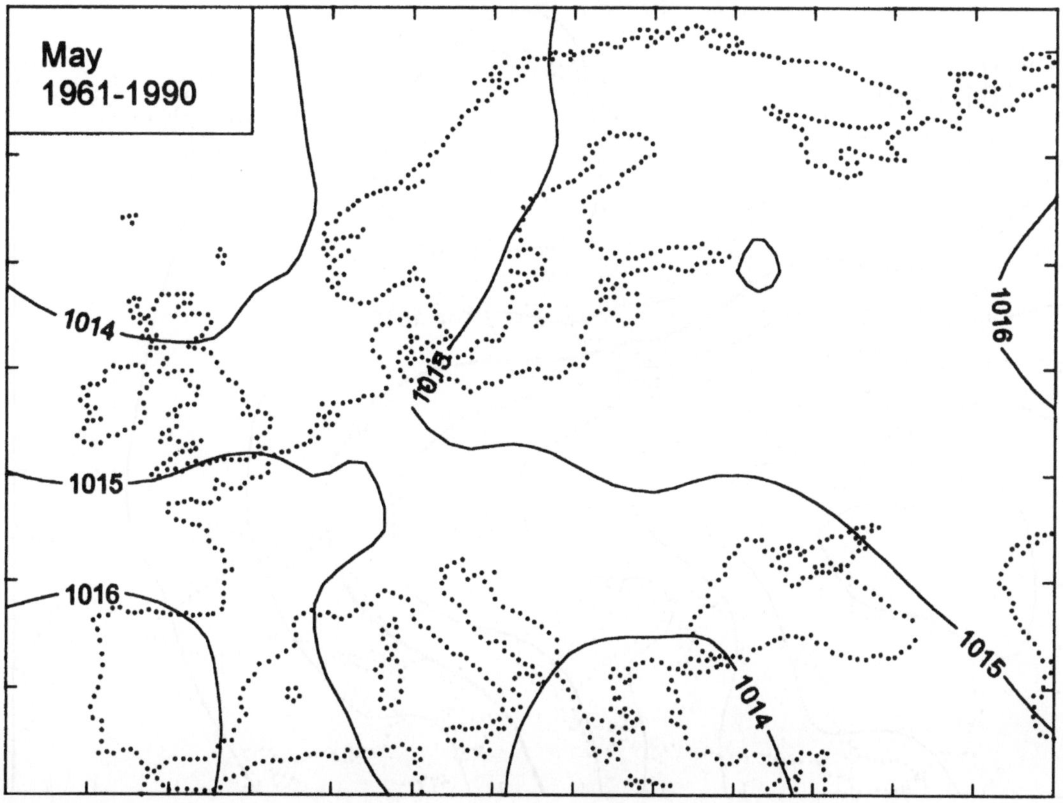

Sea level pressure, average [hpa]

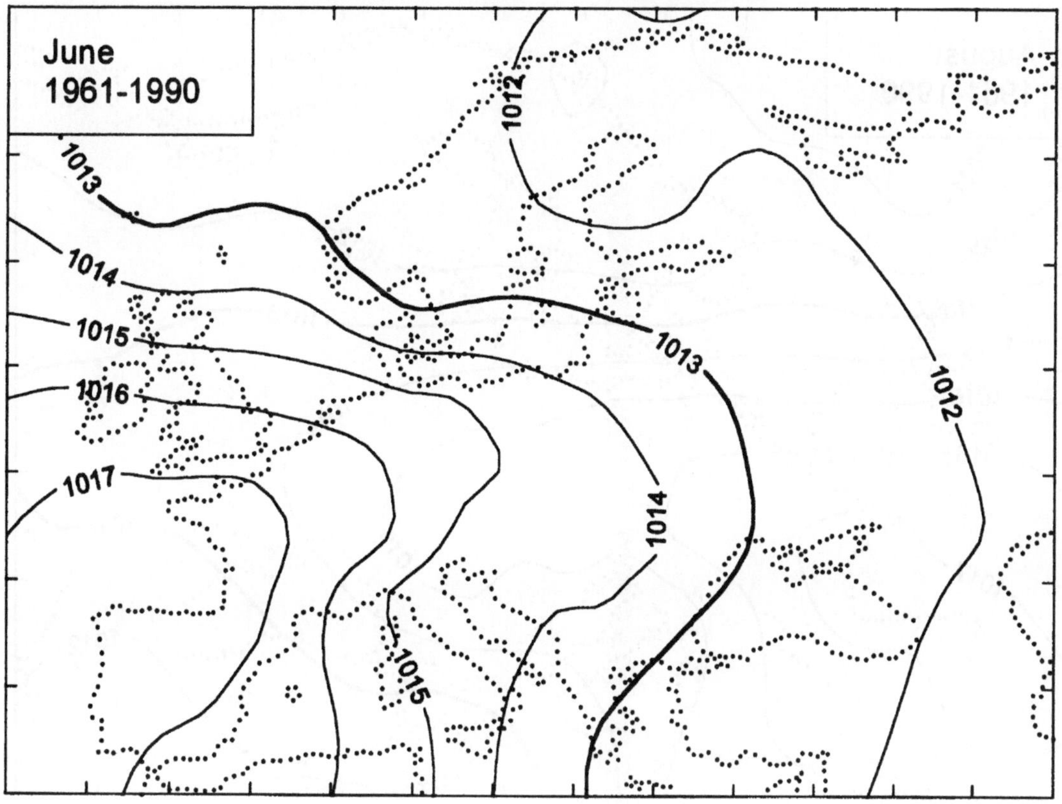

Sea level pressure, average [hpa]

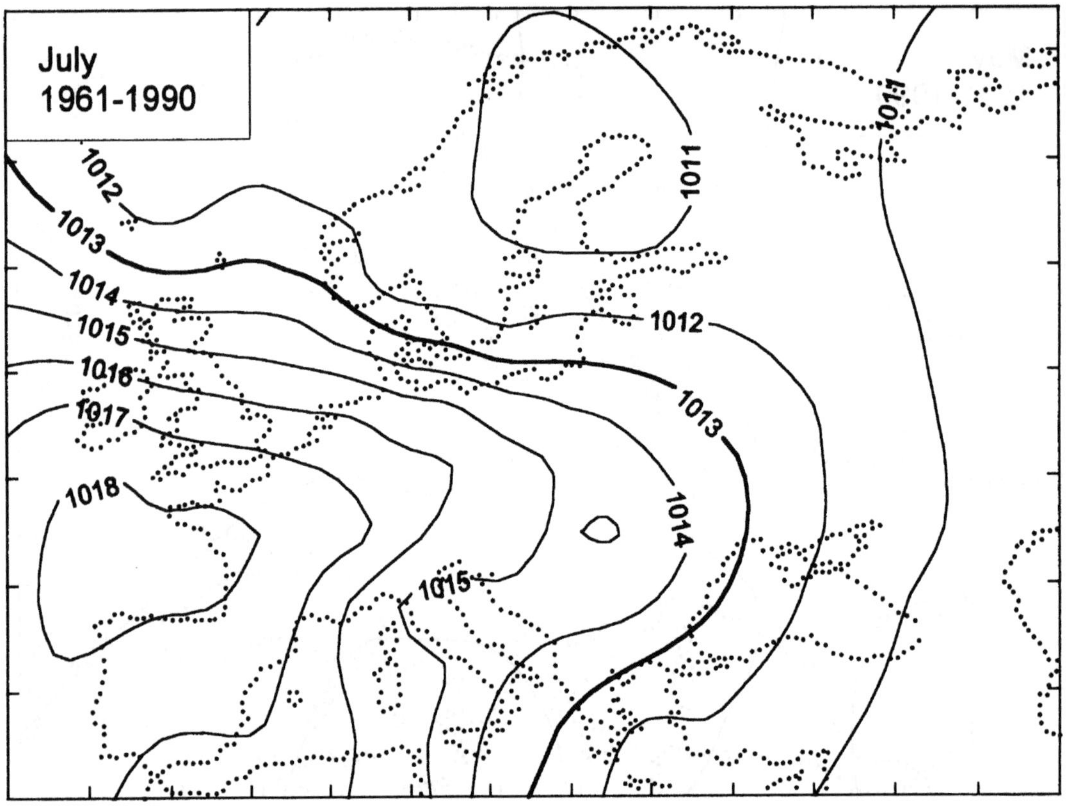

Sea level pressure, average [hpa]

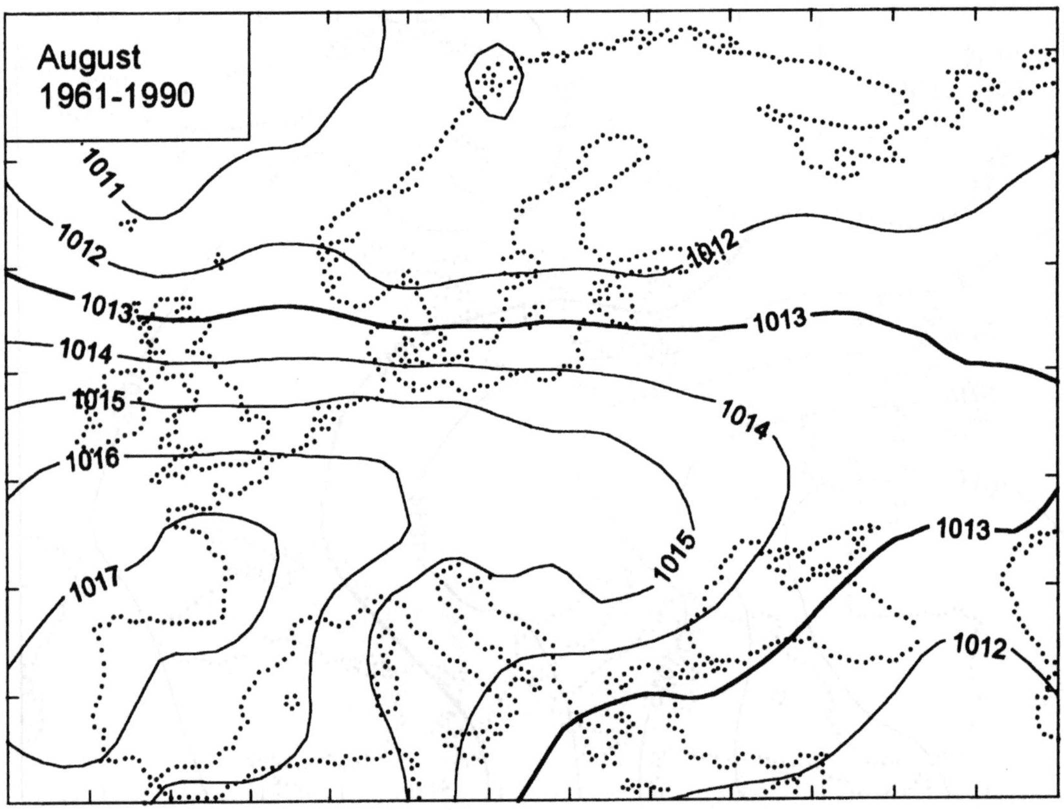

Sea level pressure, average [hpa]

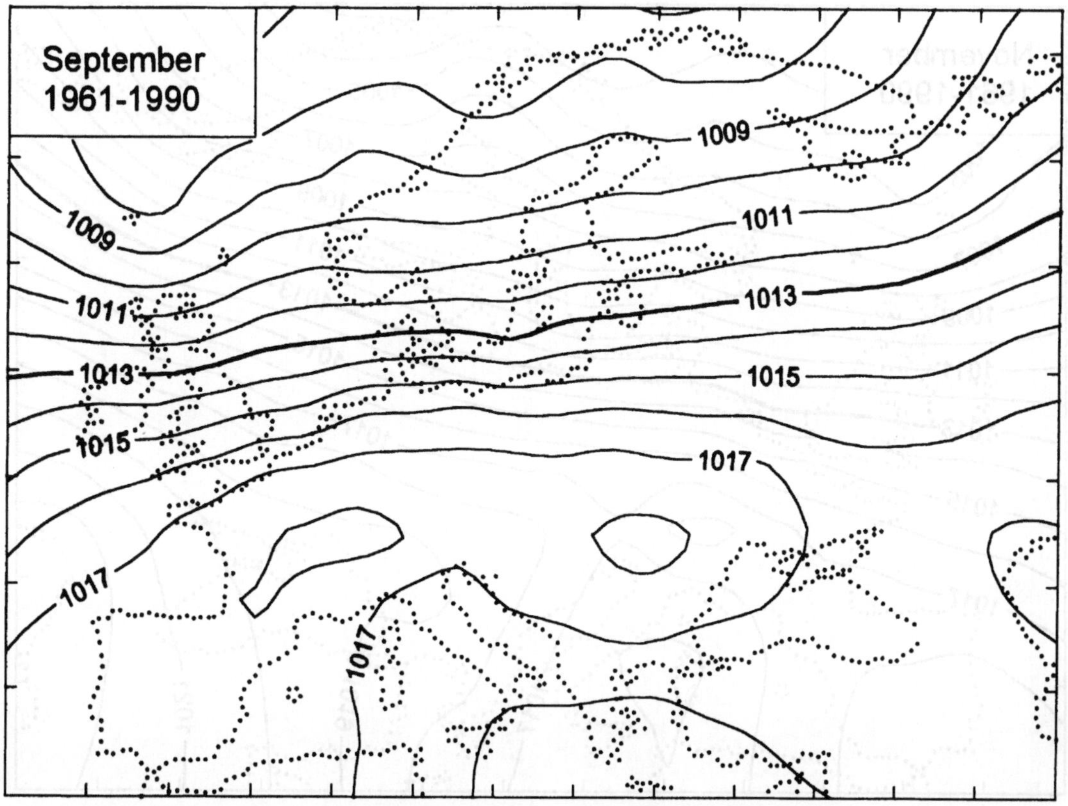

Sea level pressure, average [hpa]

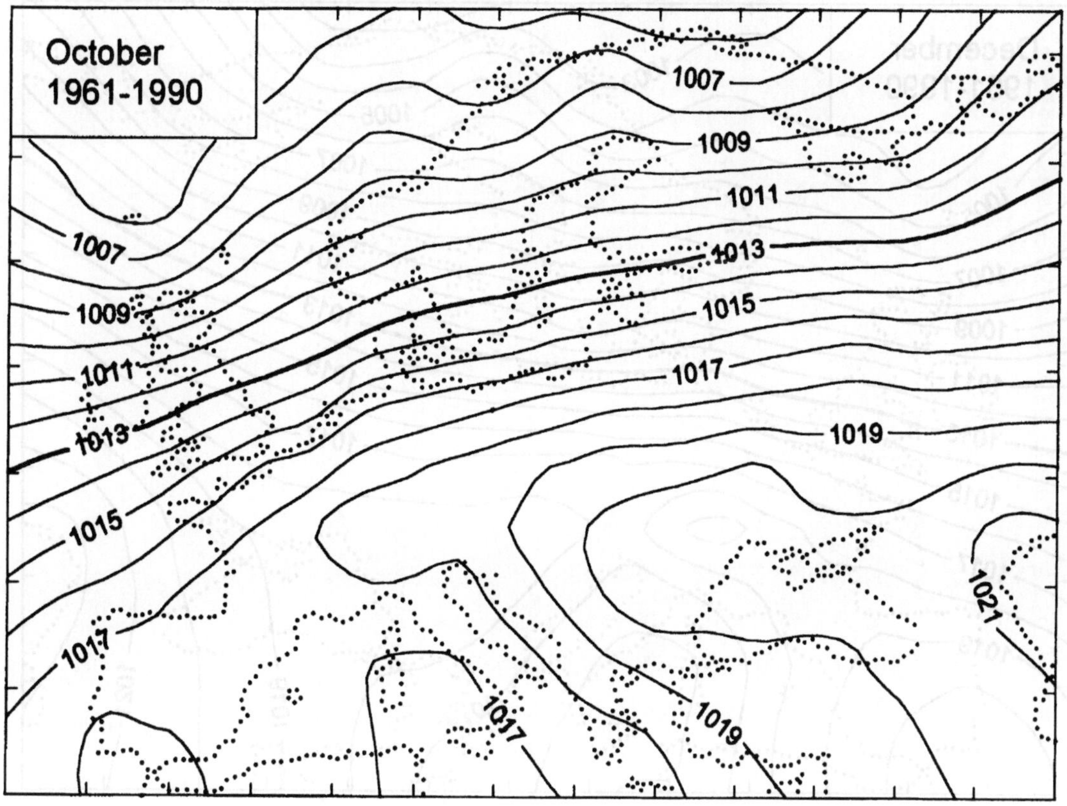

Sea level pressure, average [hpa]

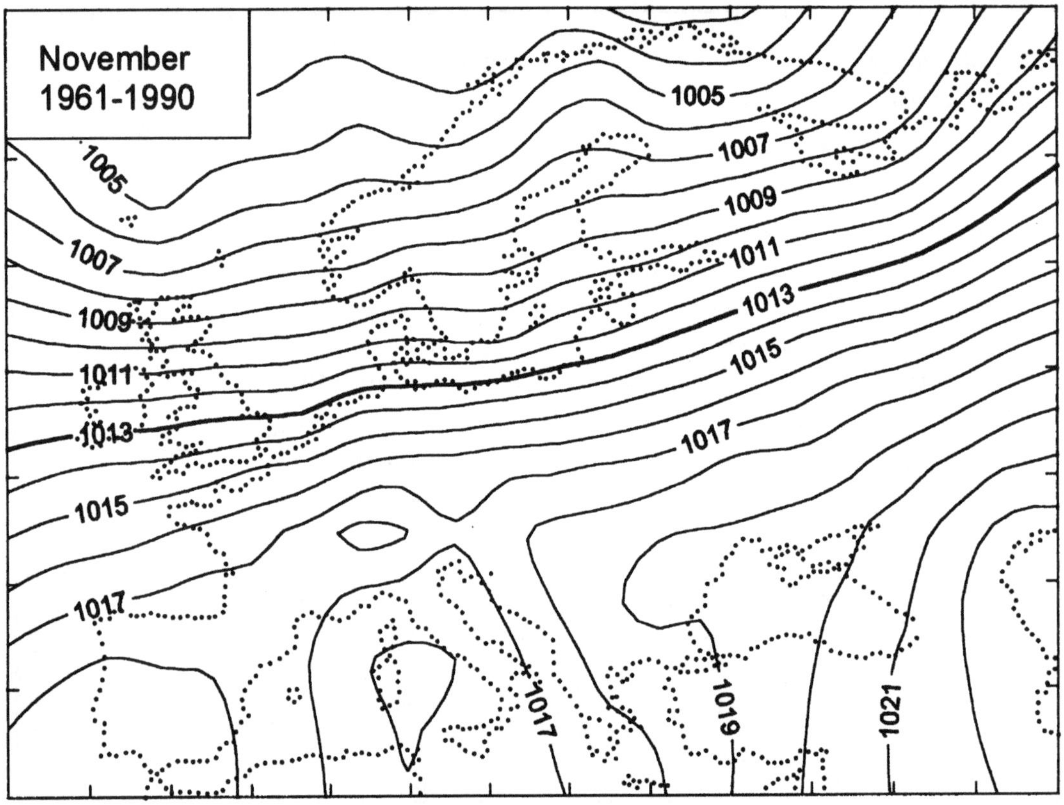

Sea level pressure, average [hpa]

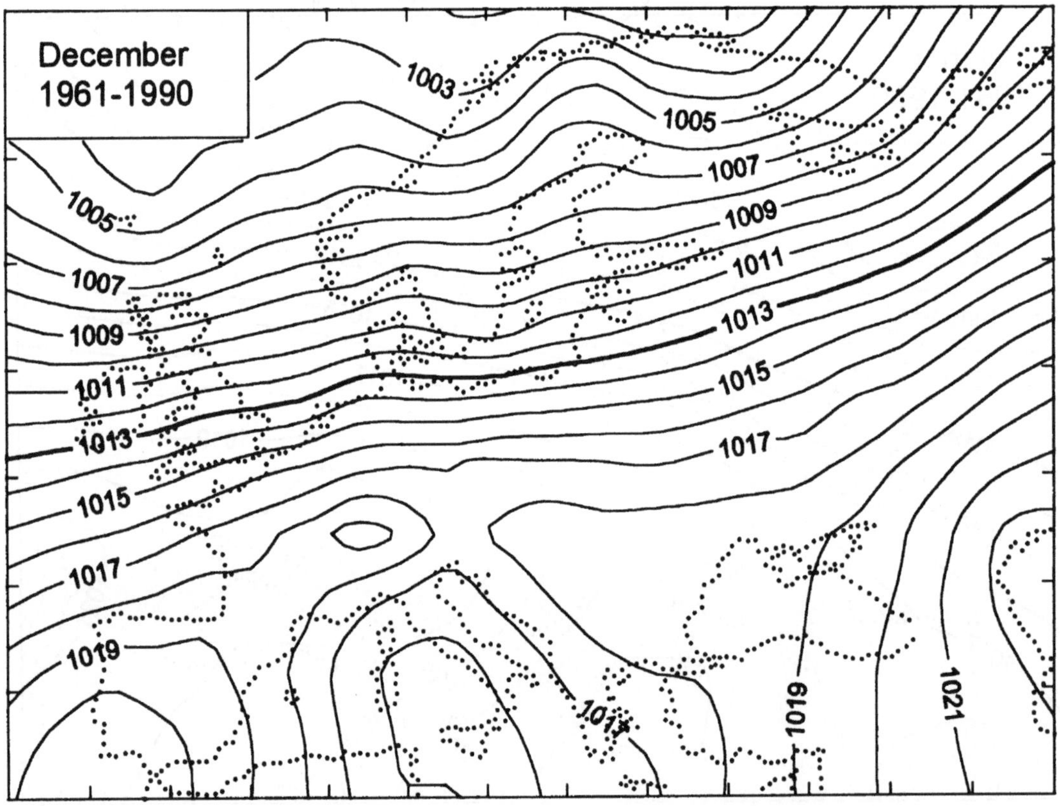

Sea level pressure, average [hpa]

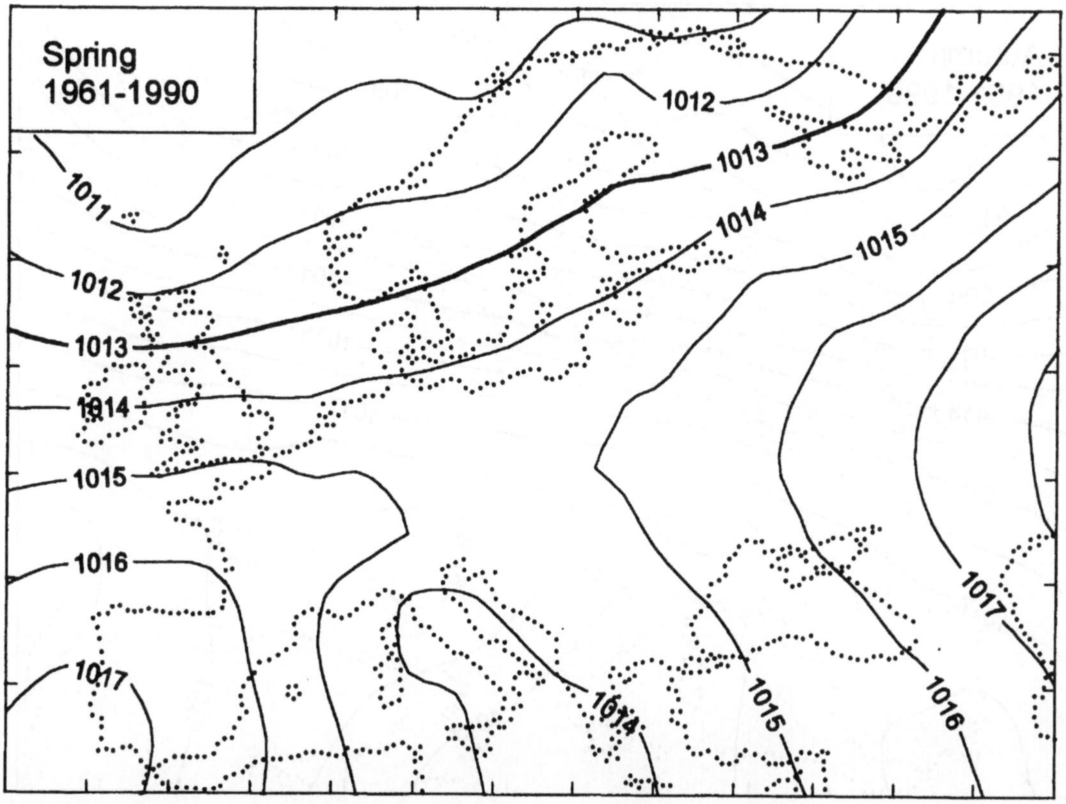

Sea level pressure, average [hpa]

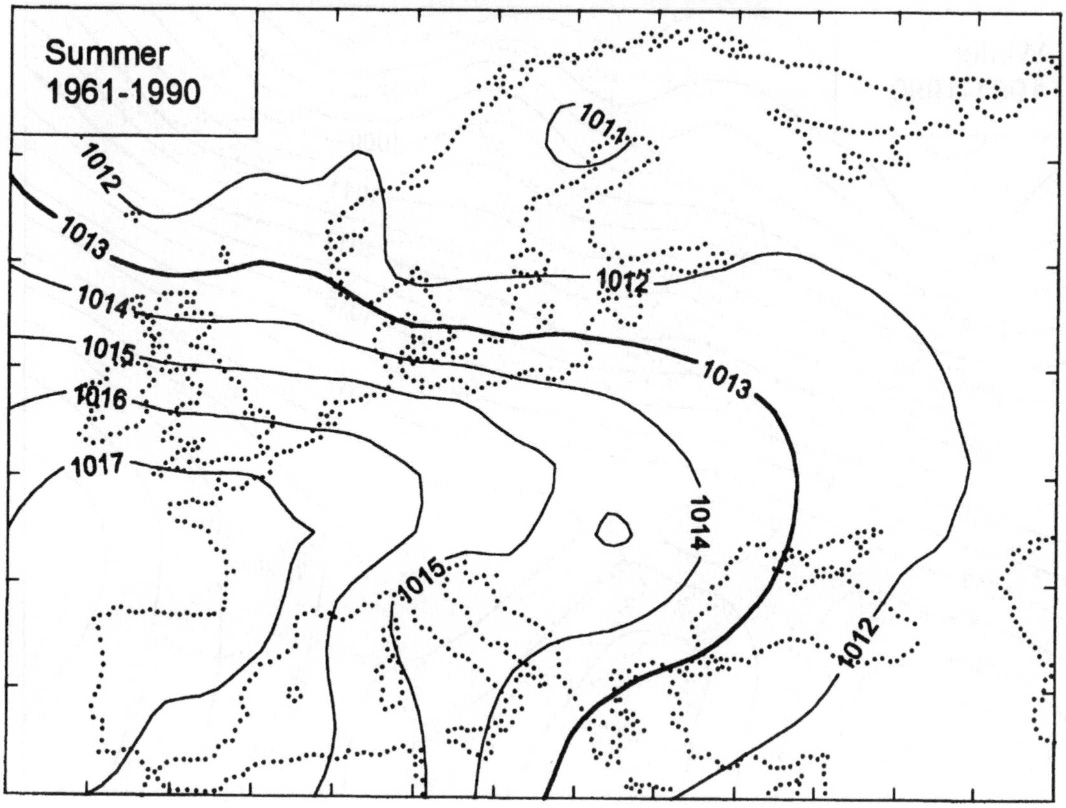

Sea level pressure, average [hpa]

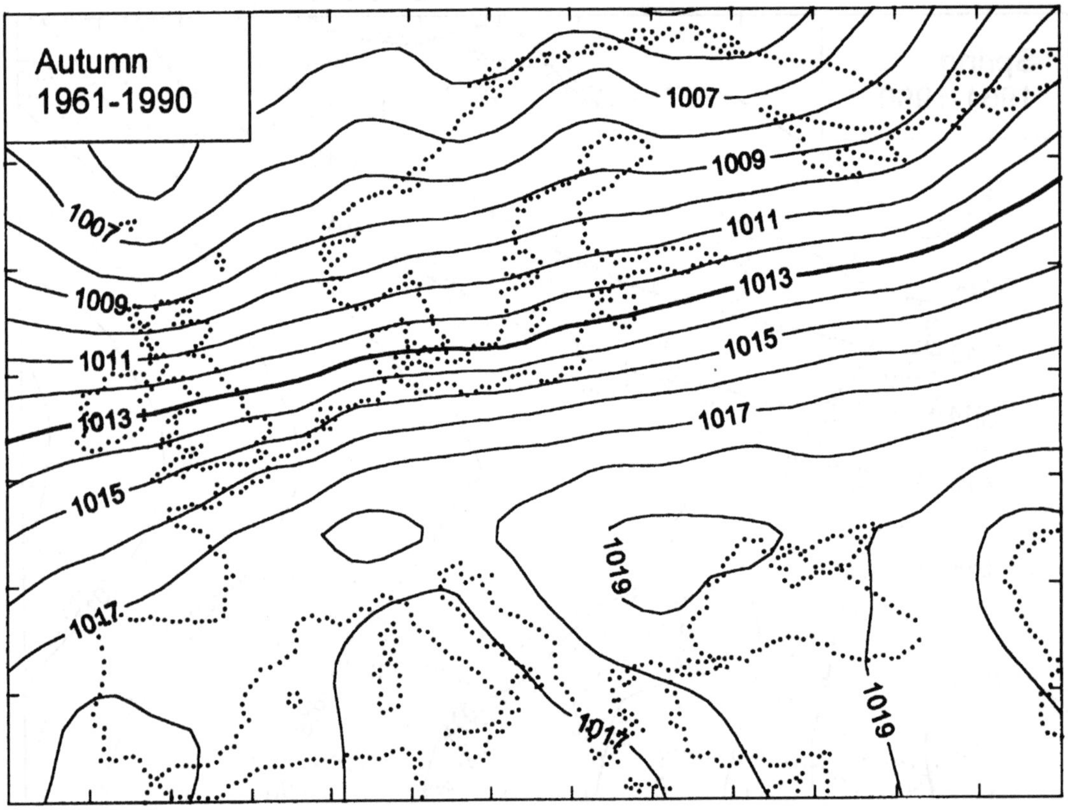

Sea level pressure, average [hpa]

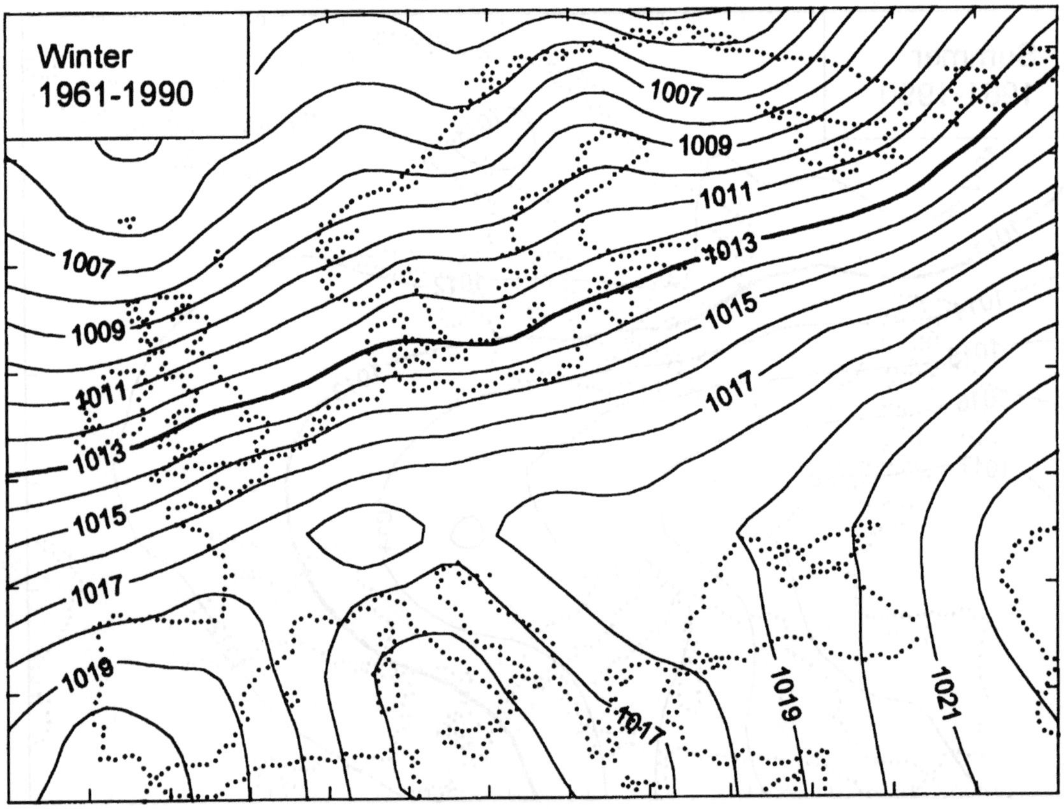

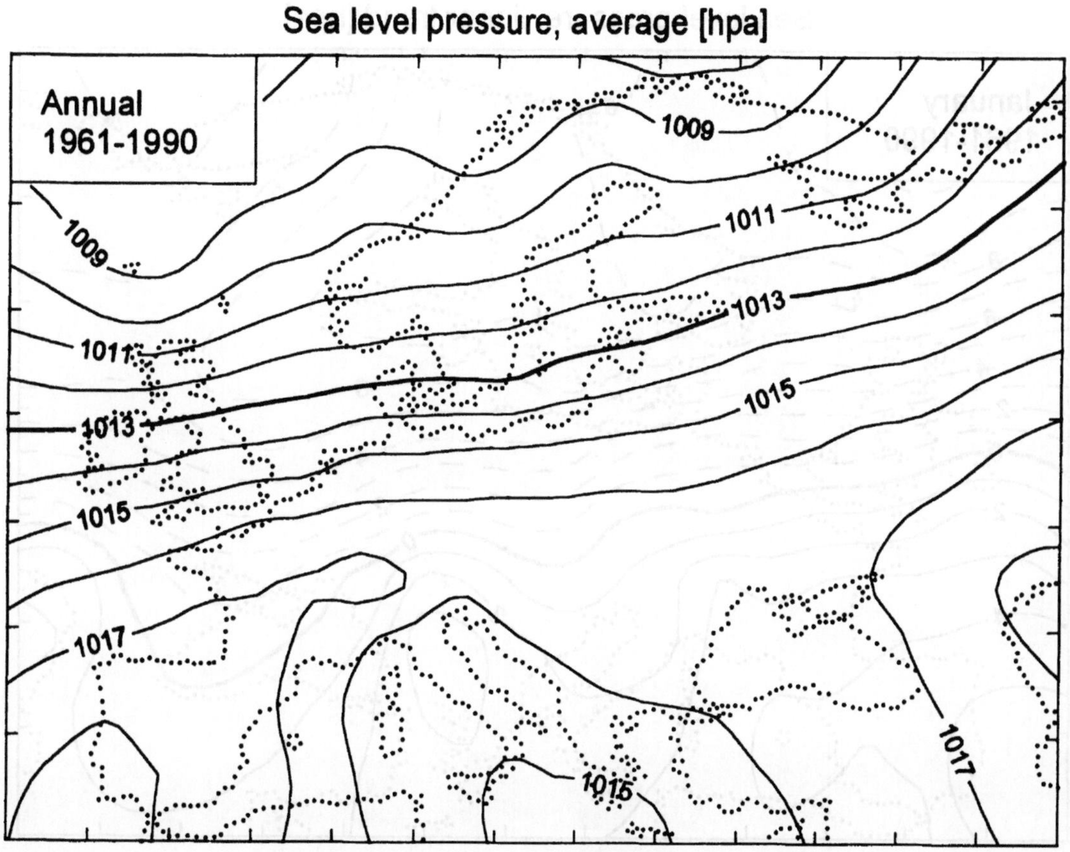
Sea level pressure, average [hpa]
Annual
1961-1990
1009
1009
1011
1011
1013
1013
1015
1015
1017
1017
1015

Sea level pressure, linear trend [hpa]

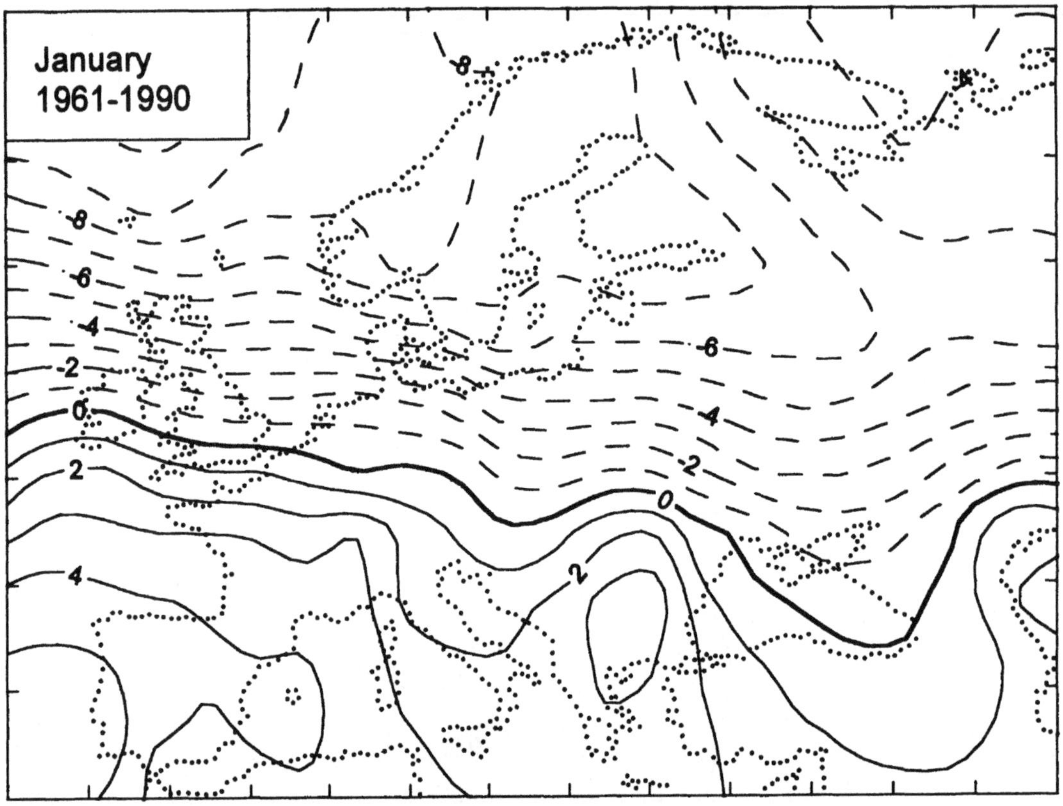

Sea level pressure, trend test

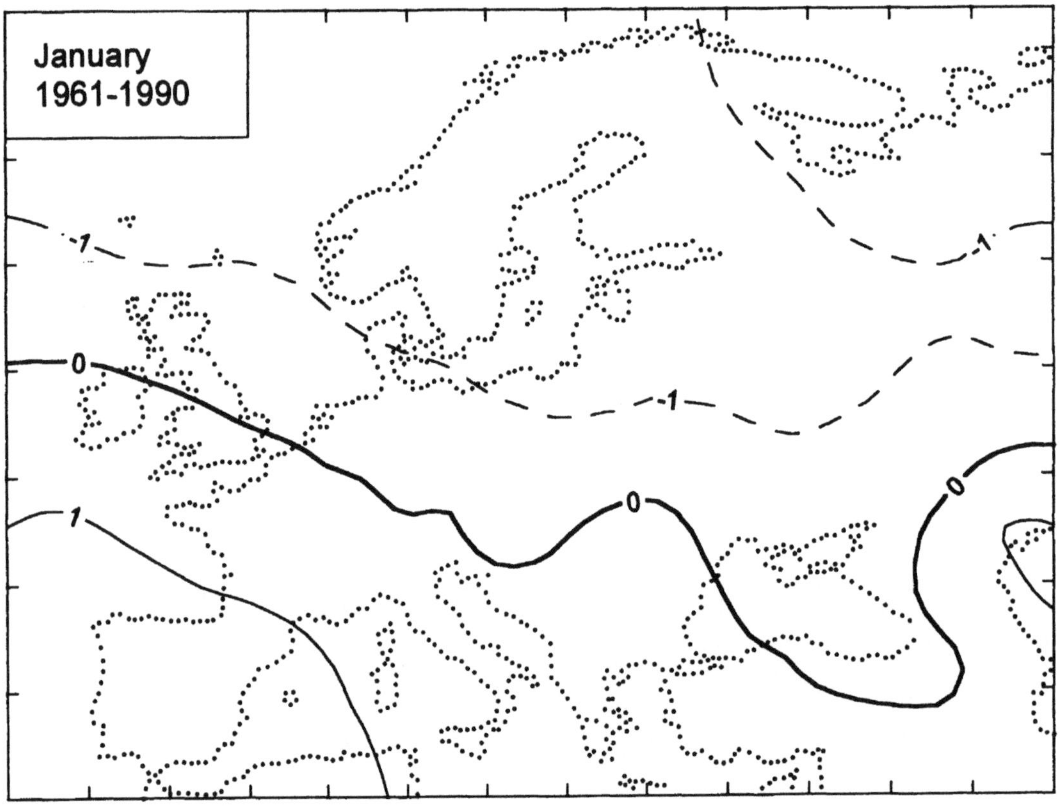

Sea level pressure, linear trend [hpa]

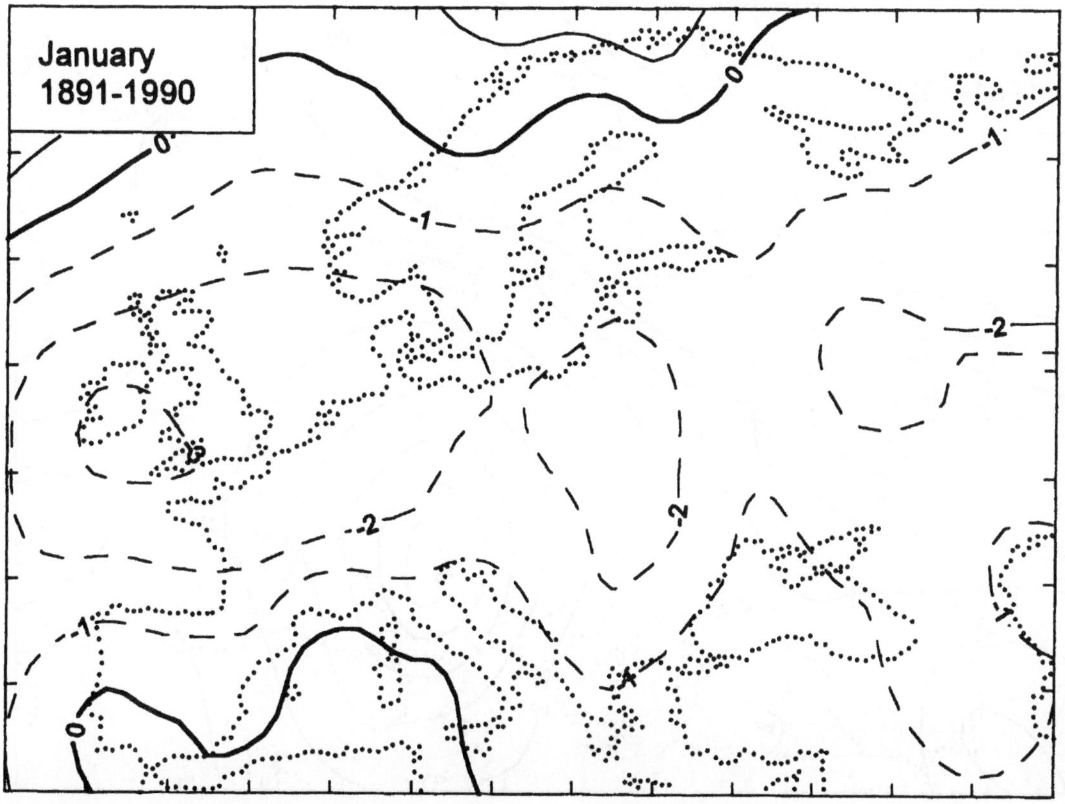

Sea level pressure, trend test

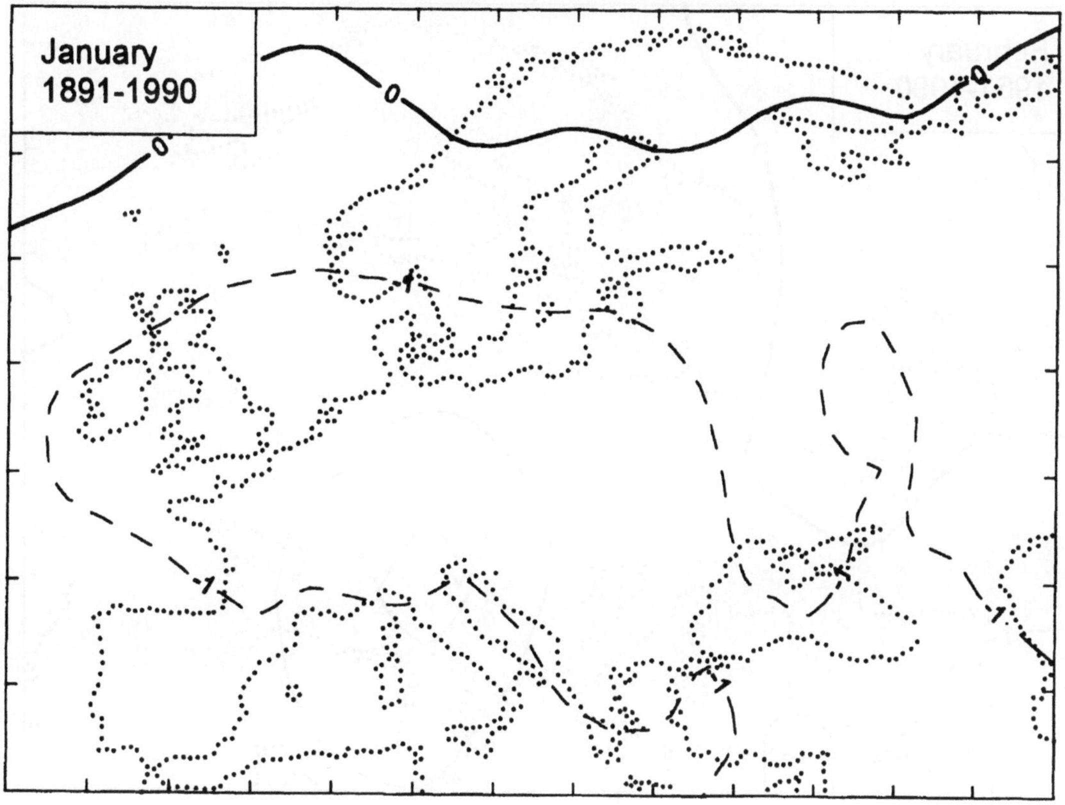

Sea level pressure, linear trend [hpa]

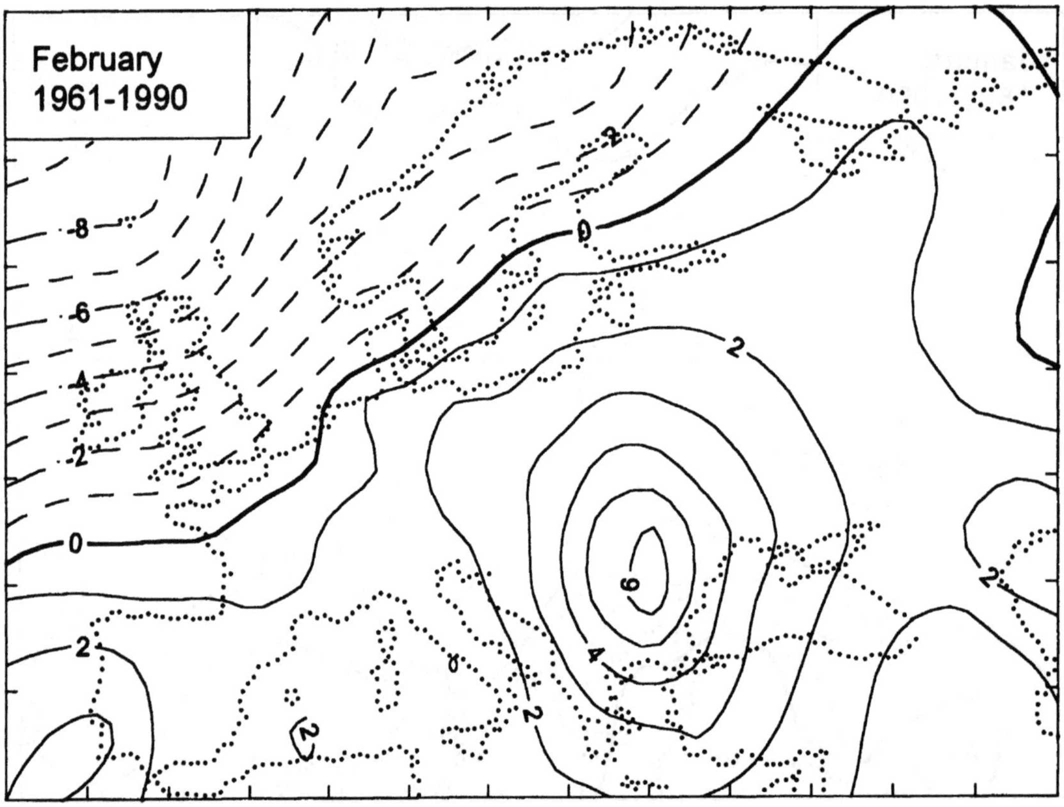

Sea level pressure, trend test

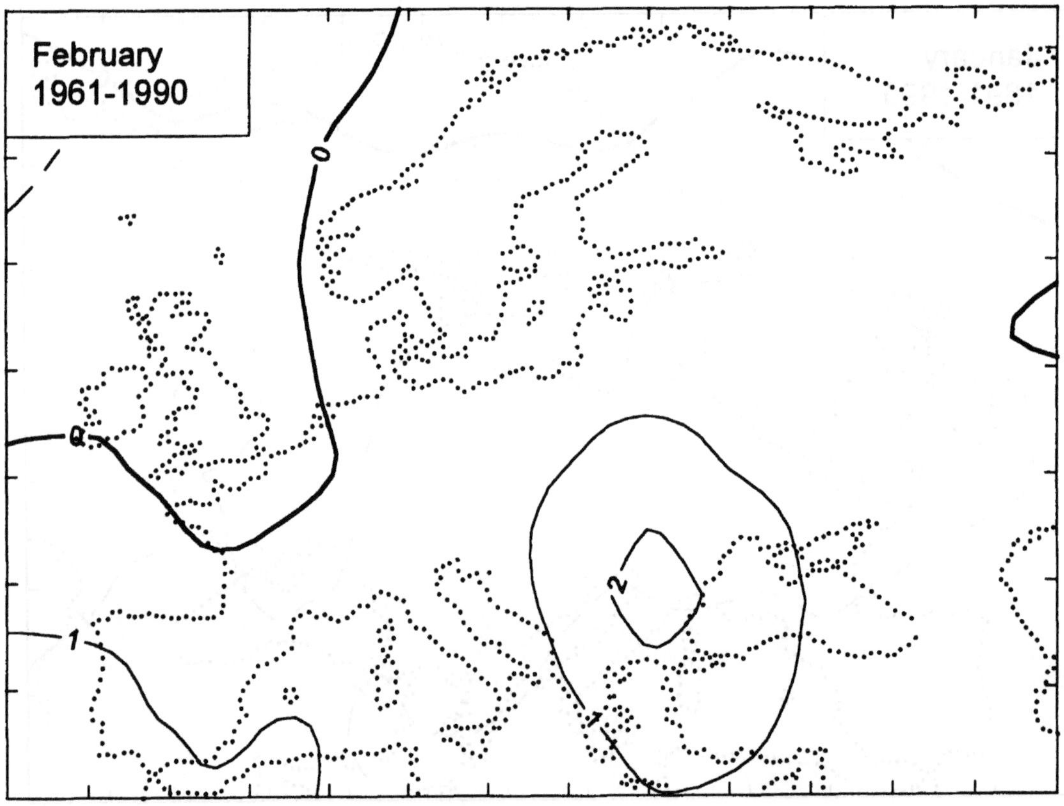

Sea level pressure, linear trend [hpa]

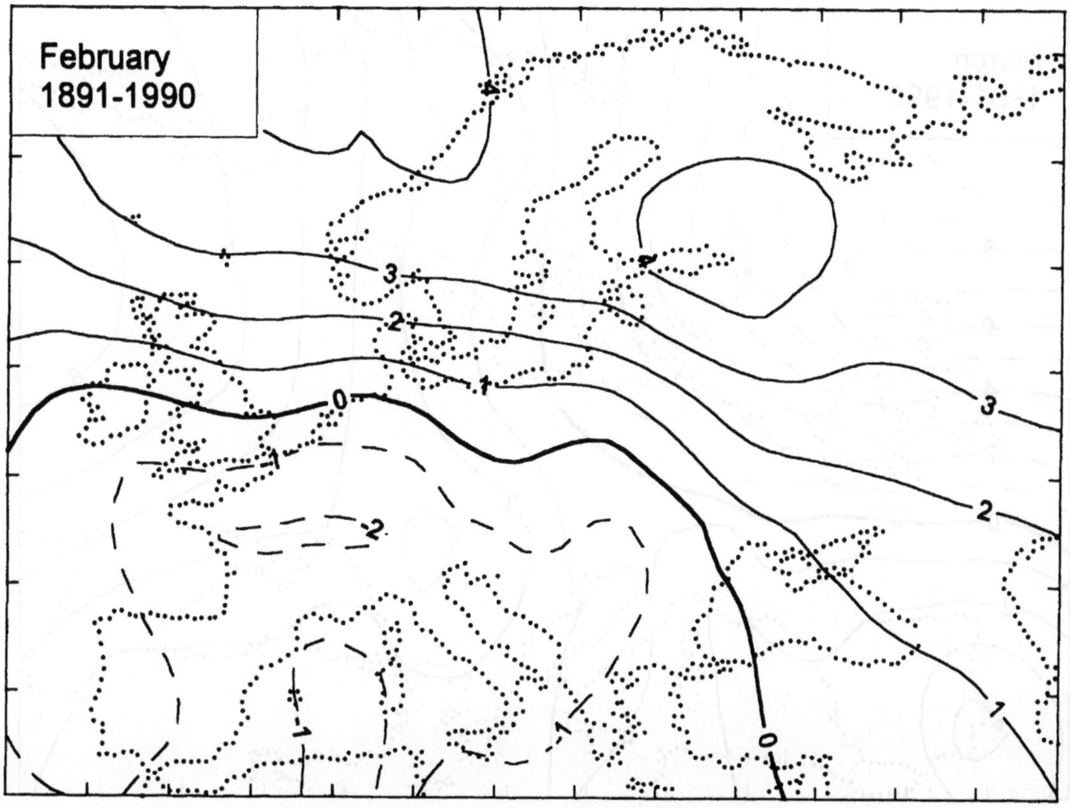

Sea level pressure, trend test

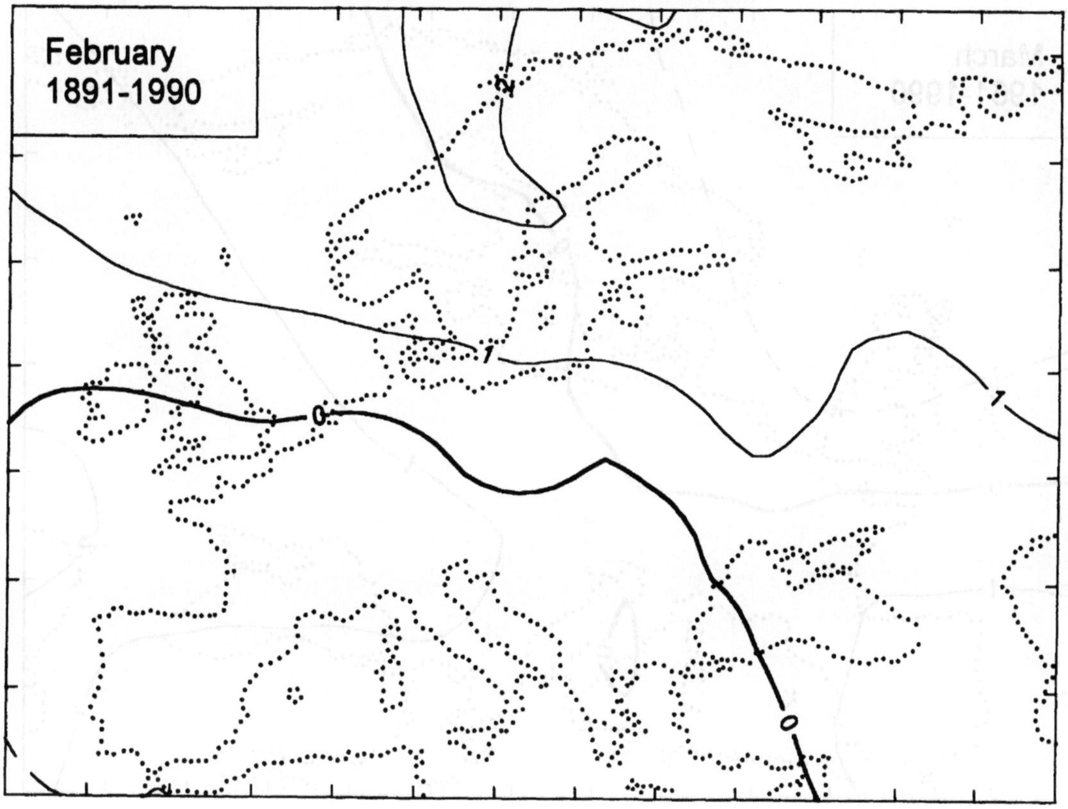

Sea level pressure, linear trend [hpa]

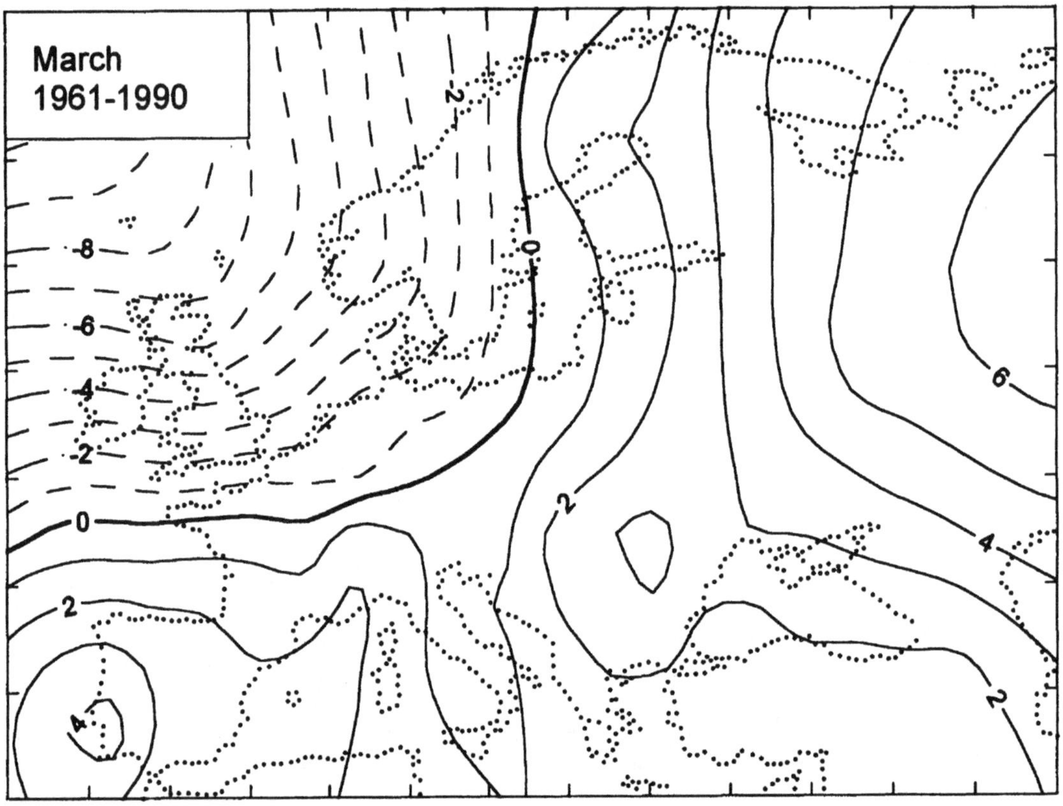

Sea level pressure, trend test

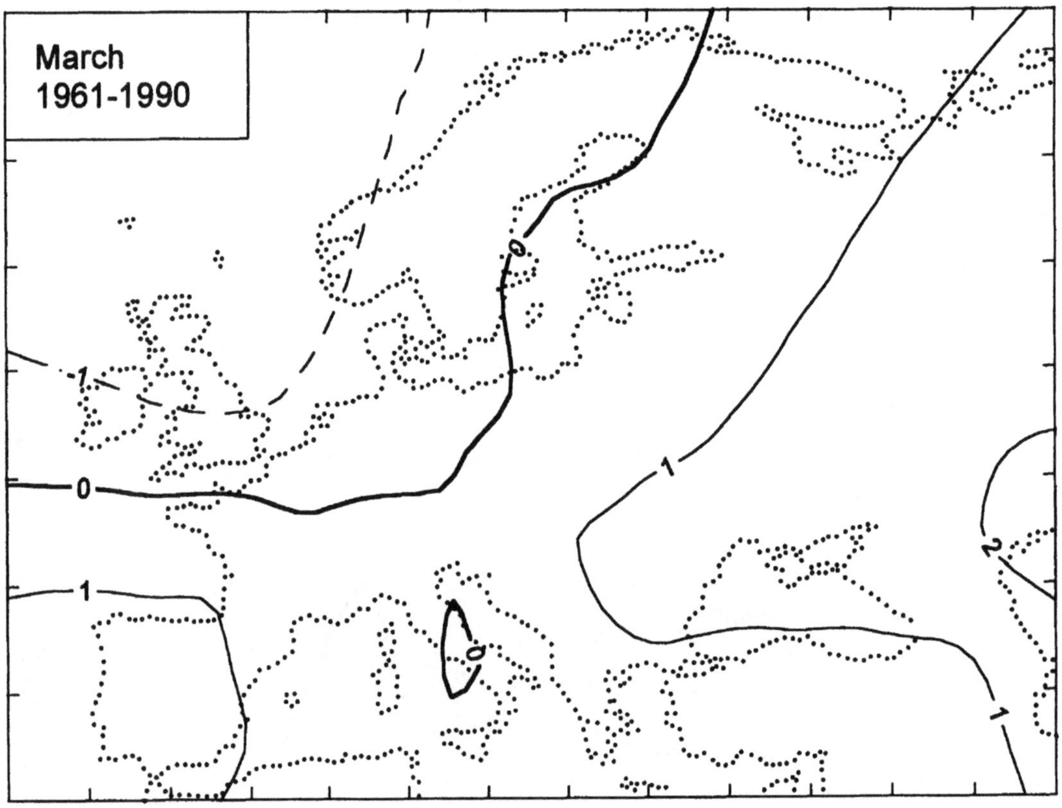

Sea level pressure, linear trend [hpa]

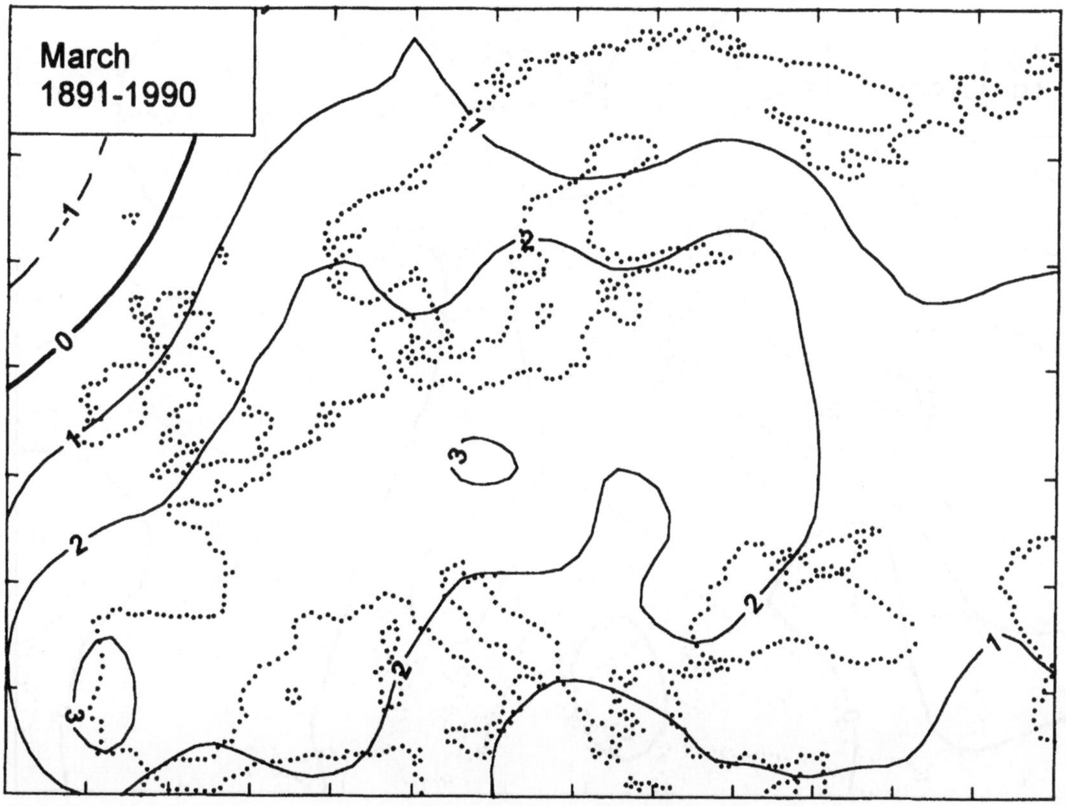

Sea level pressure, trend test

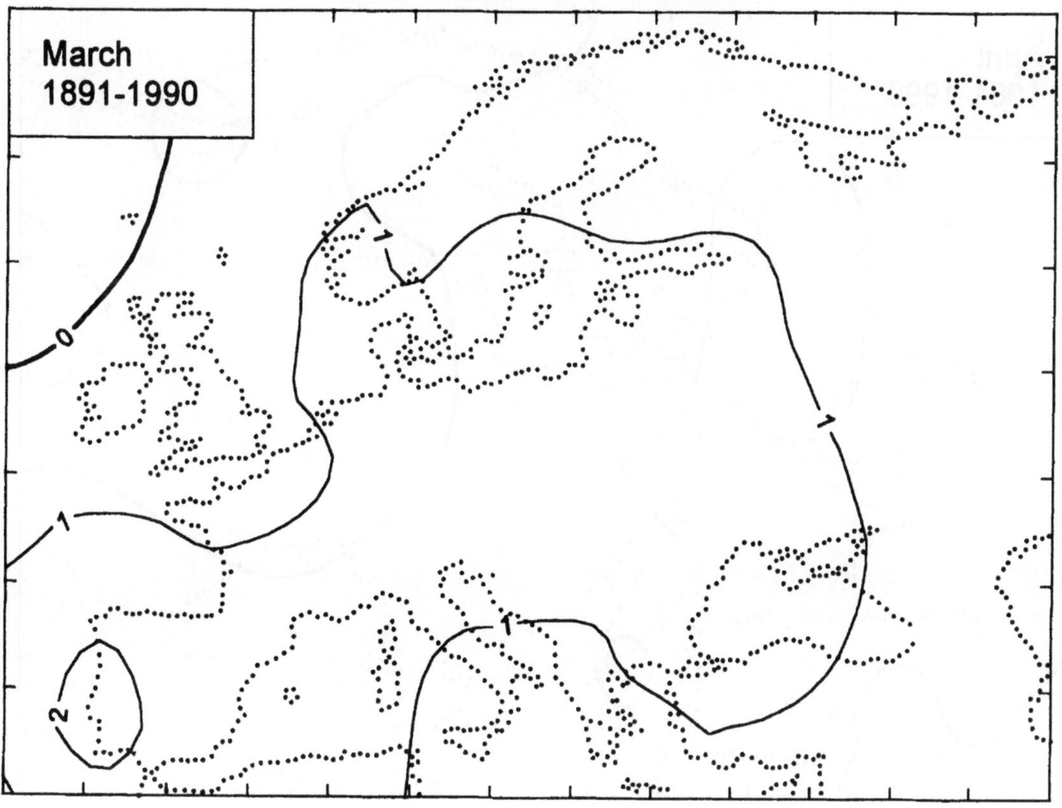

Sea level pressure, linear trend [hpa]

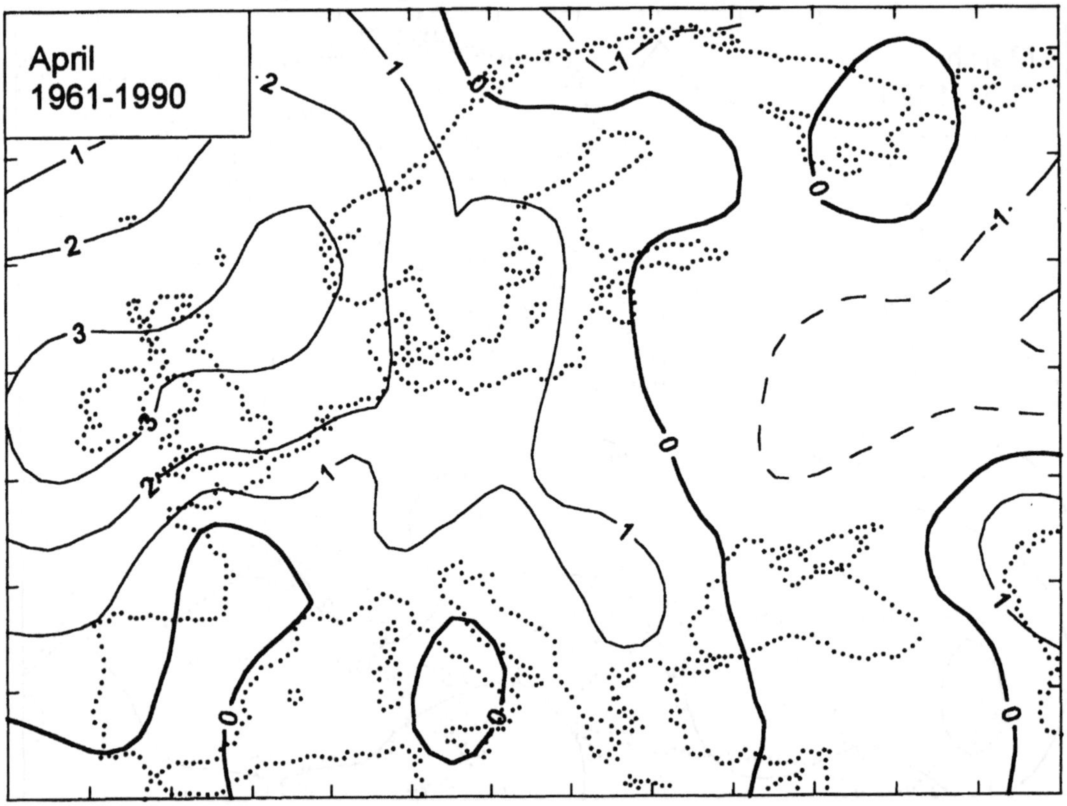

Sea level pressure, trend test

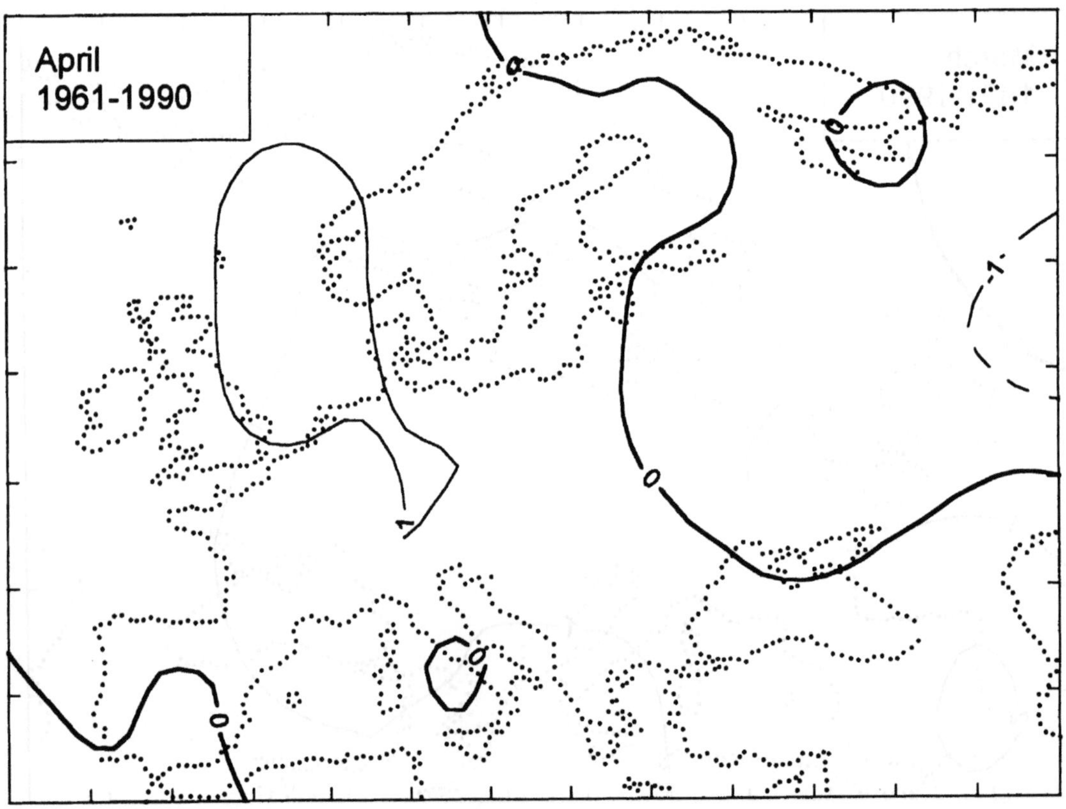

Sea level pressure, linear trend [hpa]

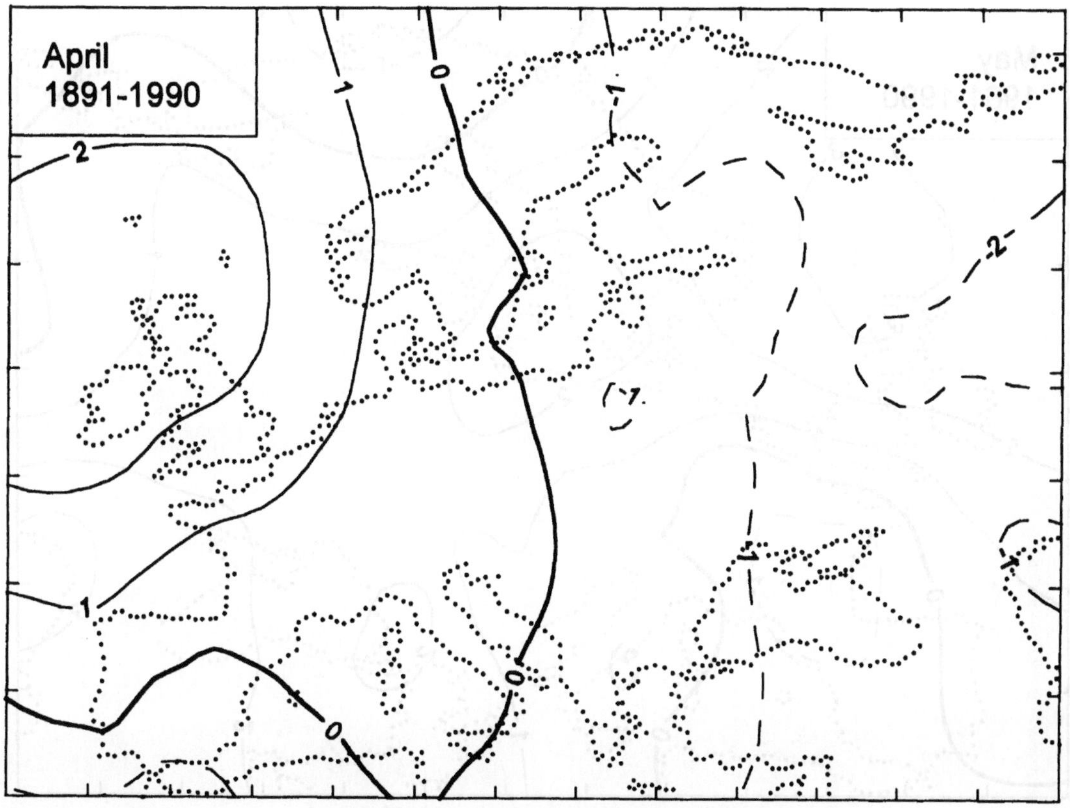

Sea level pressure, trend test

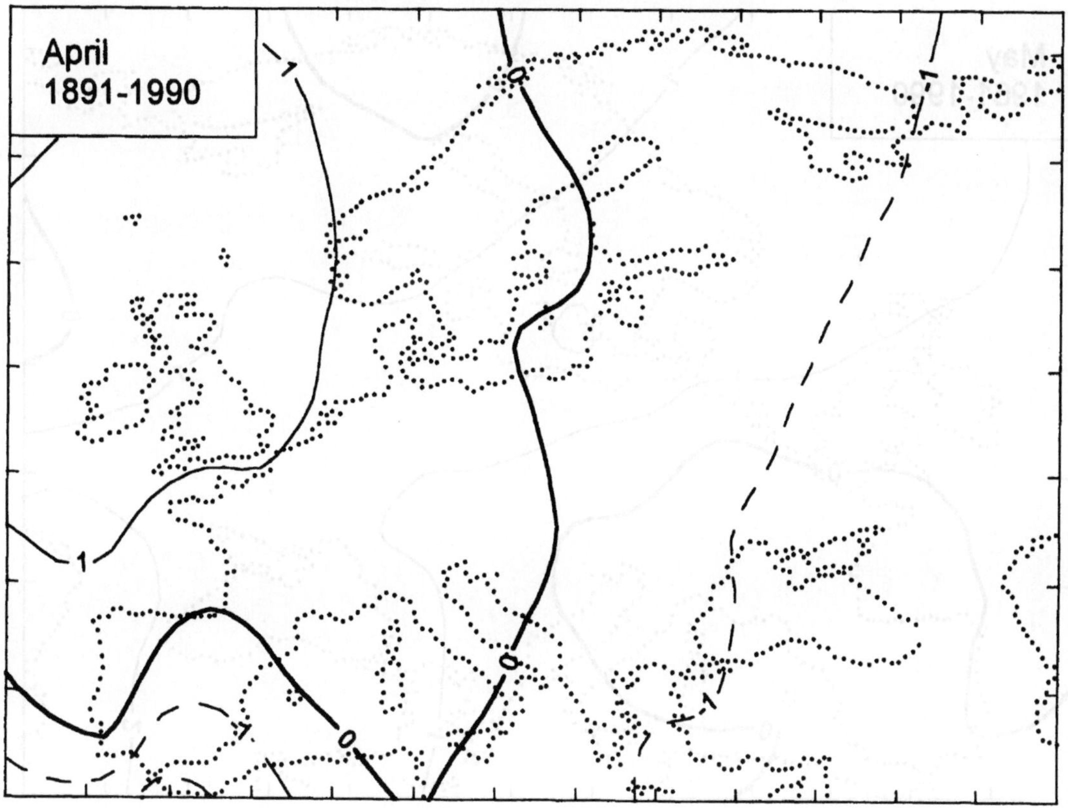

Sea level pressure, linear trend [hpa]

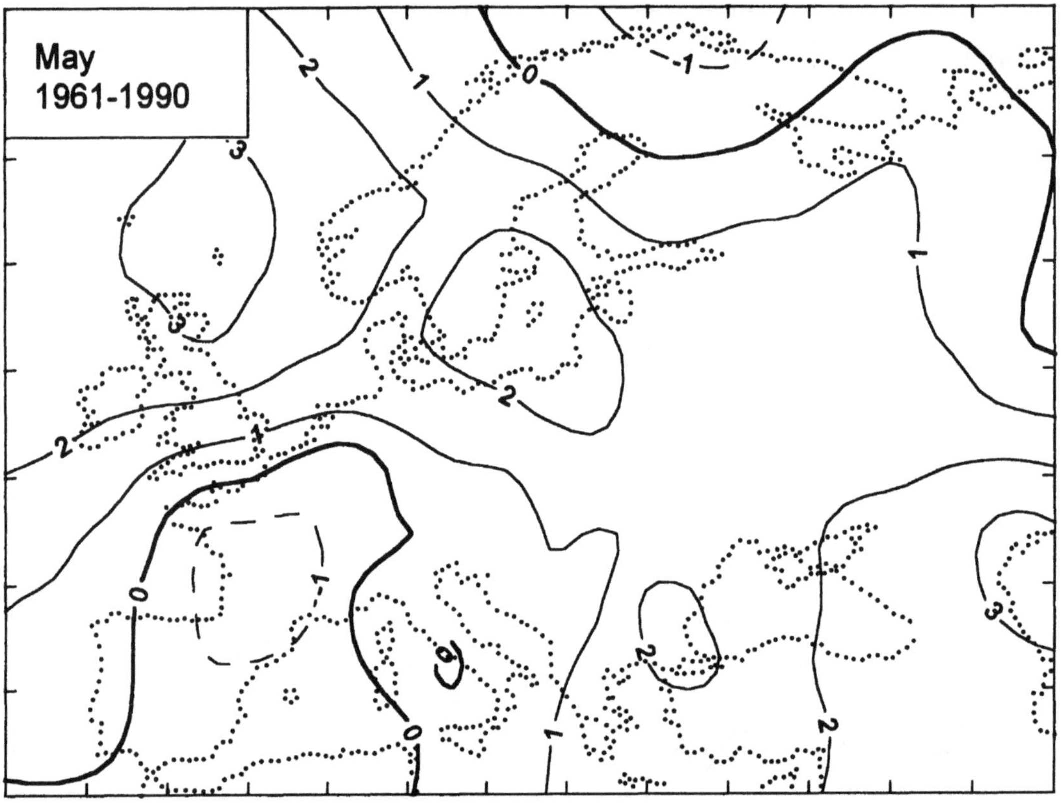

Sea level pressure, trend test

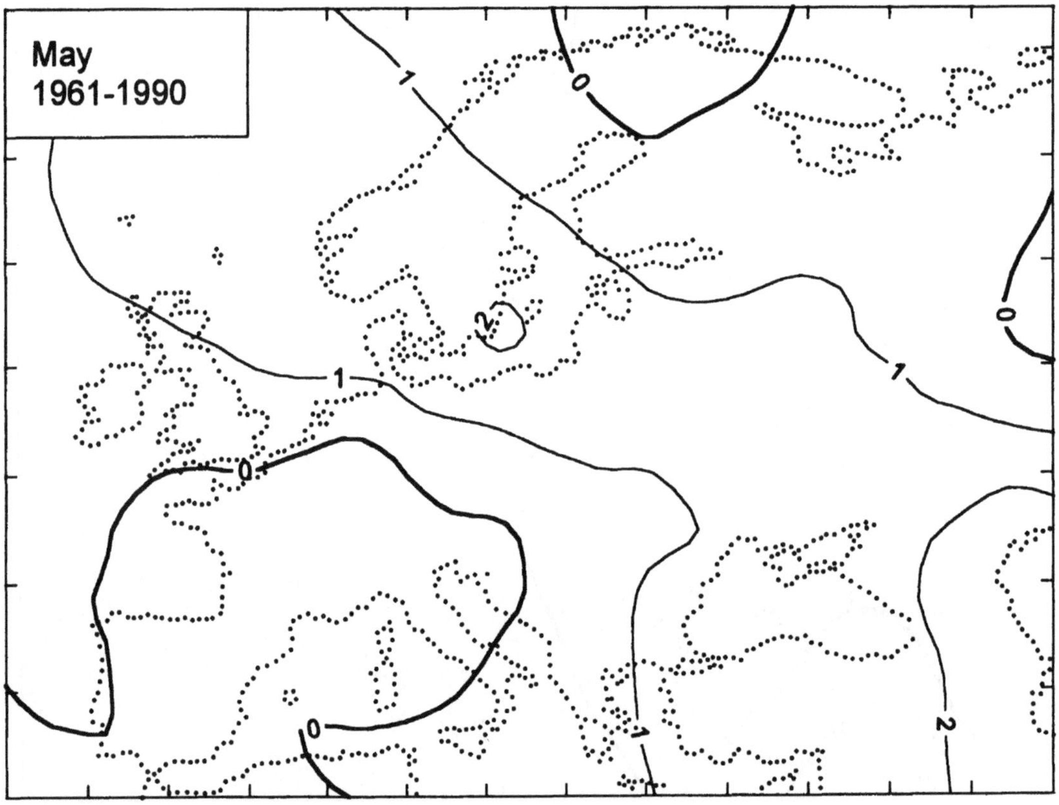

Sea level pressure, linear trend [hpa]

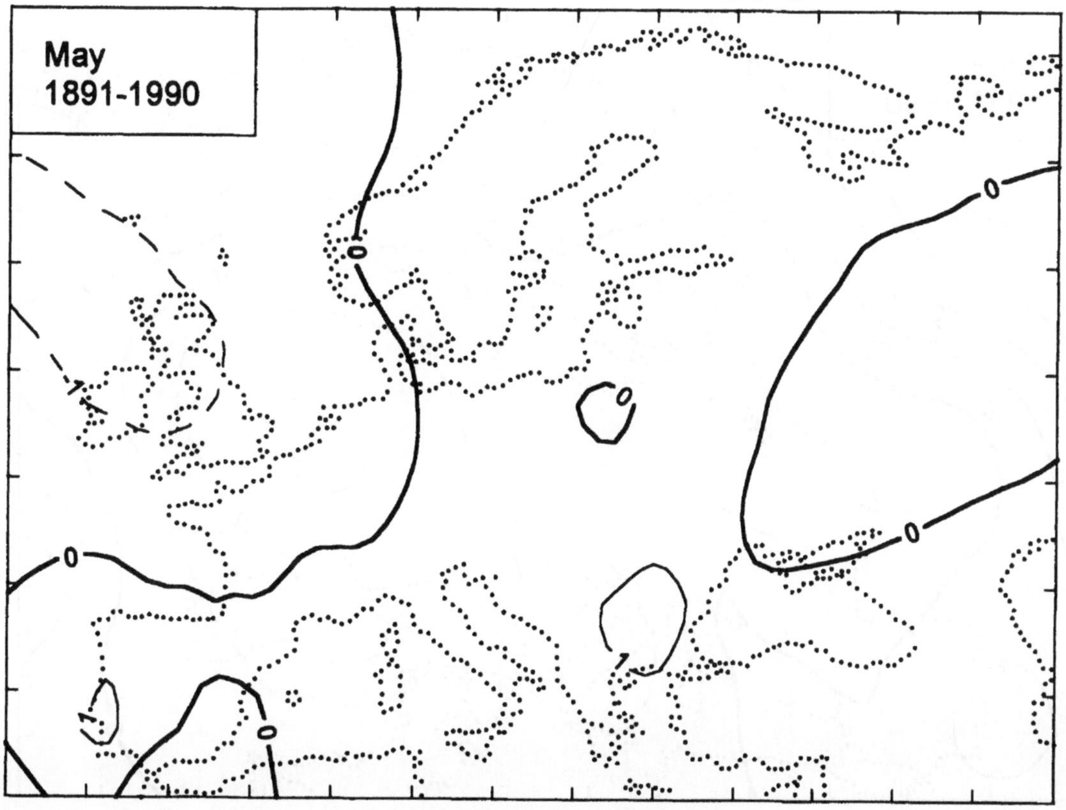

Sea level pressure, trend test

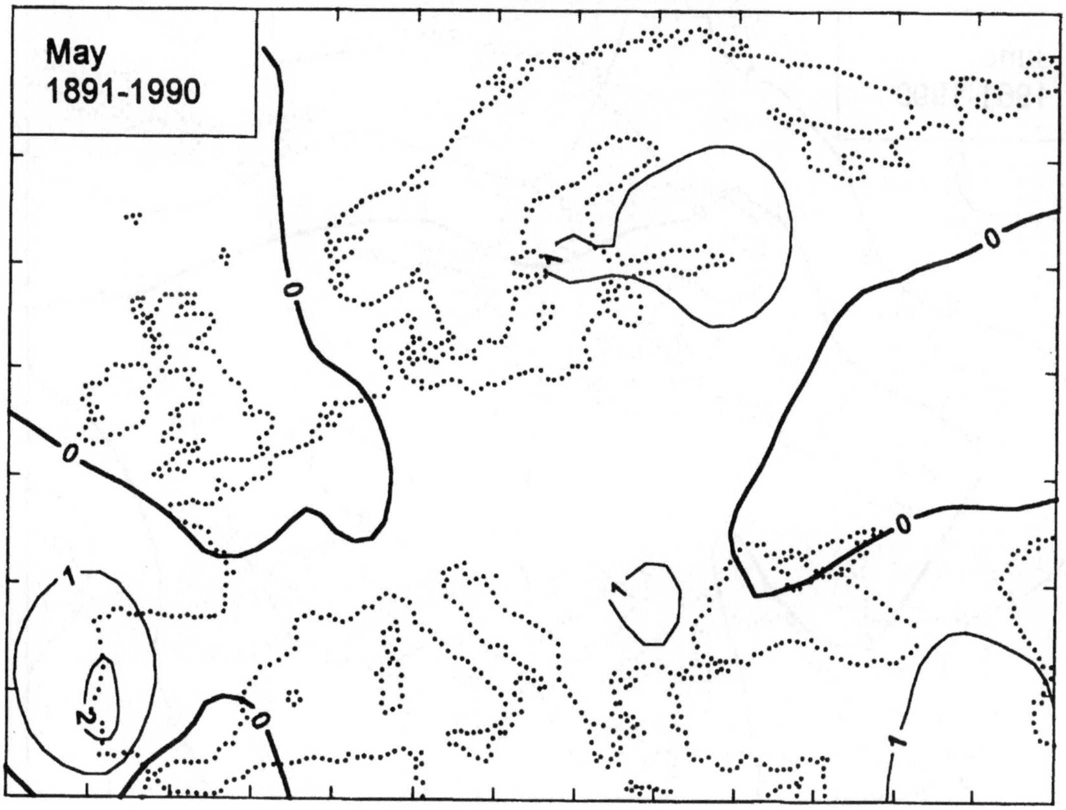

Sea level pressure, linear trend [hpa]

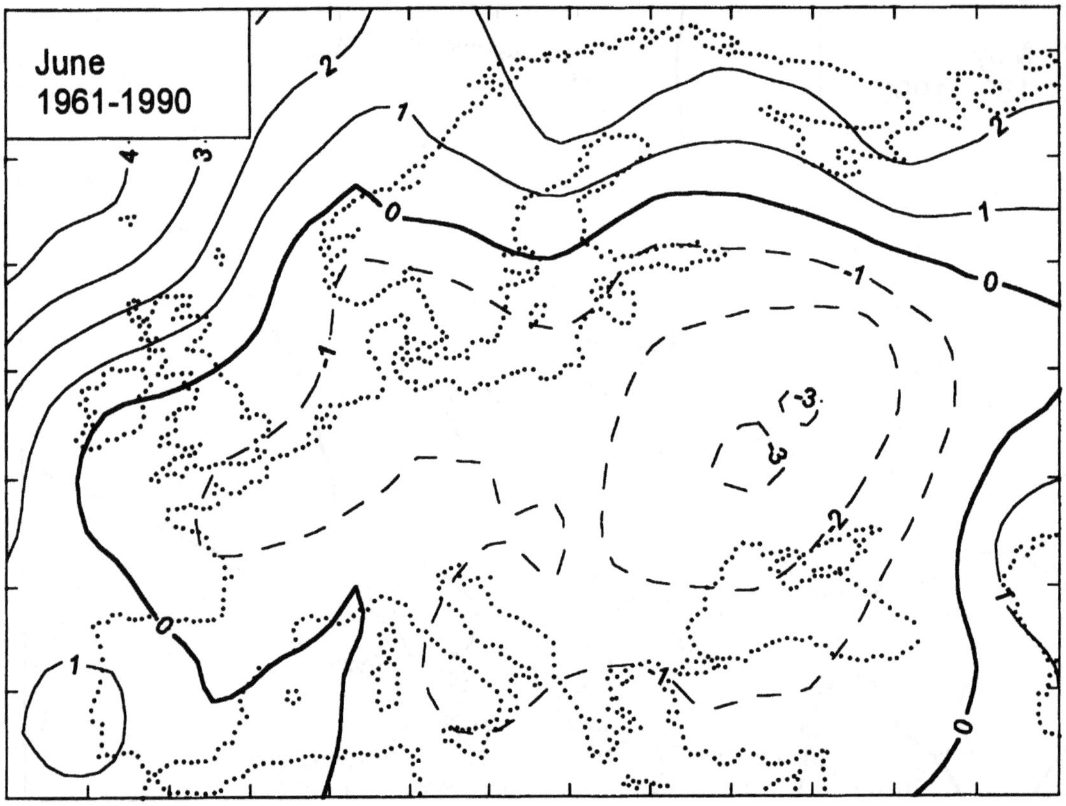

Sea level pressure, trend test

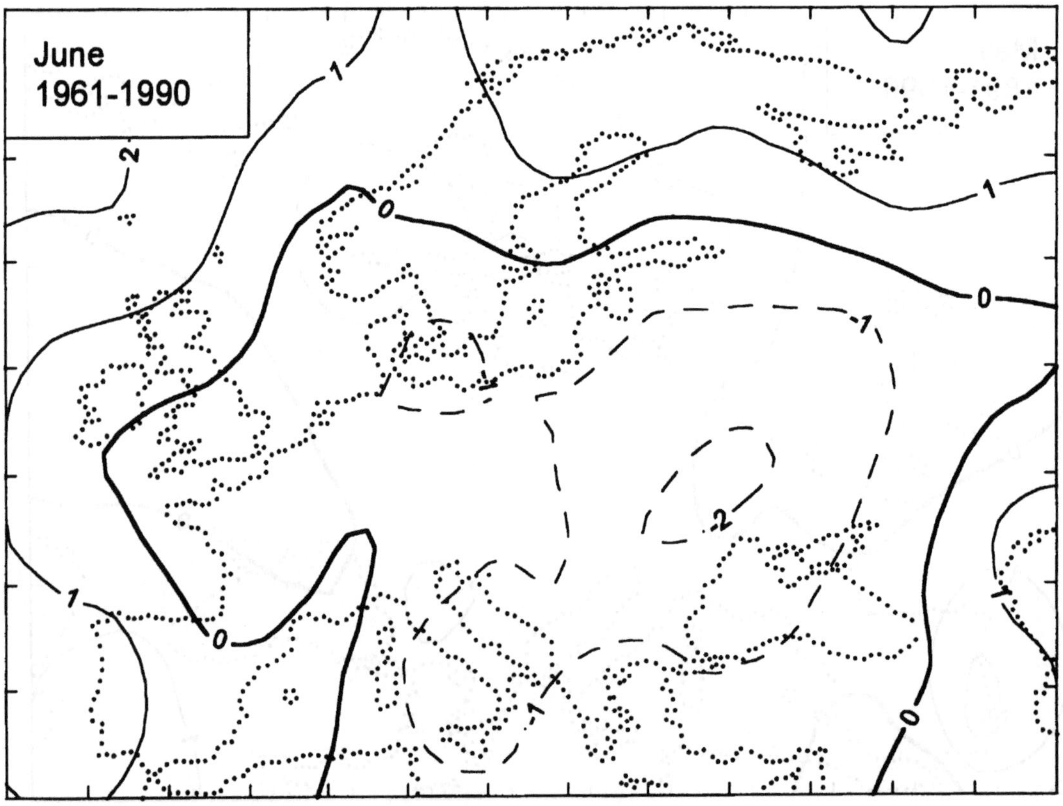

Sea level pressure, linear trend [hpa]

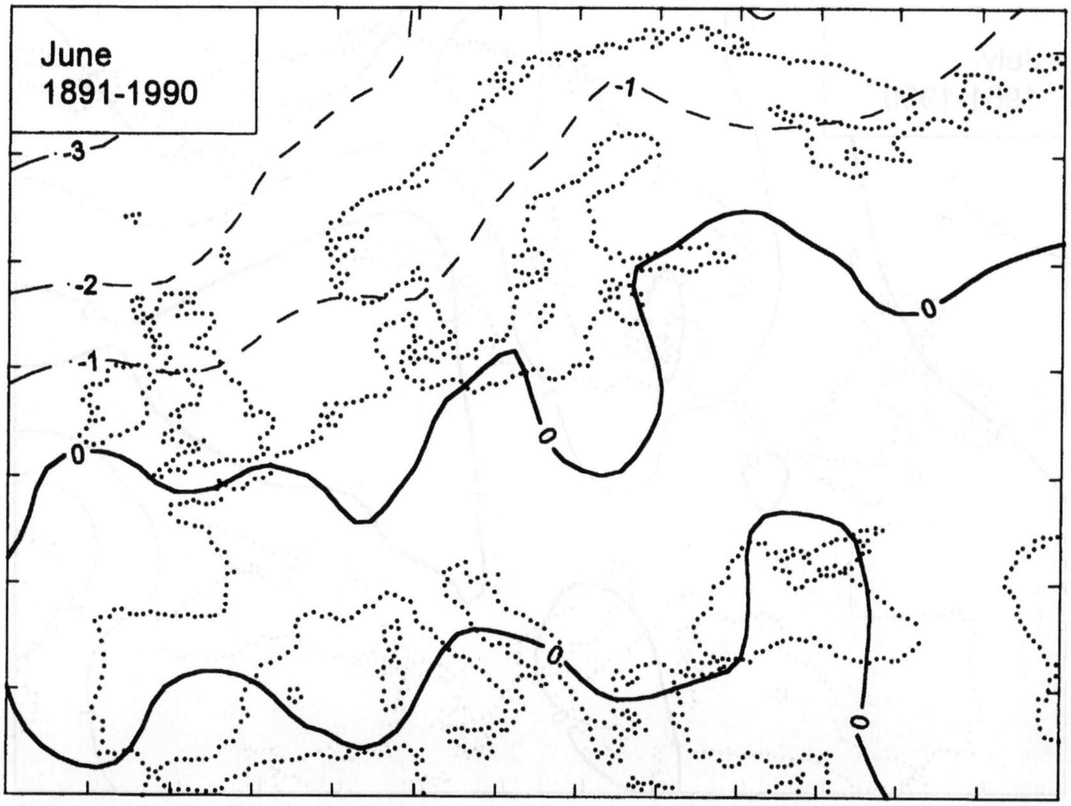

Sea level pressure, trend test

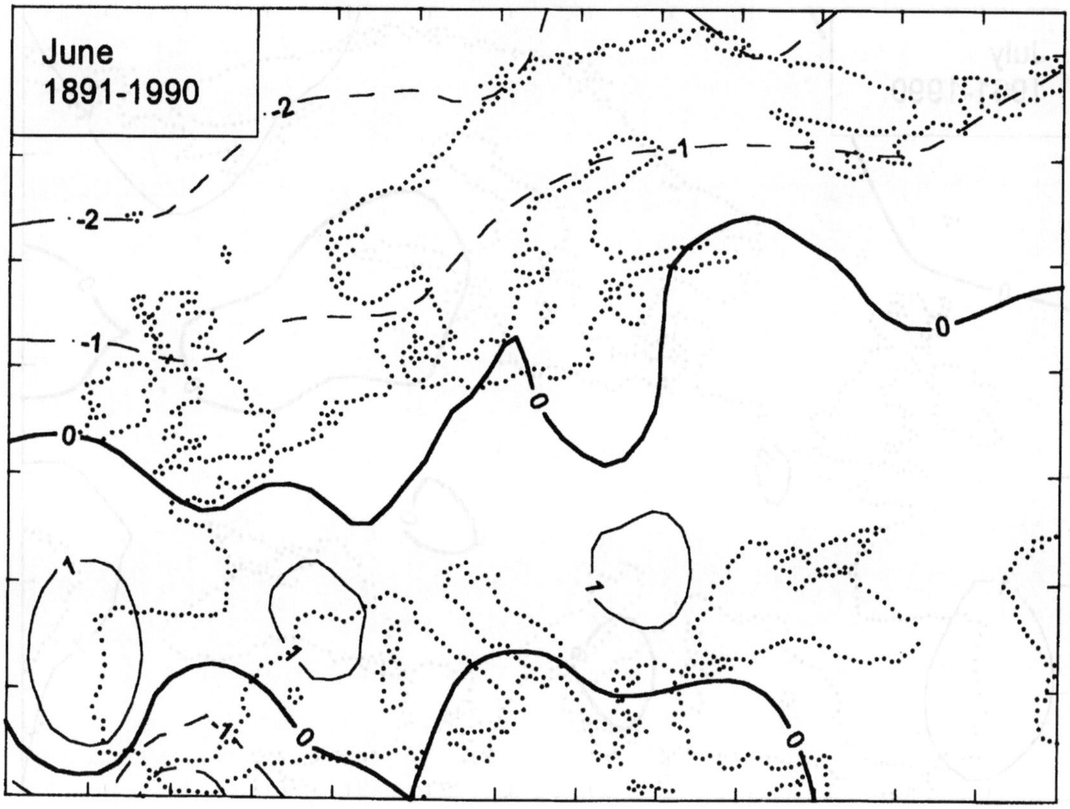

Sea level pressure, linear trend [hpa]

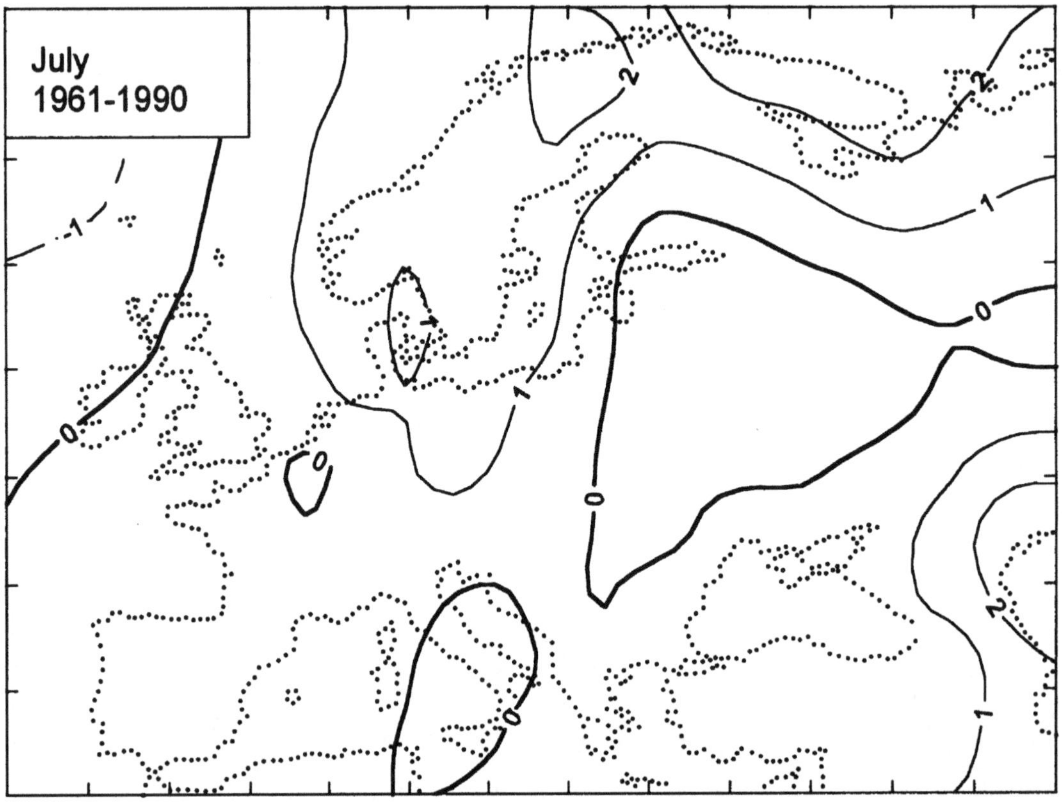

Sea level pressure, trend test

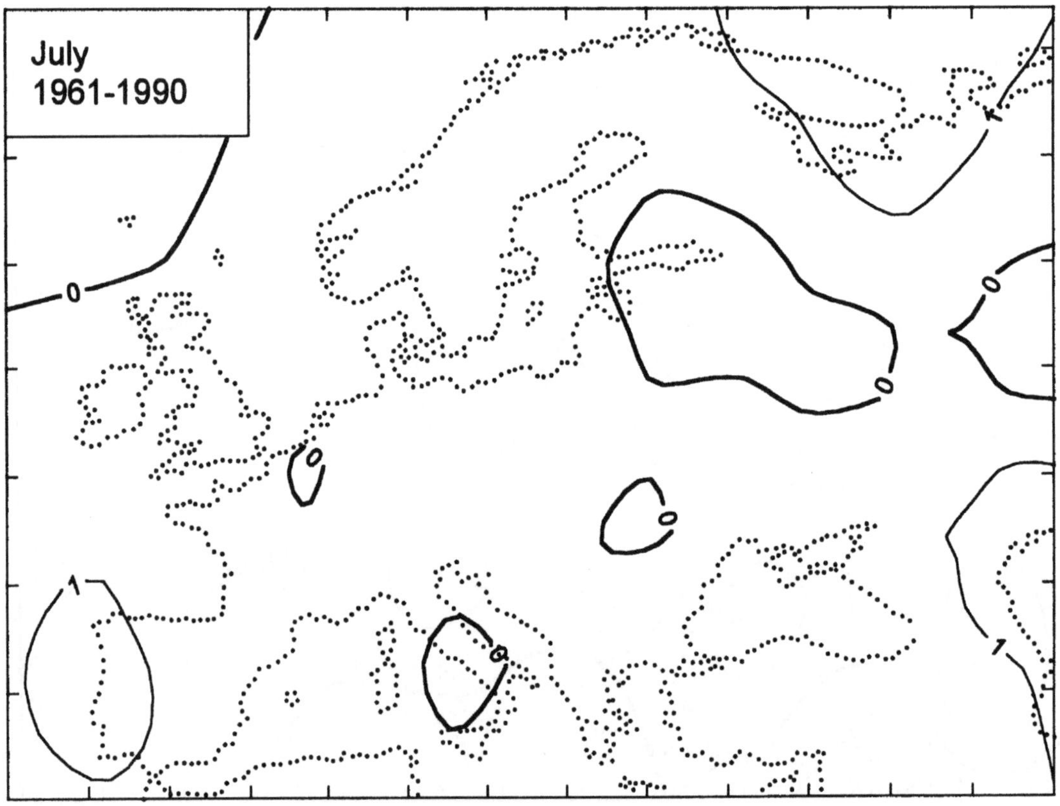

Sea level pressure, linear trend [hpa]

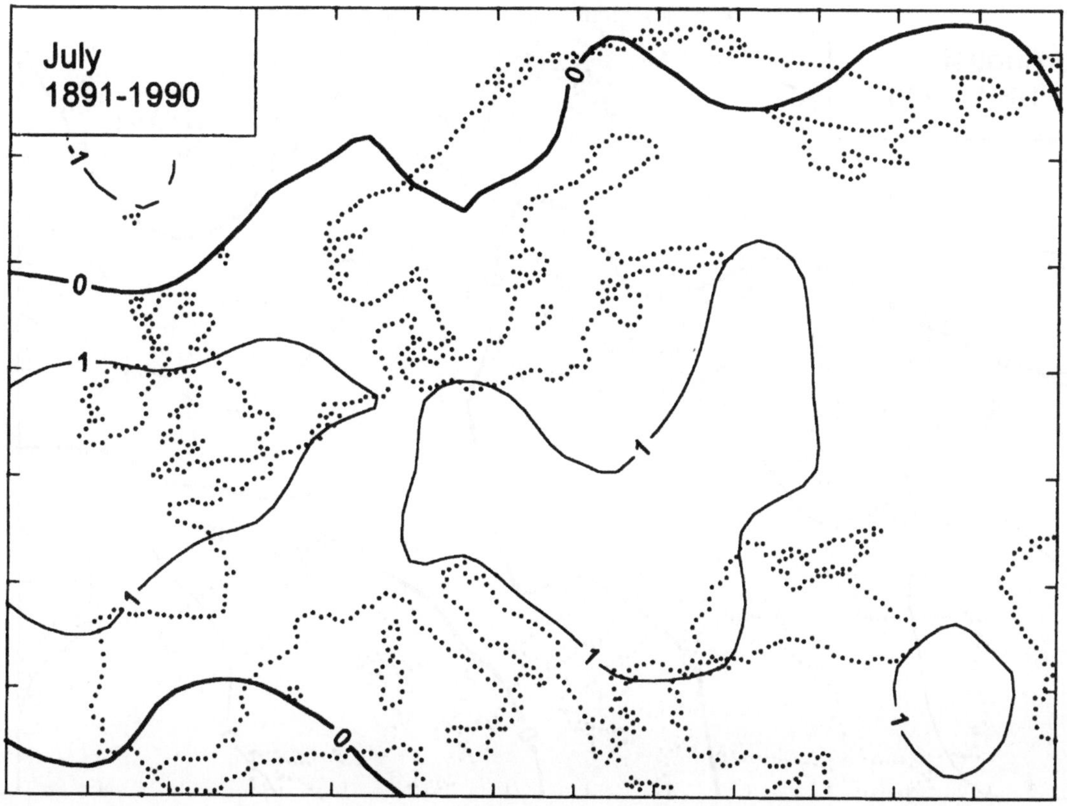

Sea level pressure, trend test

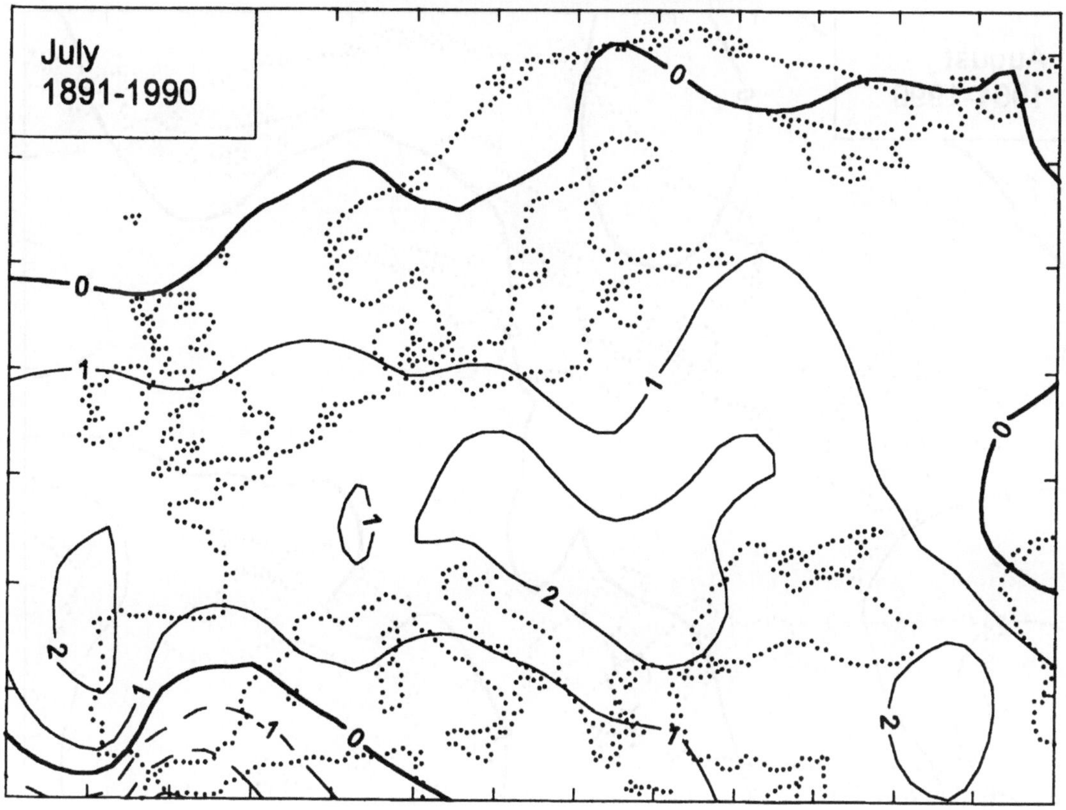

Sea level pressure, linear trend [hpa]

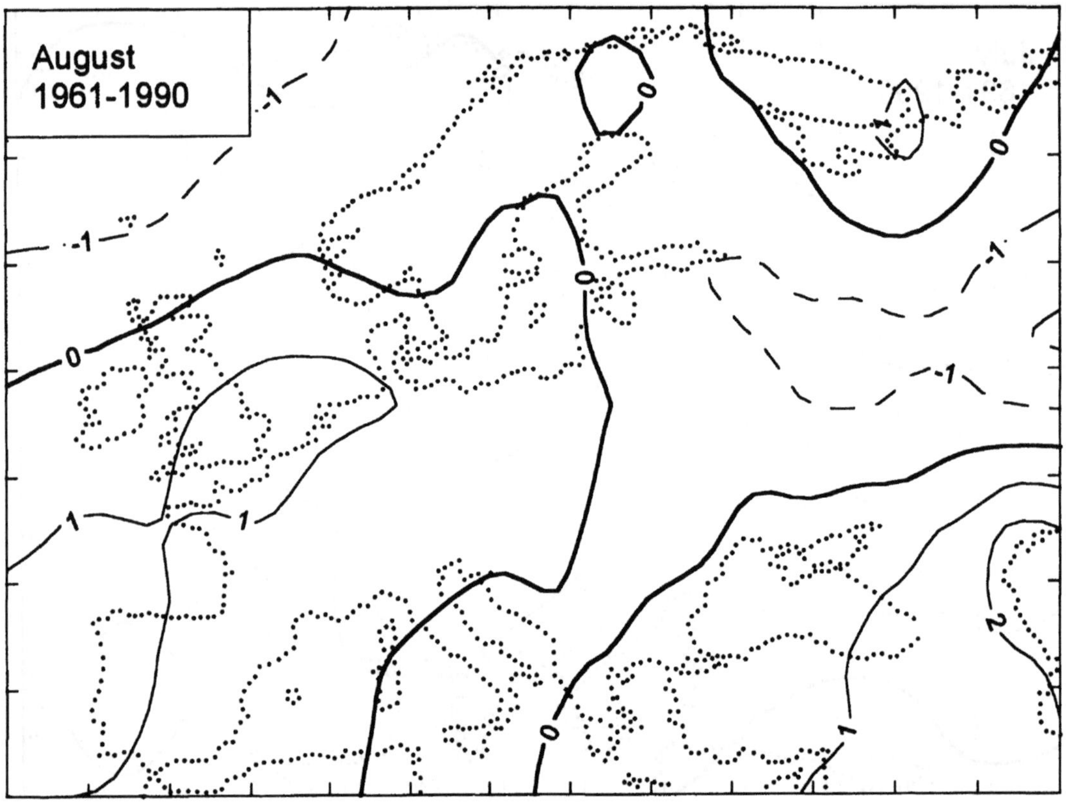

Sea level pressure, trend test

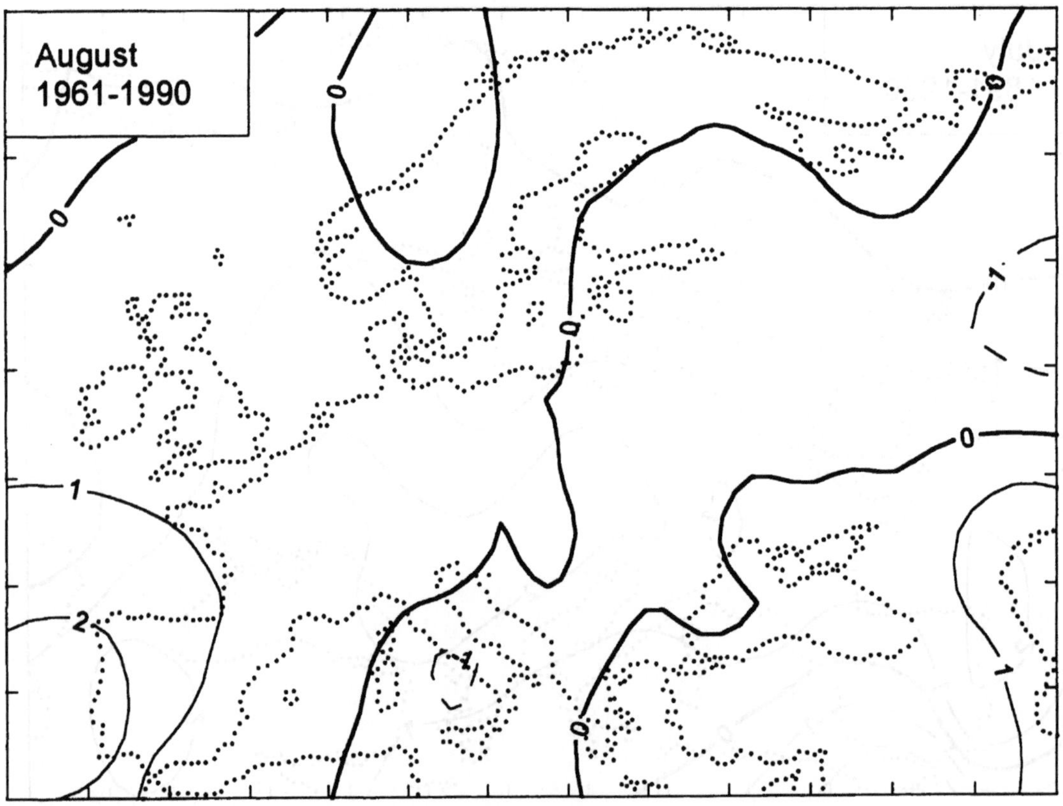

Sea level pressure, linear trend [hpa]

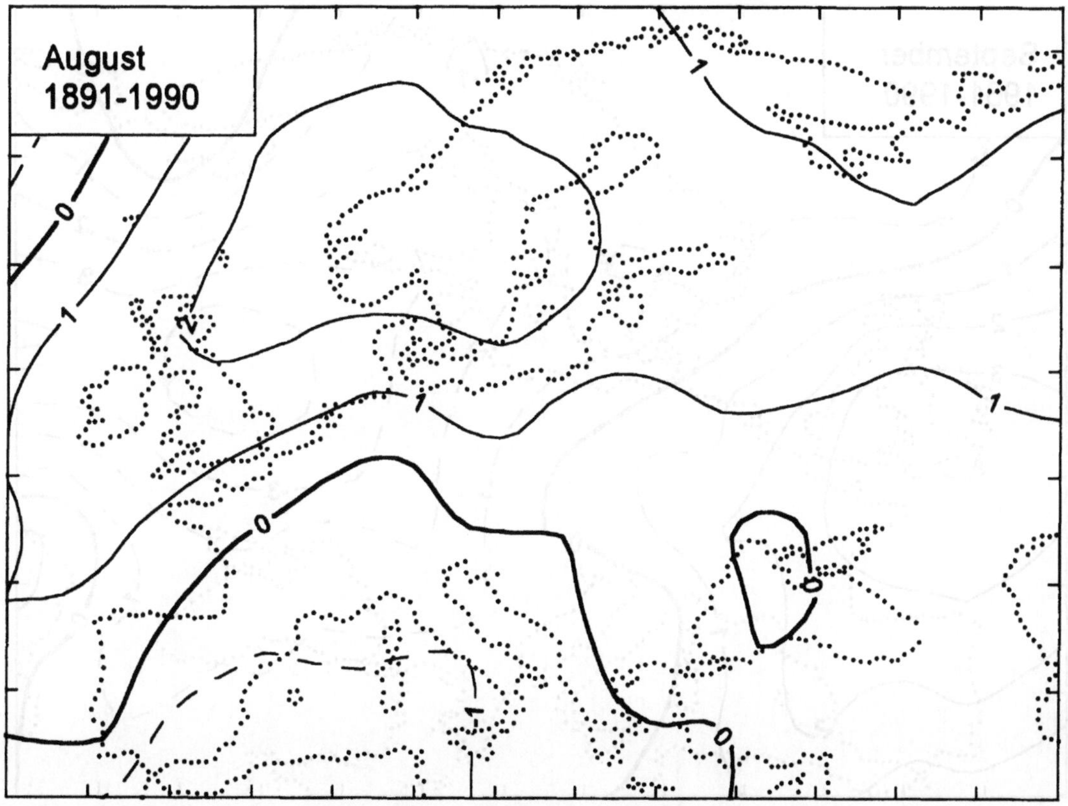

Sea level pressure, trend test

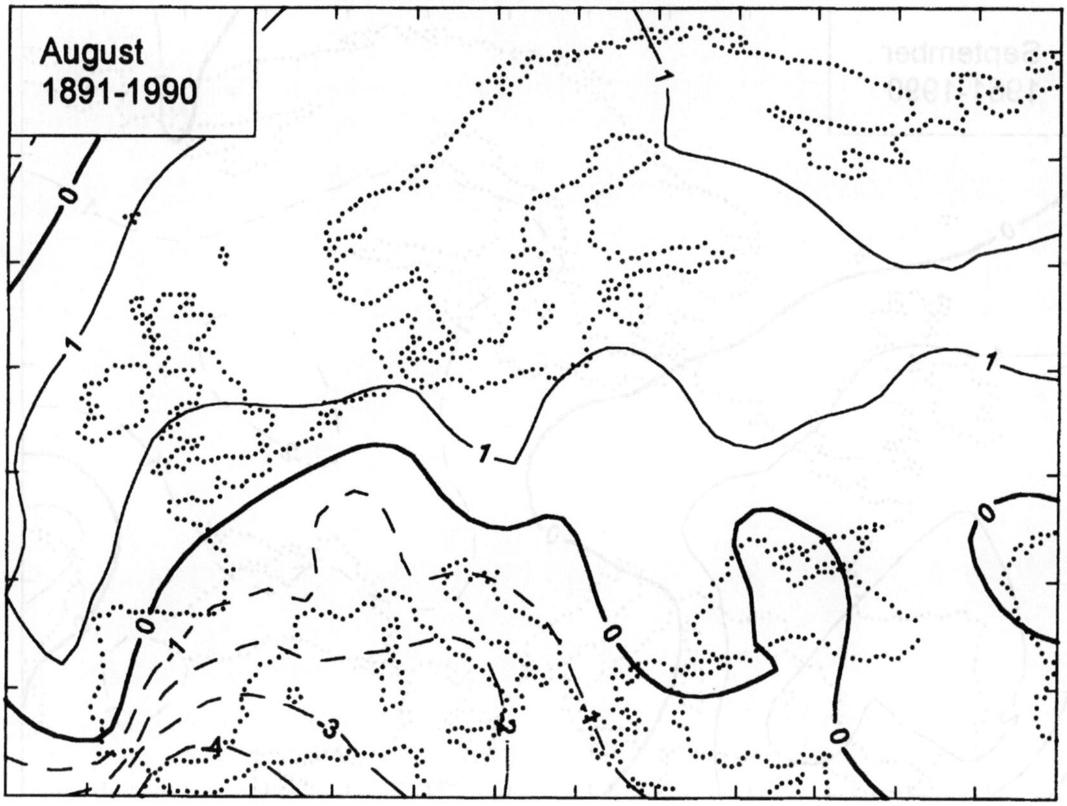

Sea level pressure, linear trend [hpa]

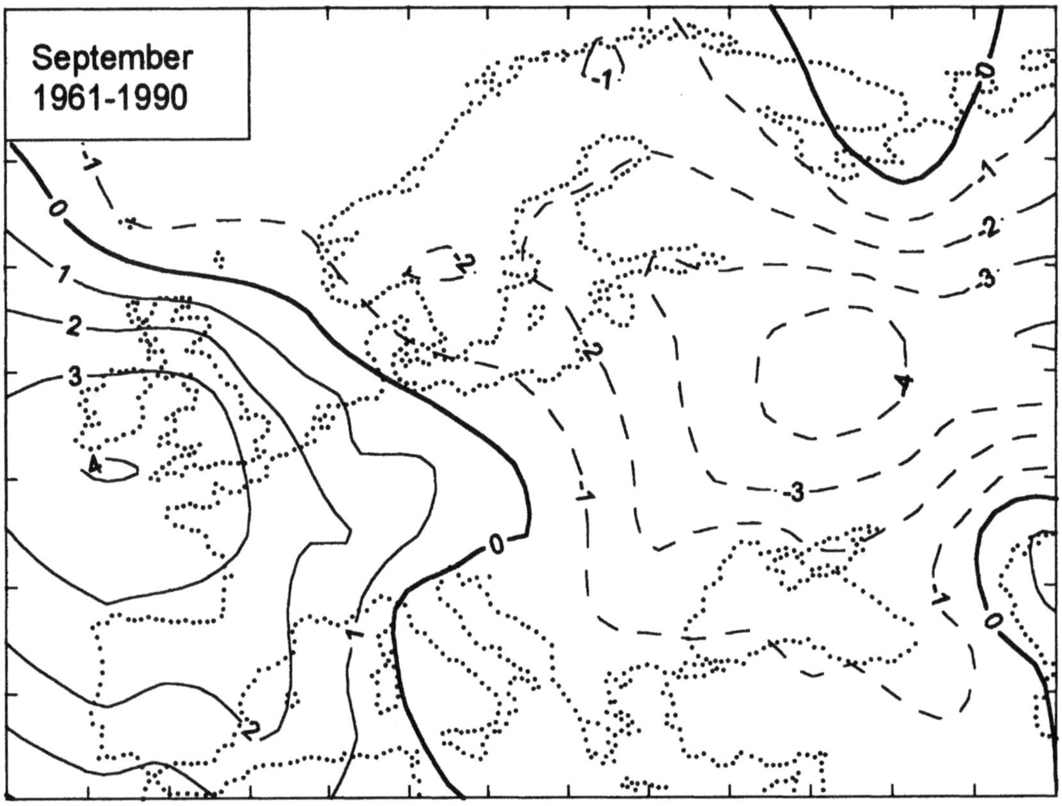

Sea level pressure, trend test

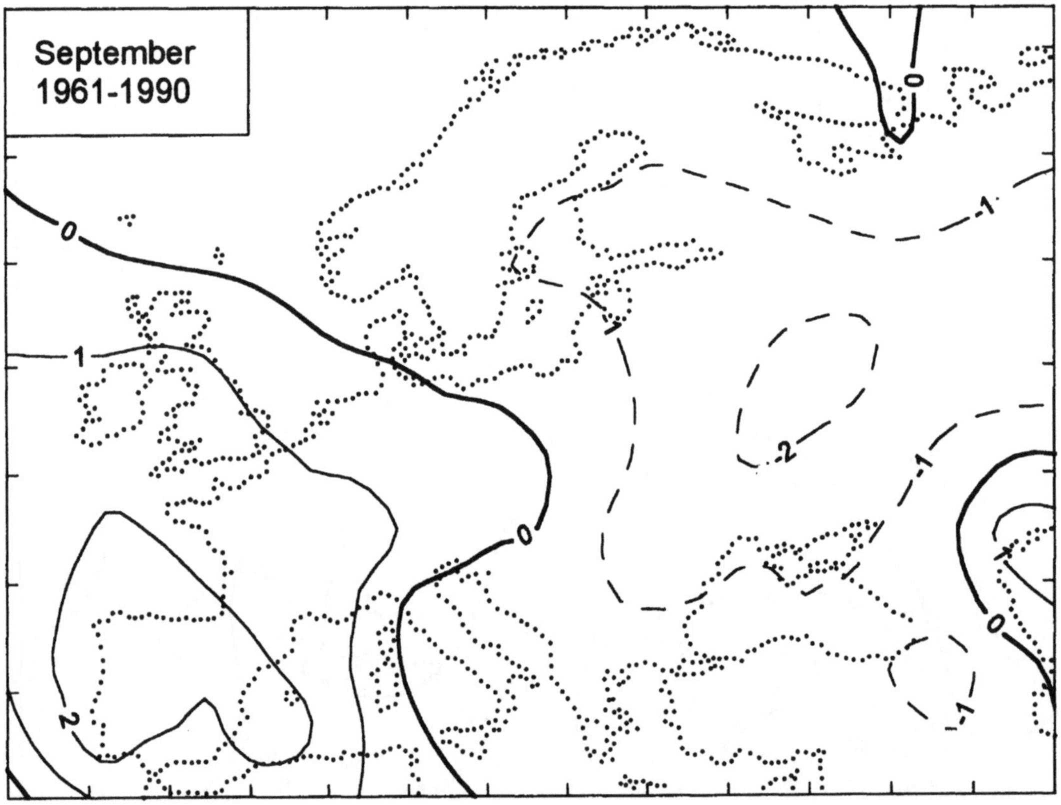

Sea level pressure, linear trend [hpa]

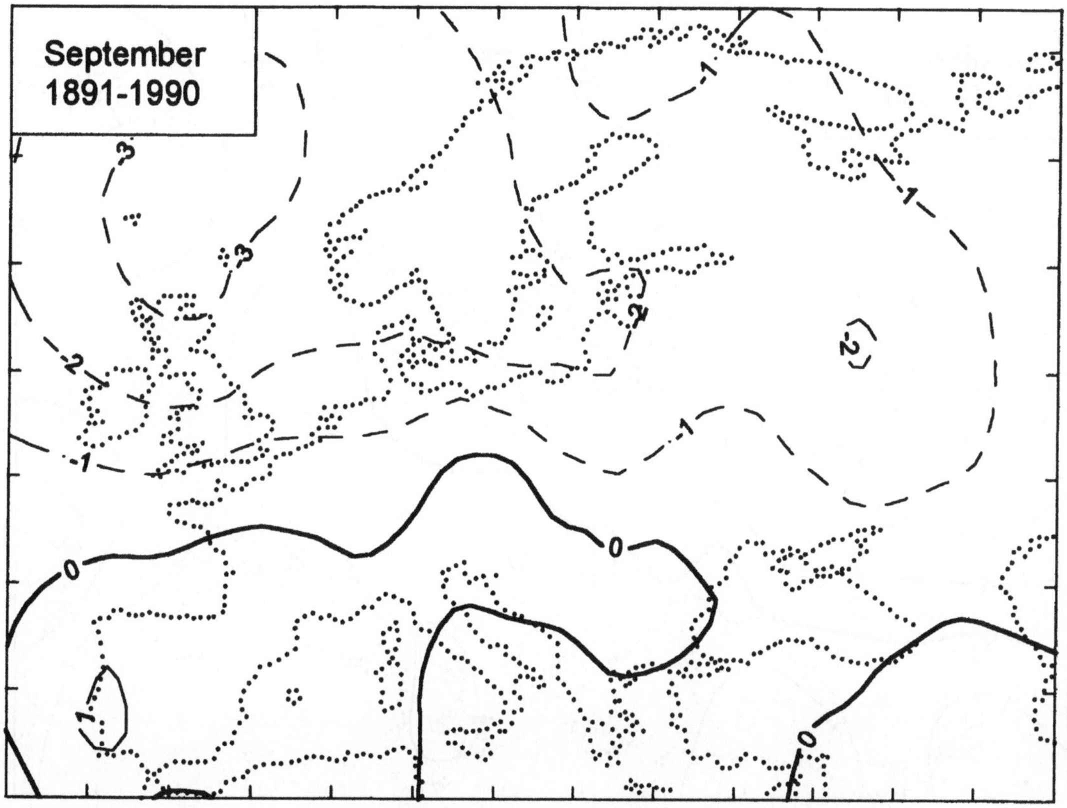

Sea level pressure, trend test

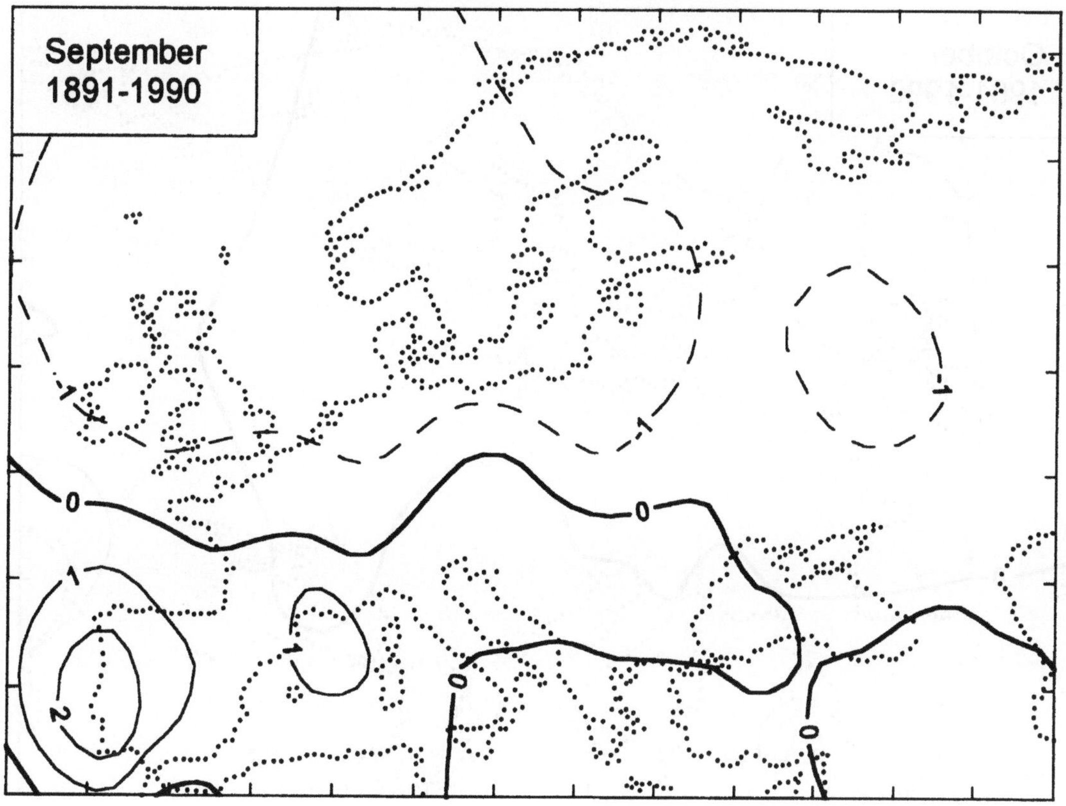

Sea level pressure, linear trend [hpa]

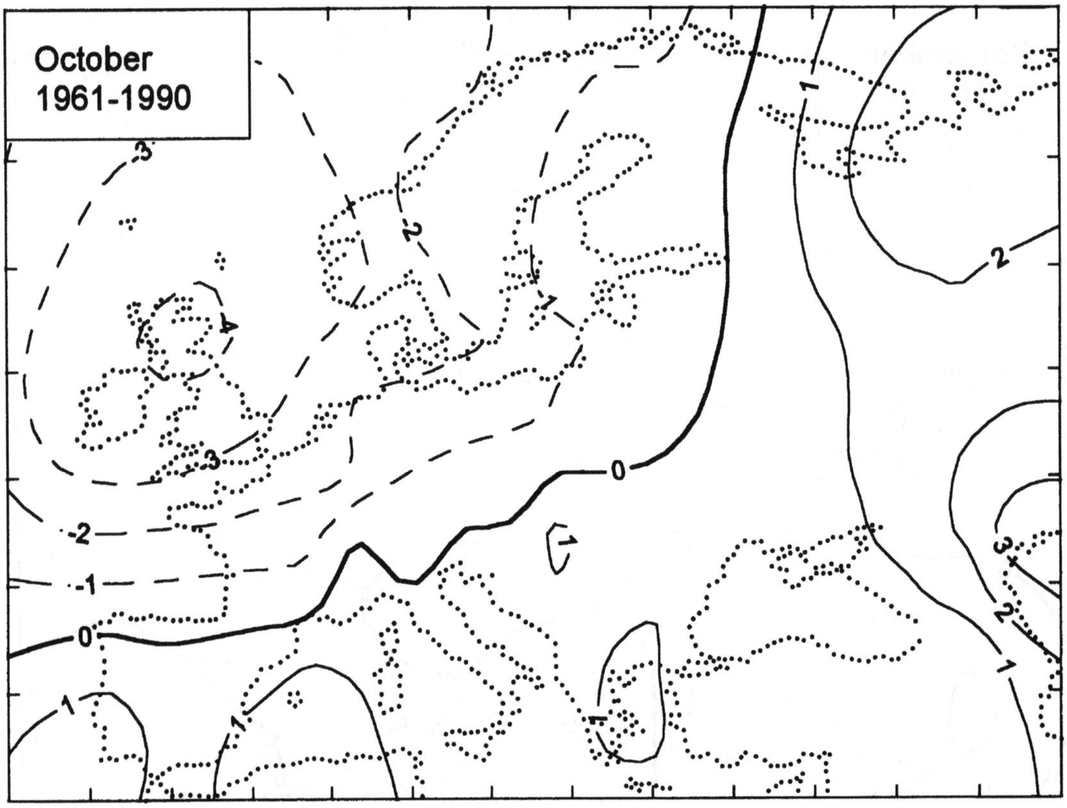

Sea level pressure, trend test

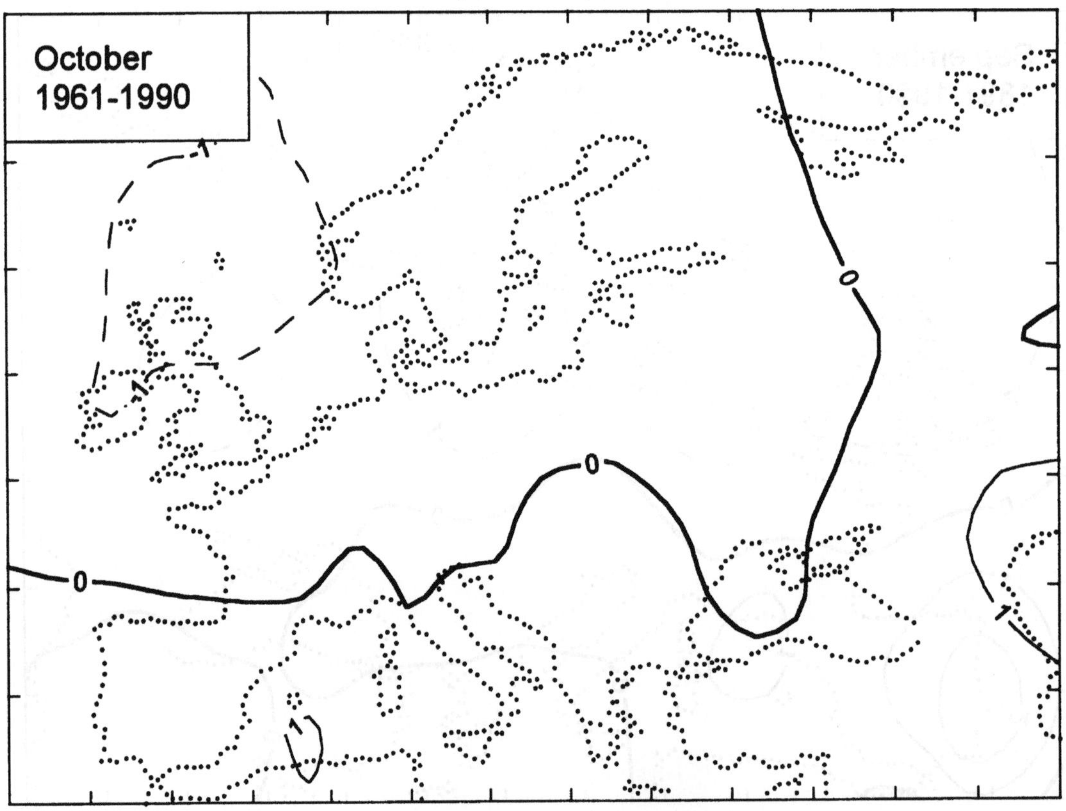

Sea level pressure, linear trend [hpa]

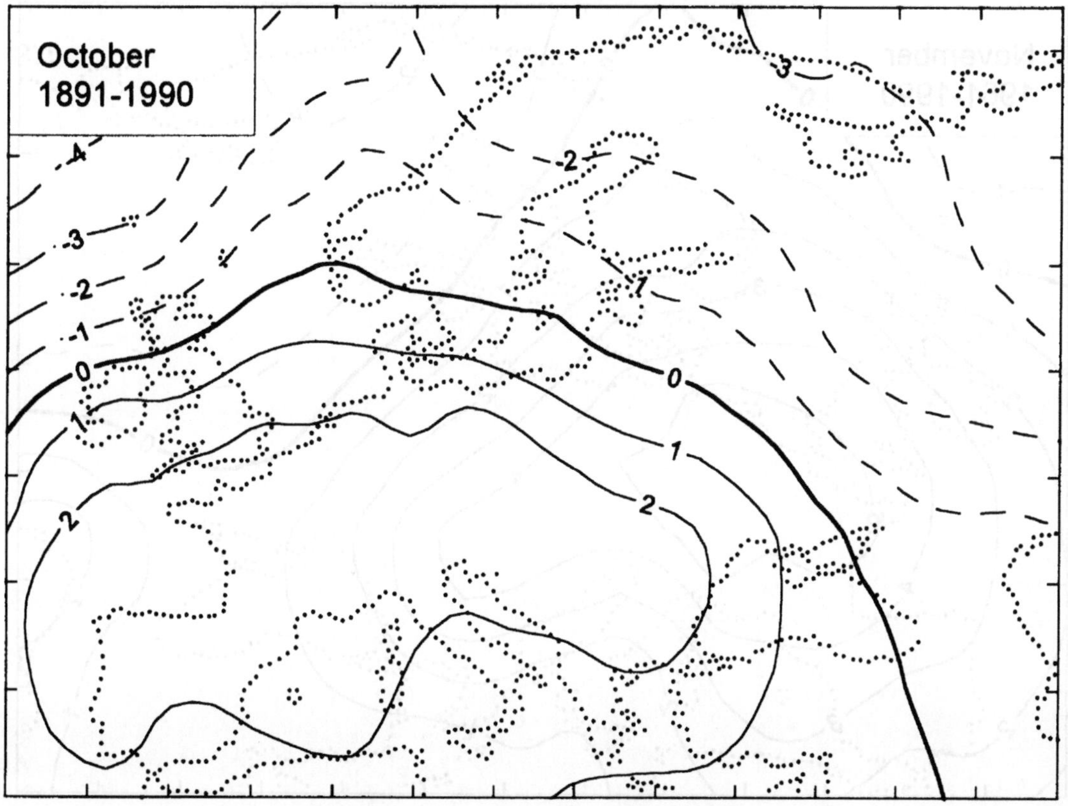

Sea level pressure, trend test

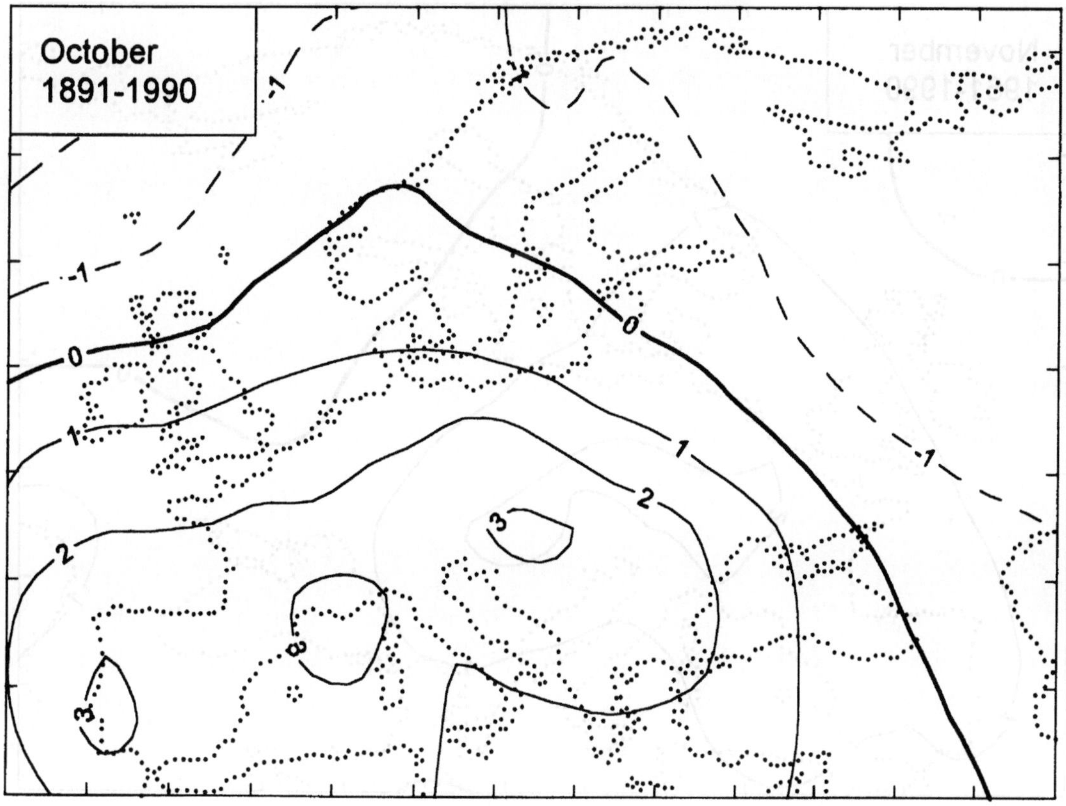

Sea level pressure, linear trend [hpa]

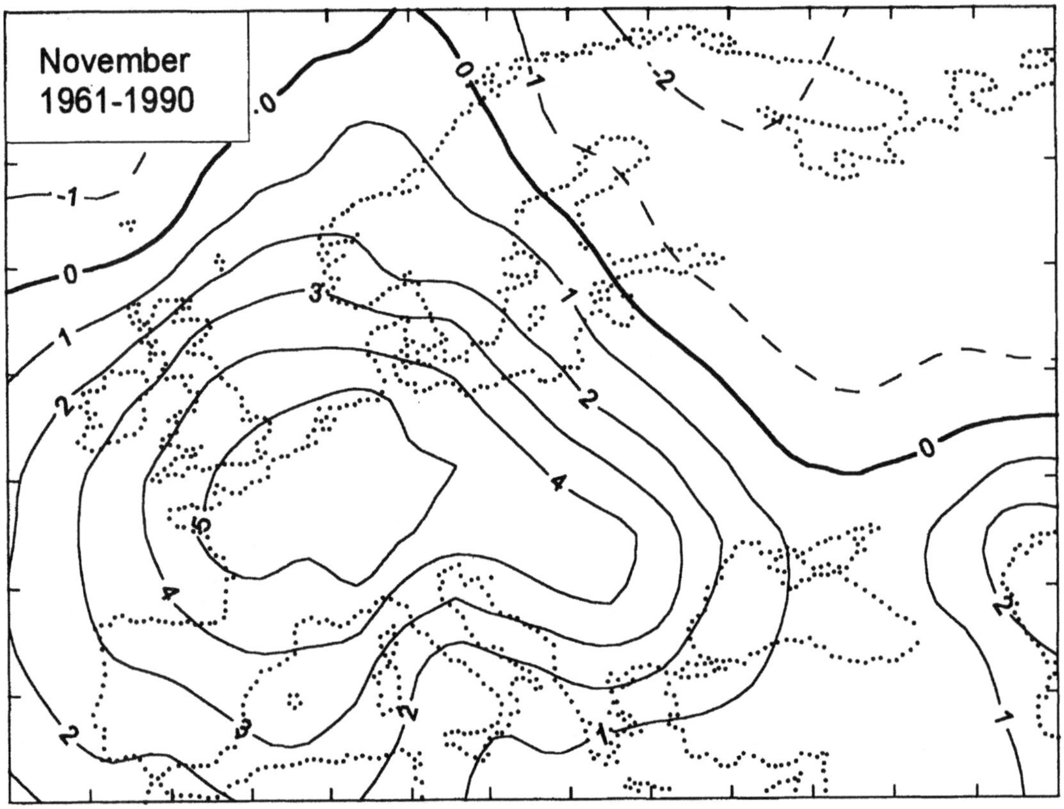

Sea level pressure, trend test

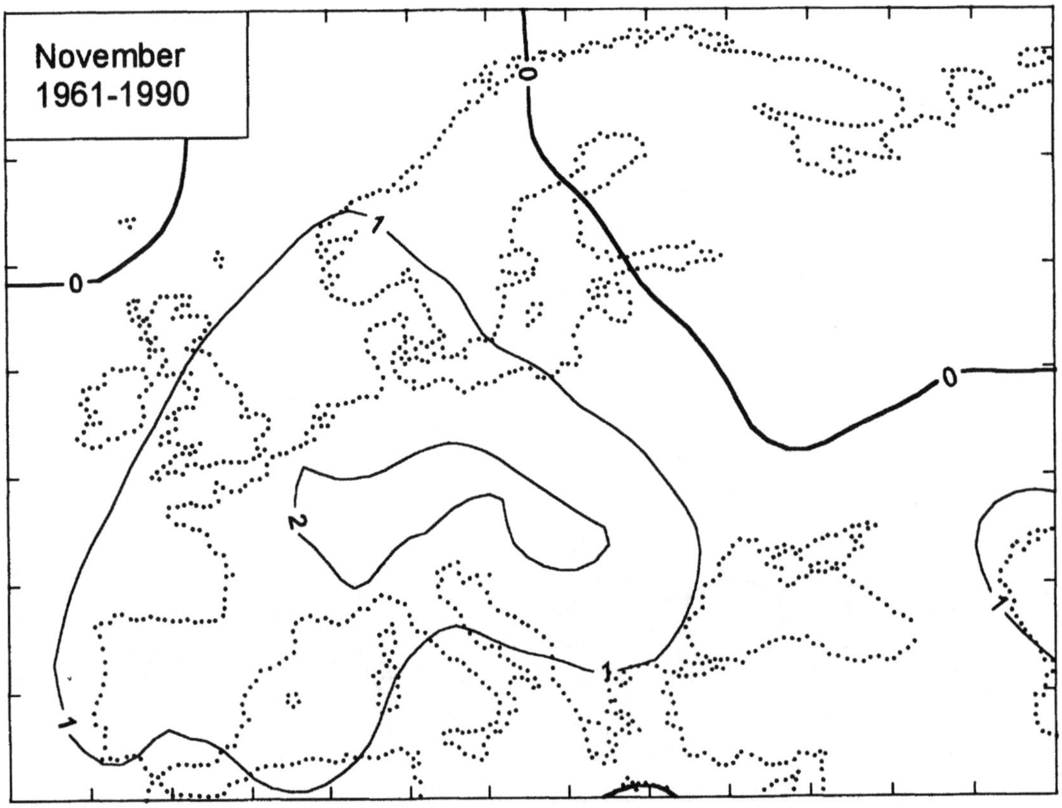

Sea level pressure, linear trend [hpa]

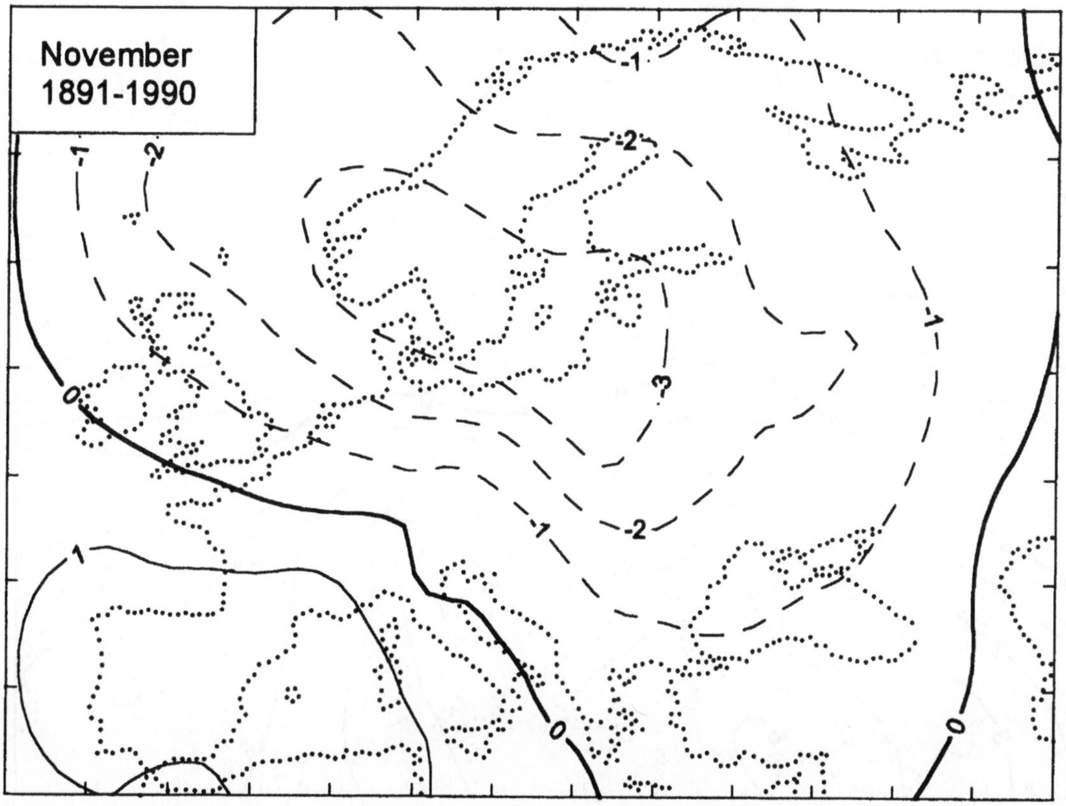

Sea level pressure, trend test

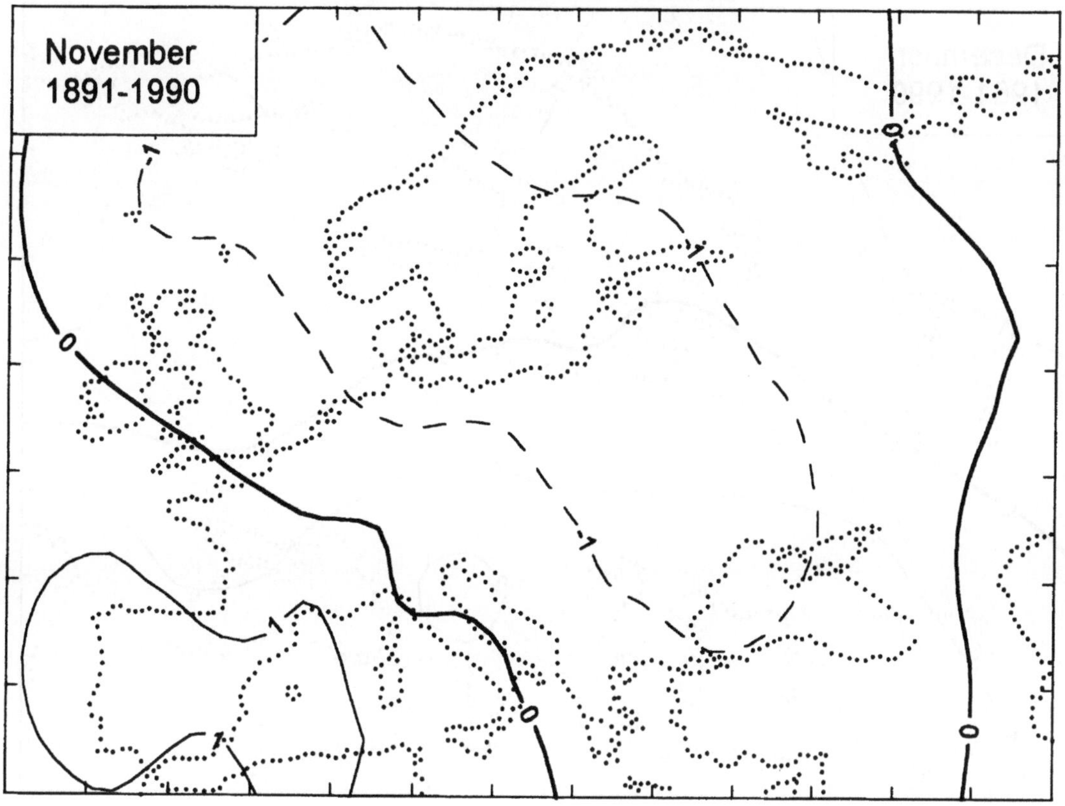

Sea level pressure, linear trend [hpa]

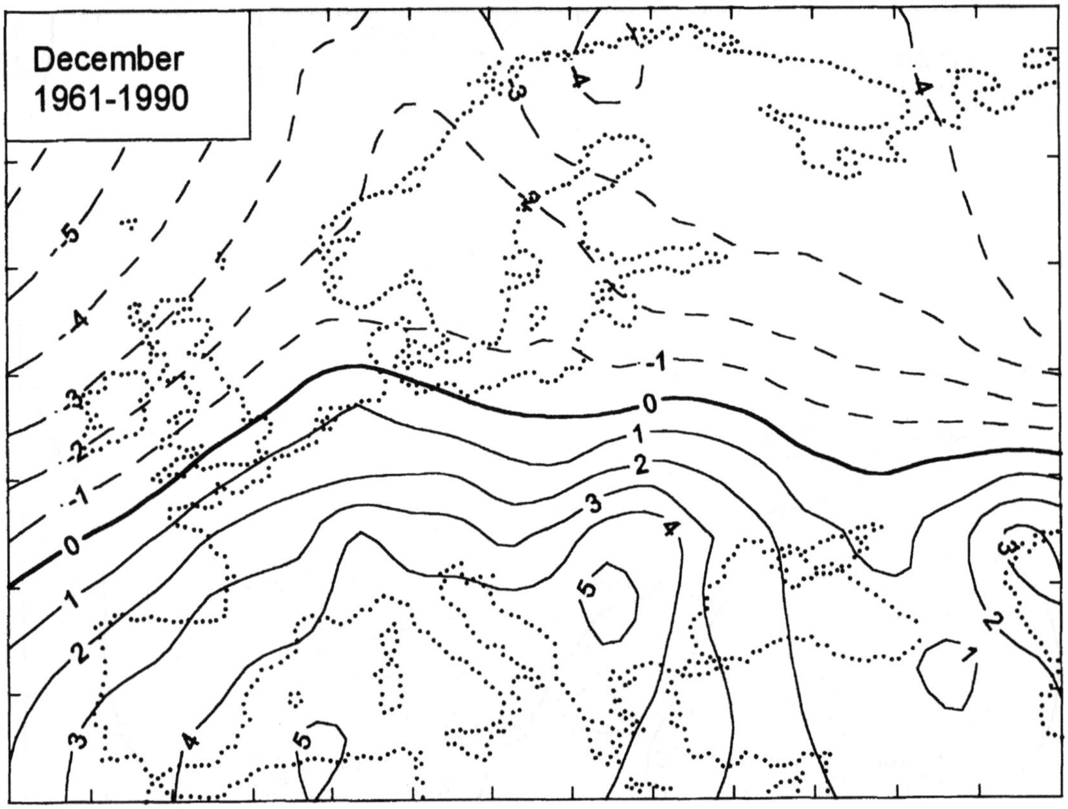

Sea level pressure, trend test

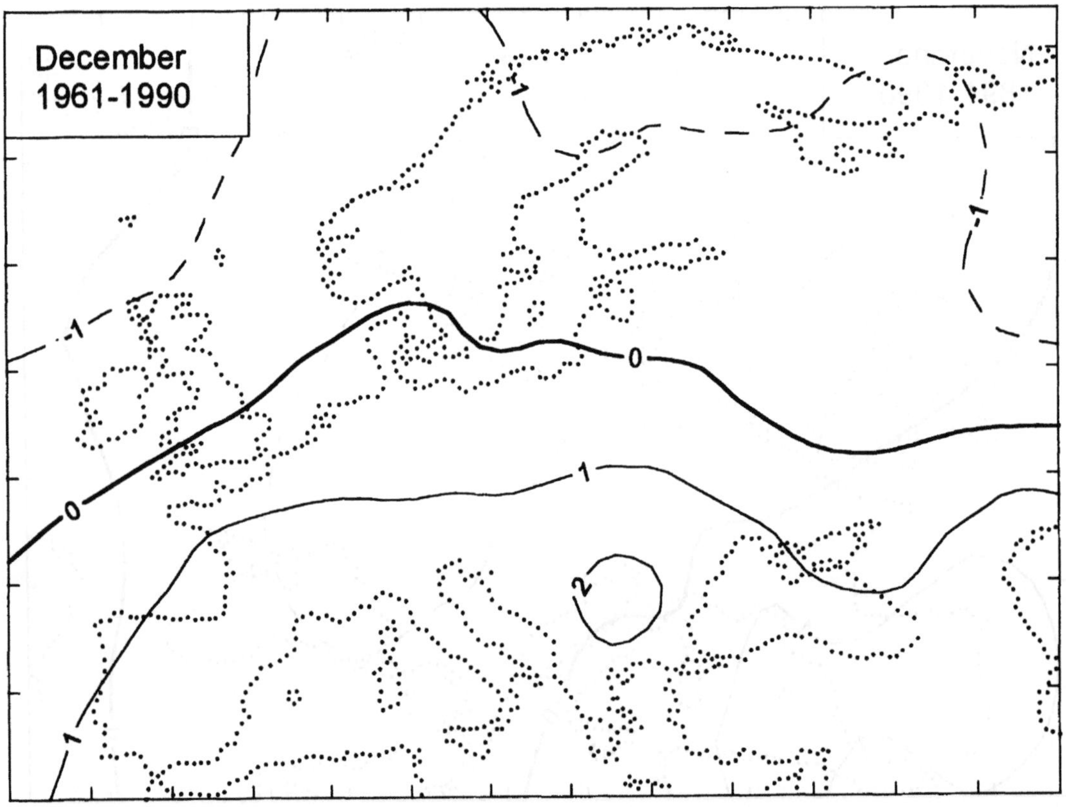

Sea level pressure, linear trend [hpa]

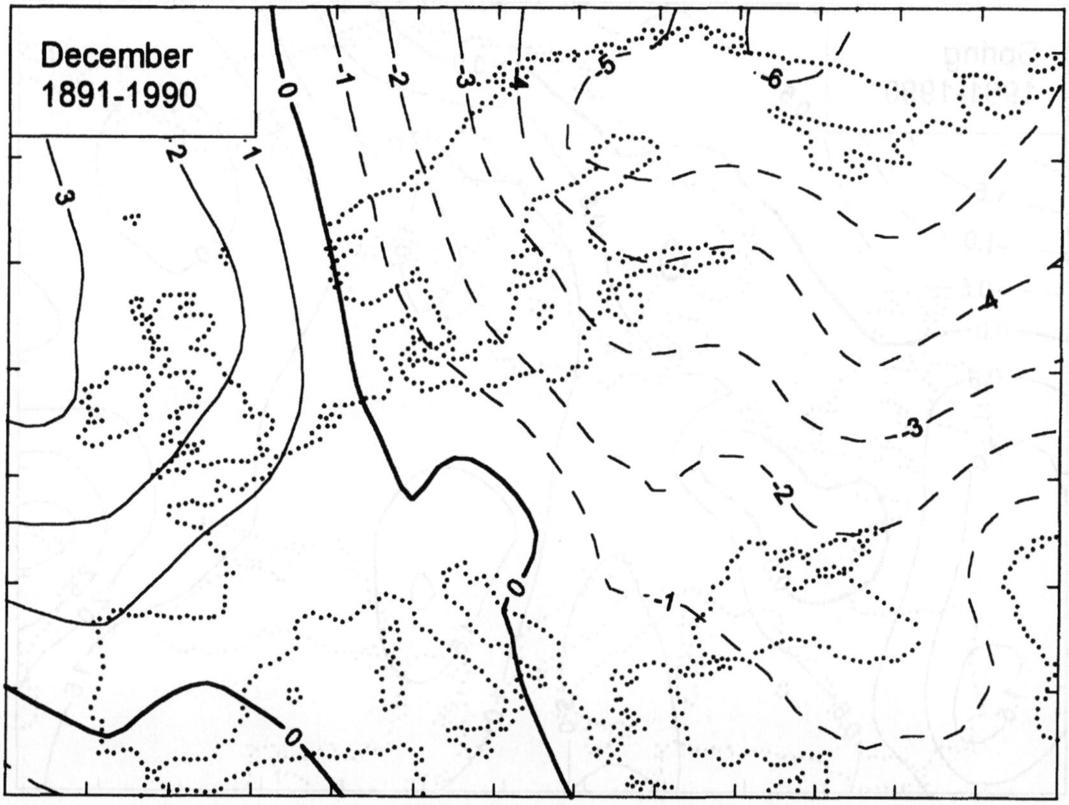

Sea level pressure, trend test

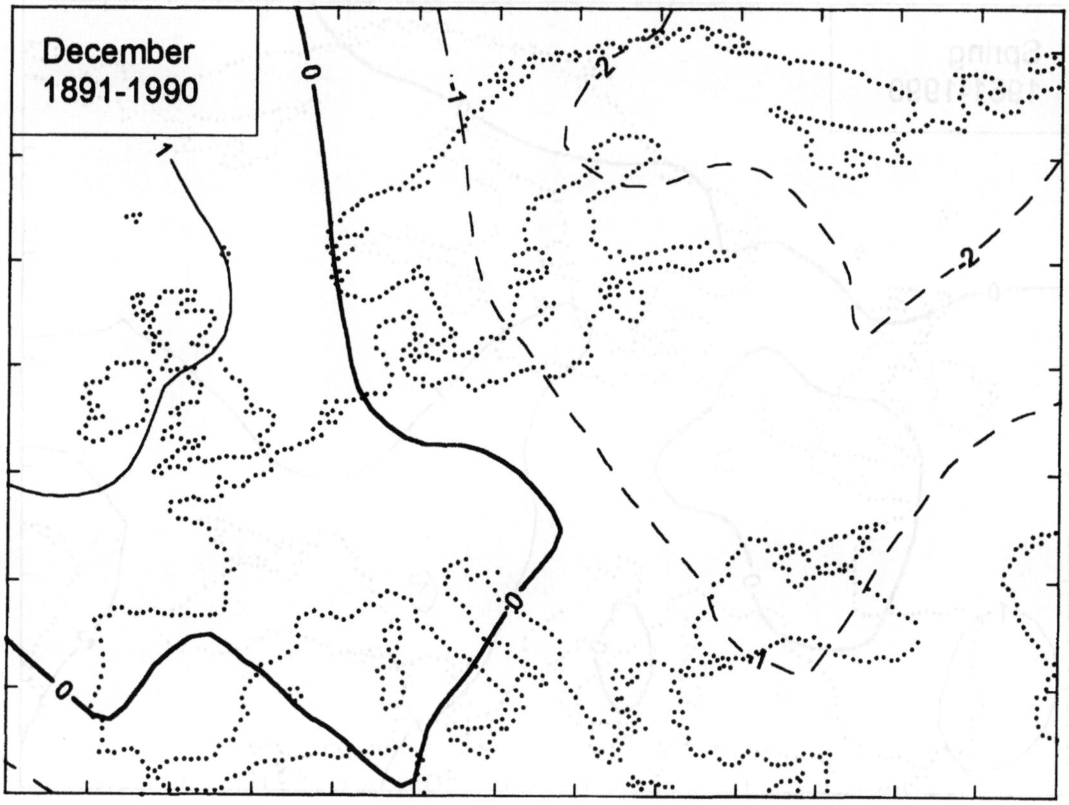

Sea level pressure, linear trend [hpa]

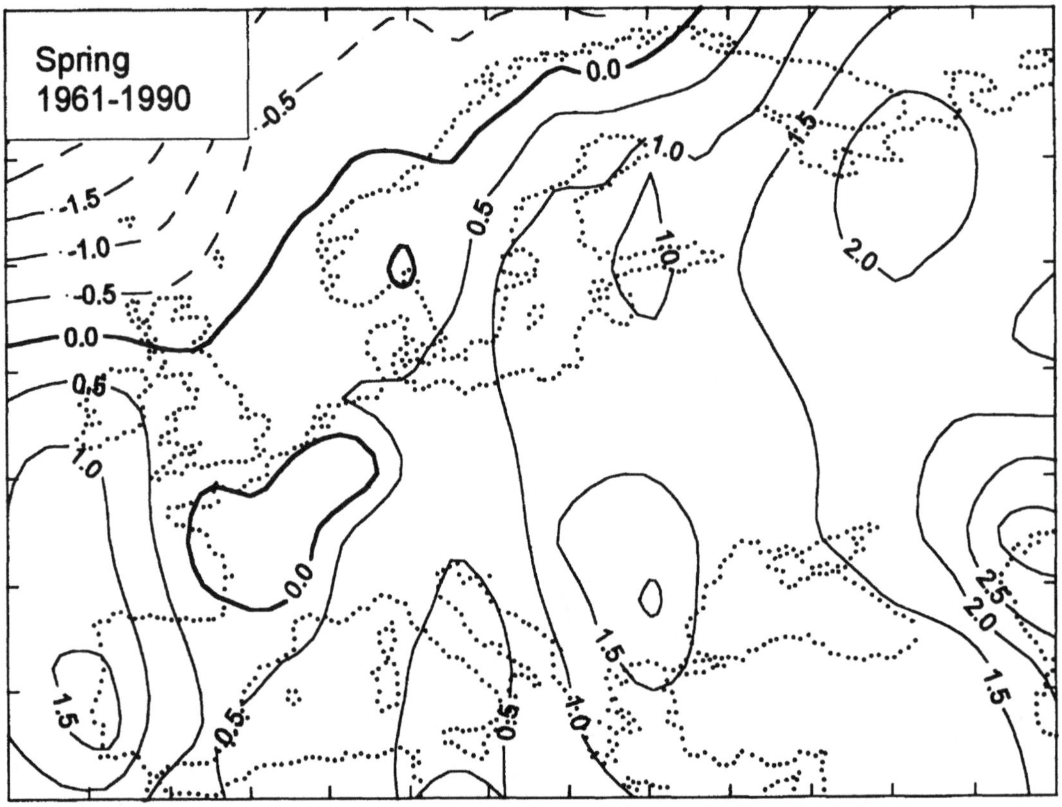

Sea level pressure, trend test

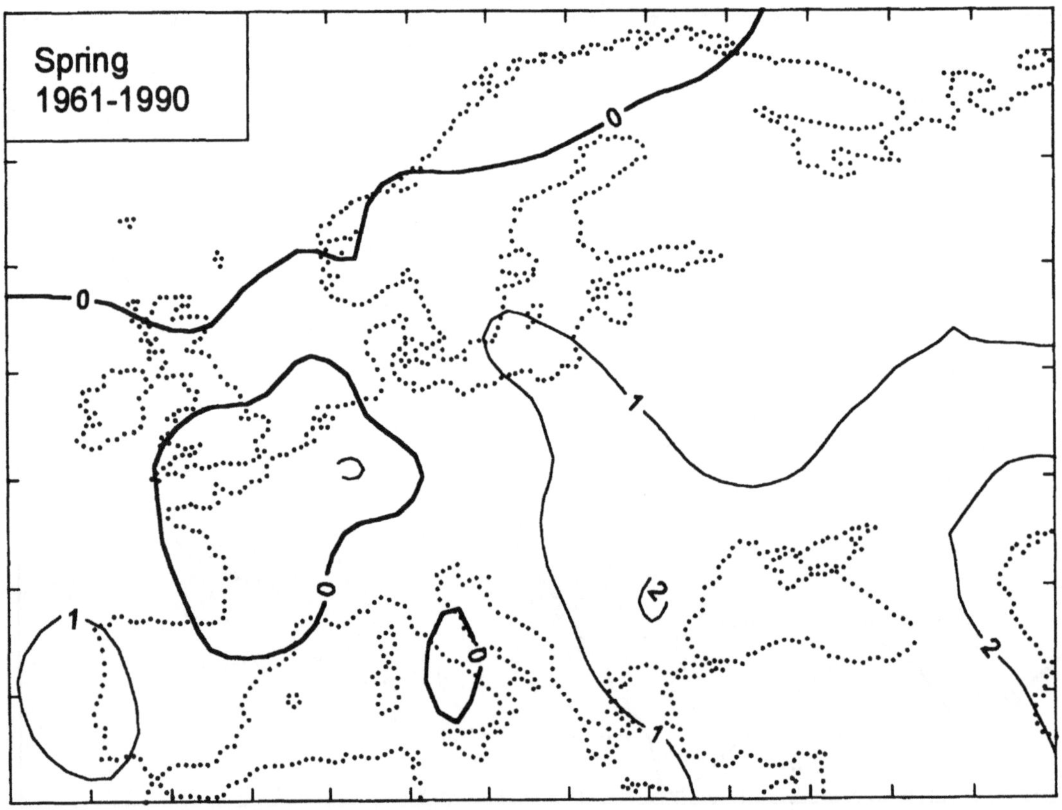

Sea level pressure, linear trend [hpa]

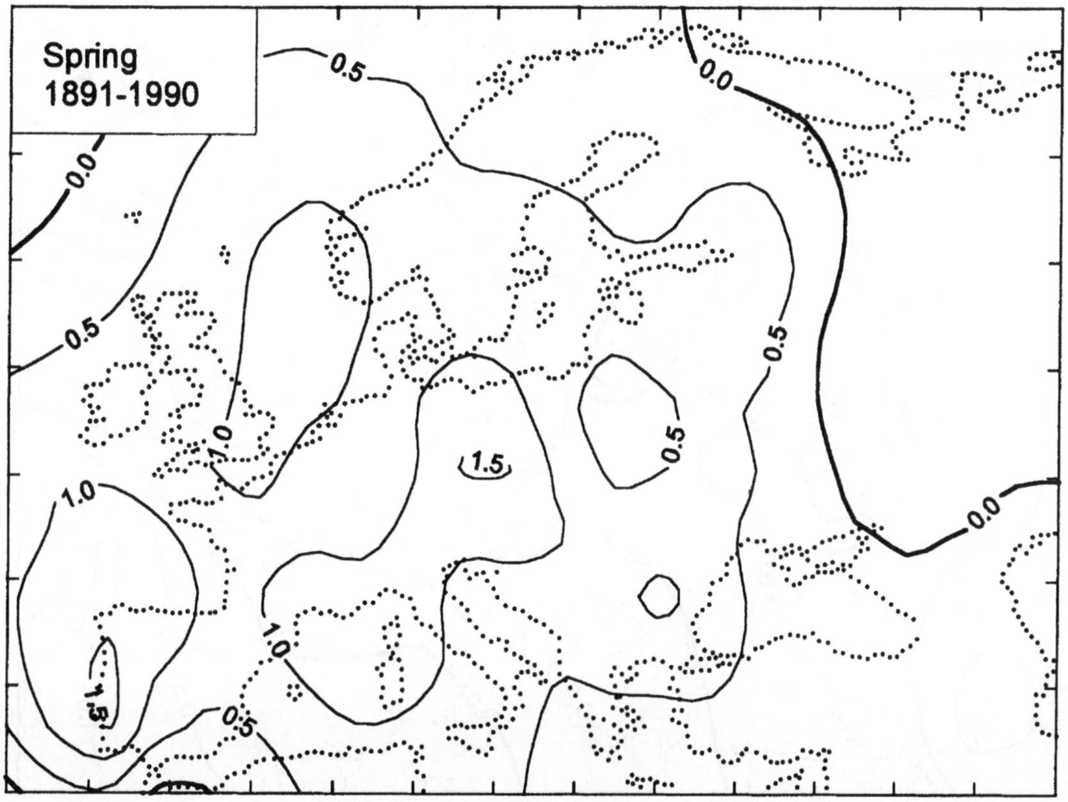

Sea level pressure, trend test

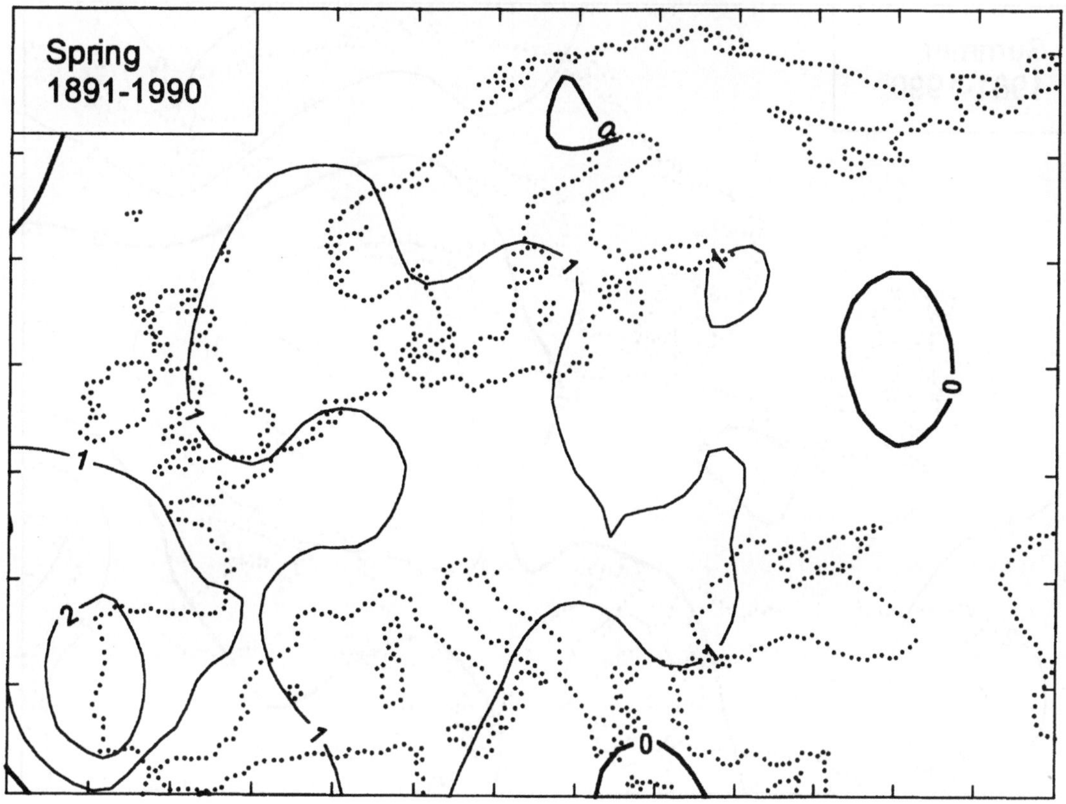

Sea level pressure, linear trend [hpa]

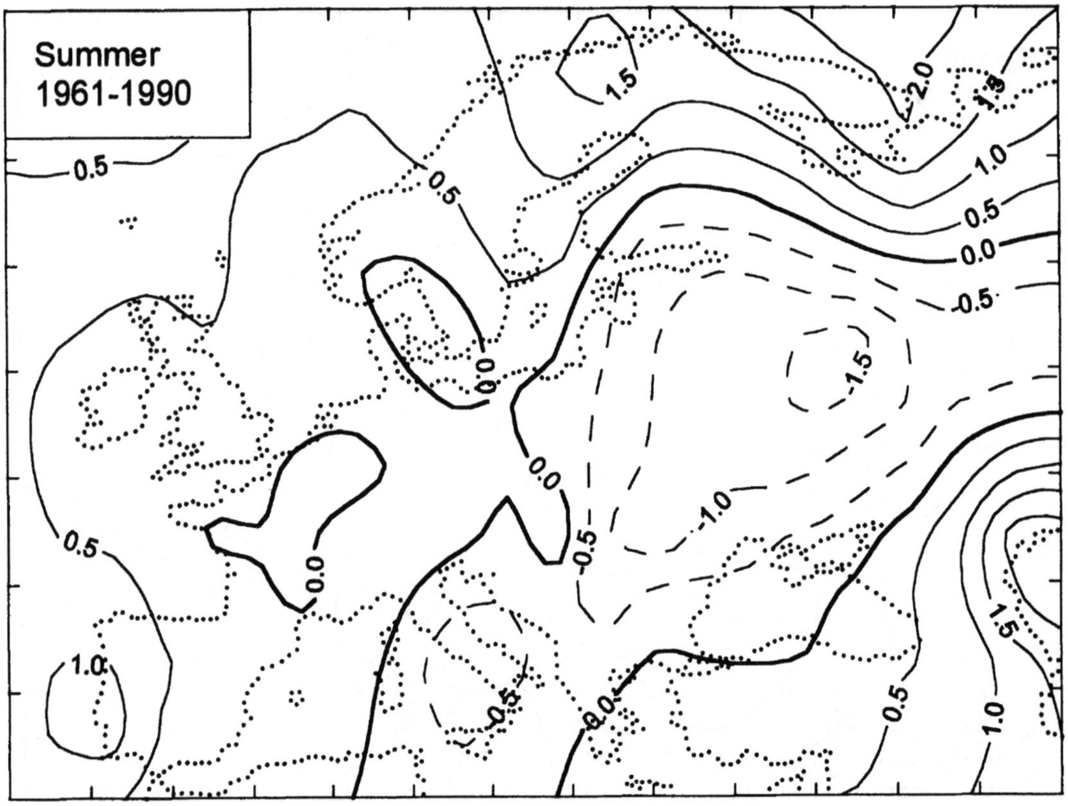

Sea level pressure, trend test

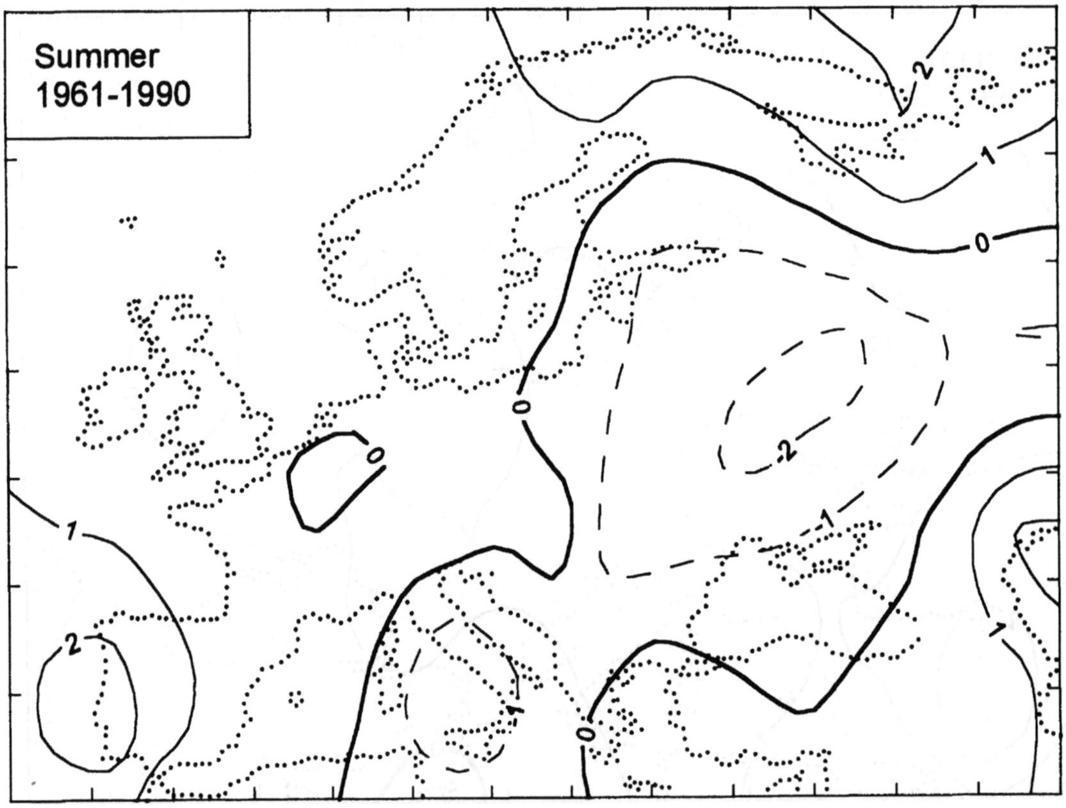

Sea level pressure, linear trend [hpa]

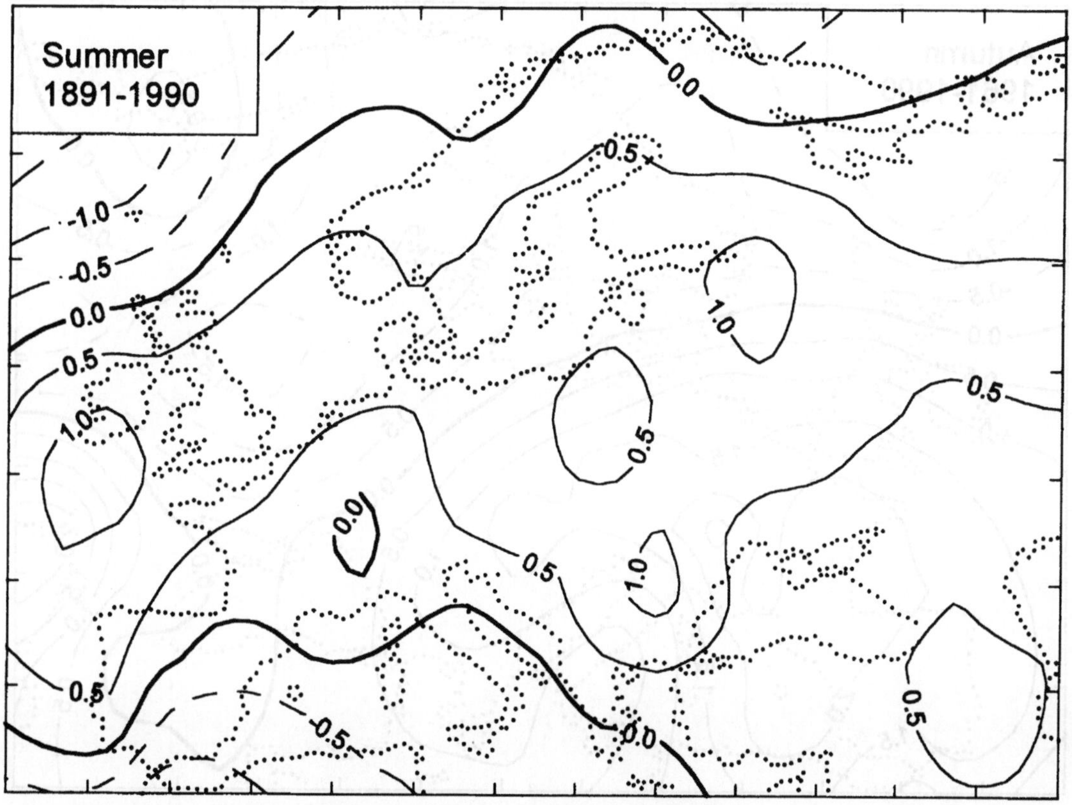

Sea level pressure, trend test

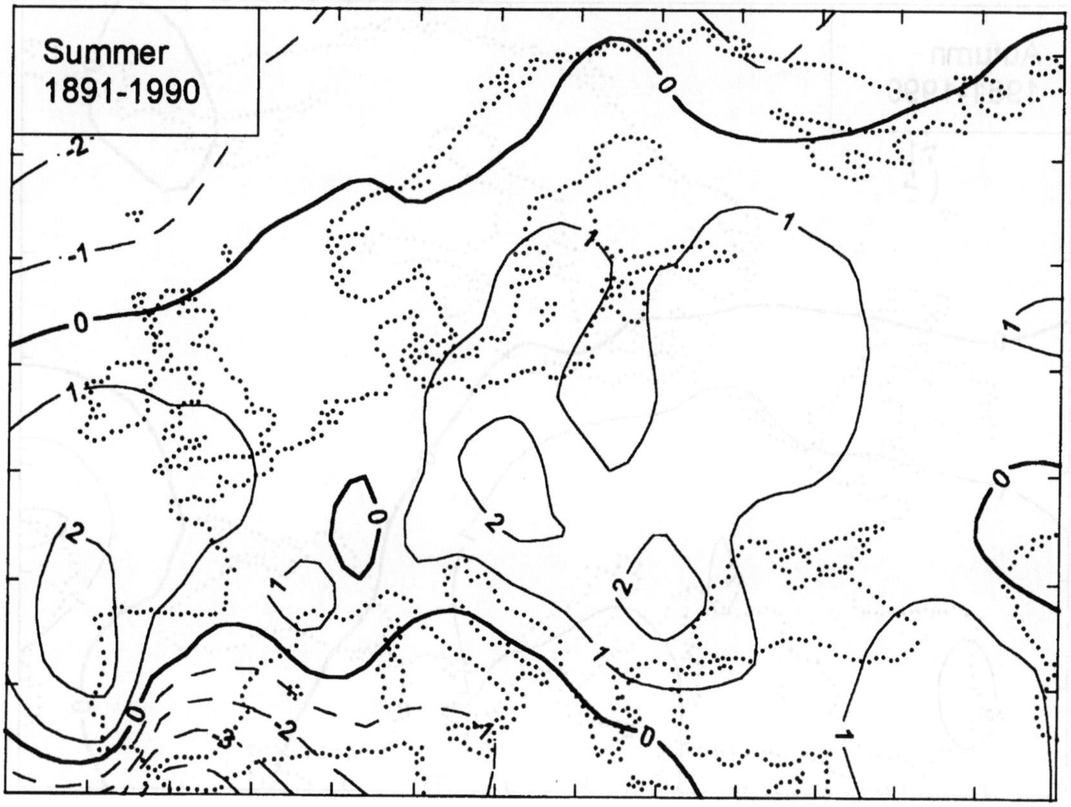

Sea level pressure, linear trend [hpa]

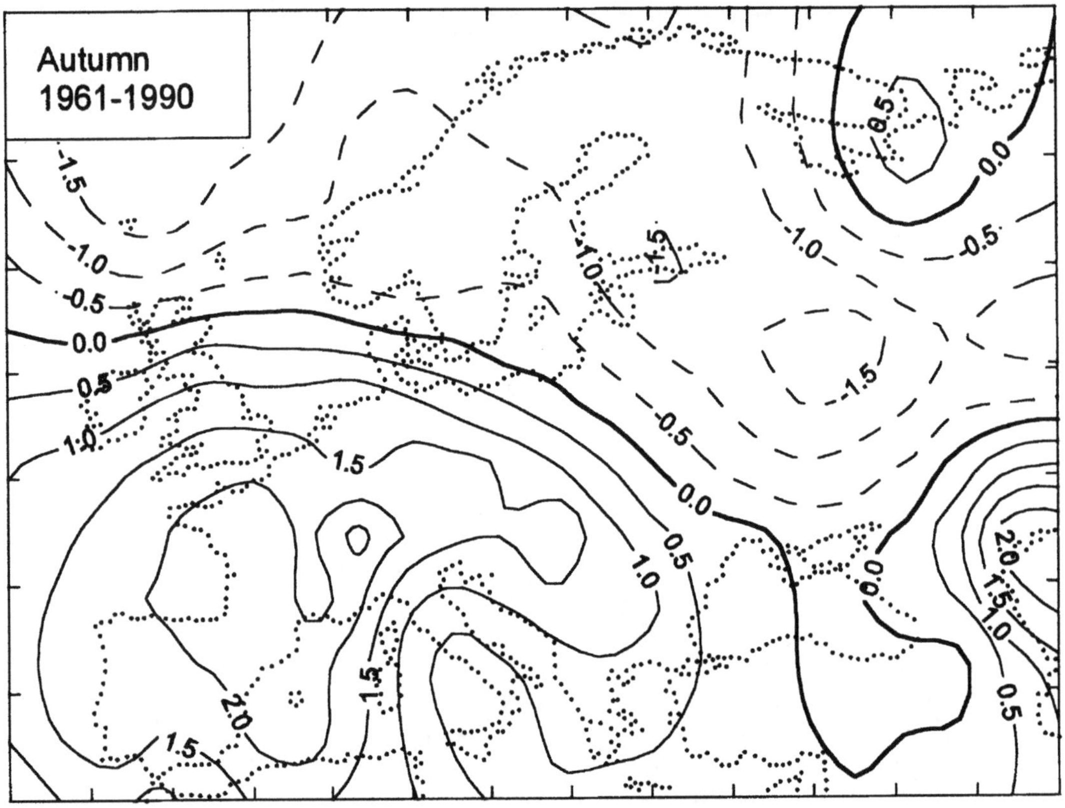

Sea level pressure, trend test

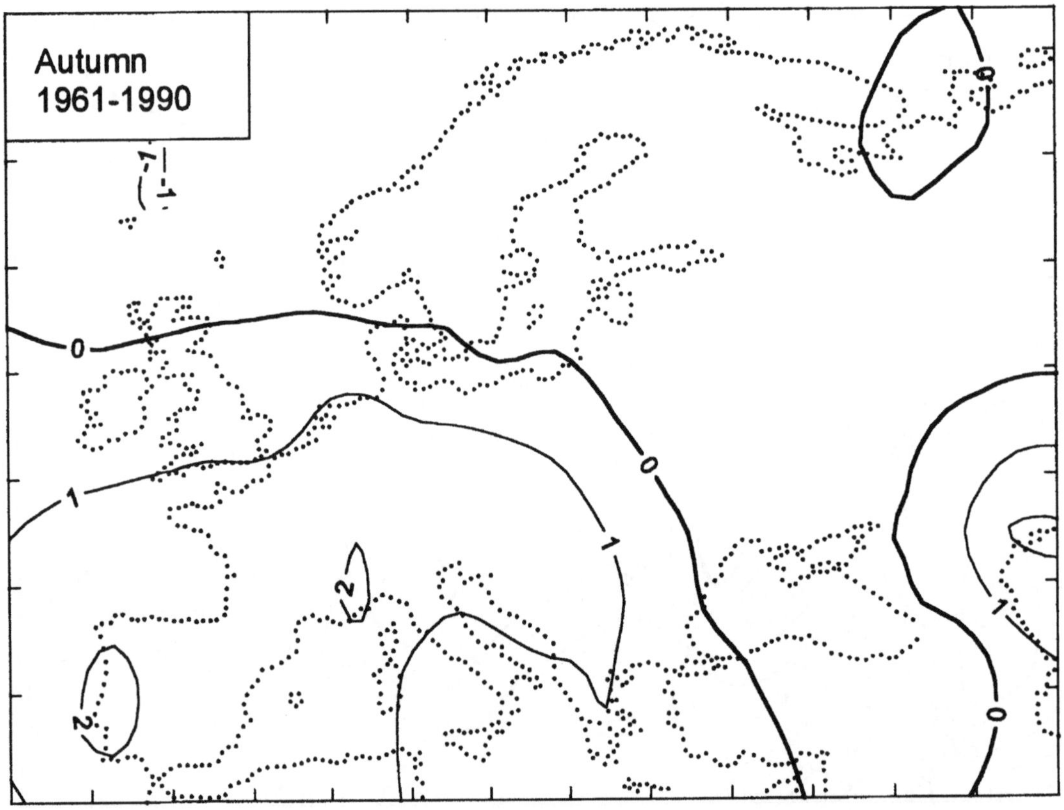

Sea level pressure, linear trend [hpa]

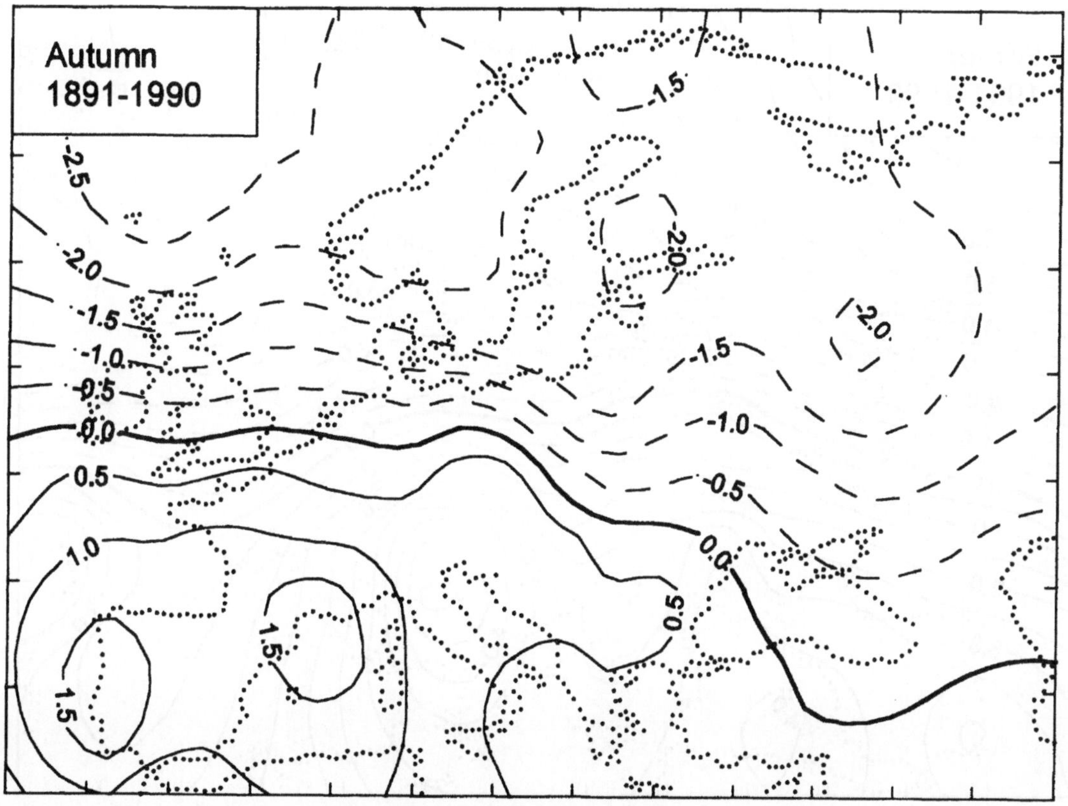

Sea level pressure, trend test

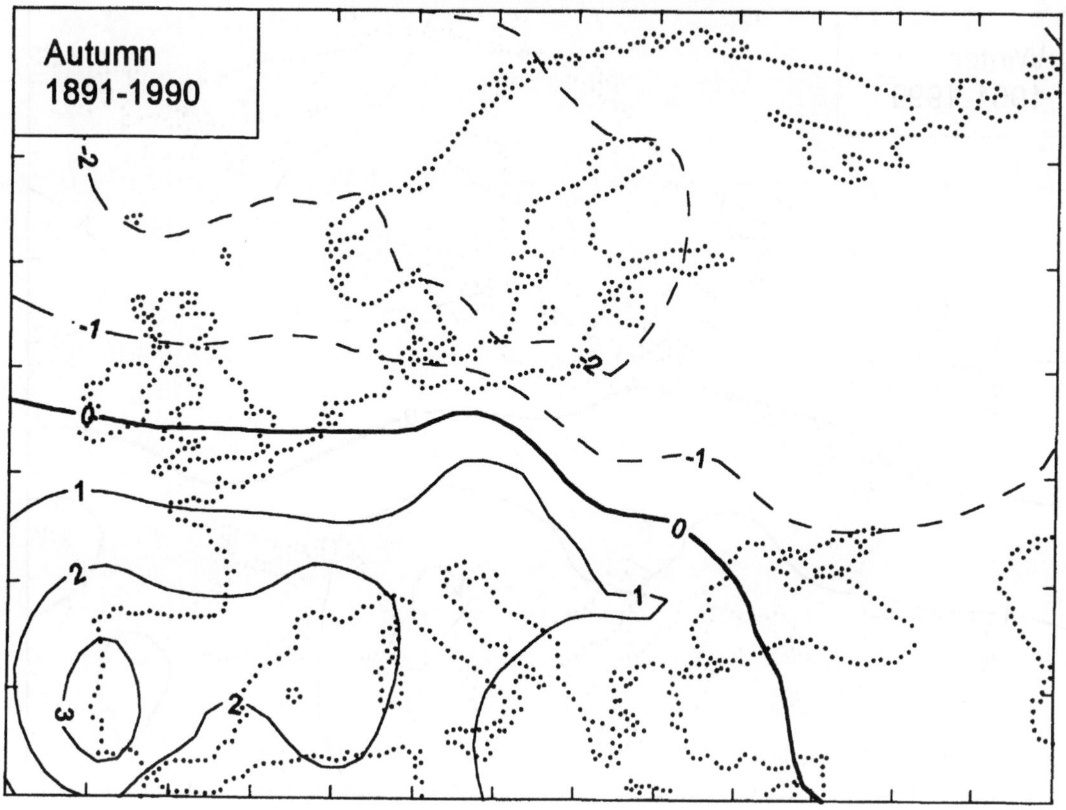

Sea level pressure, linear trend [hpa]

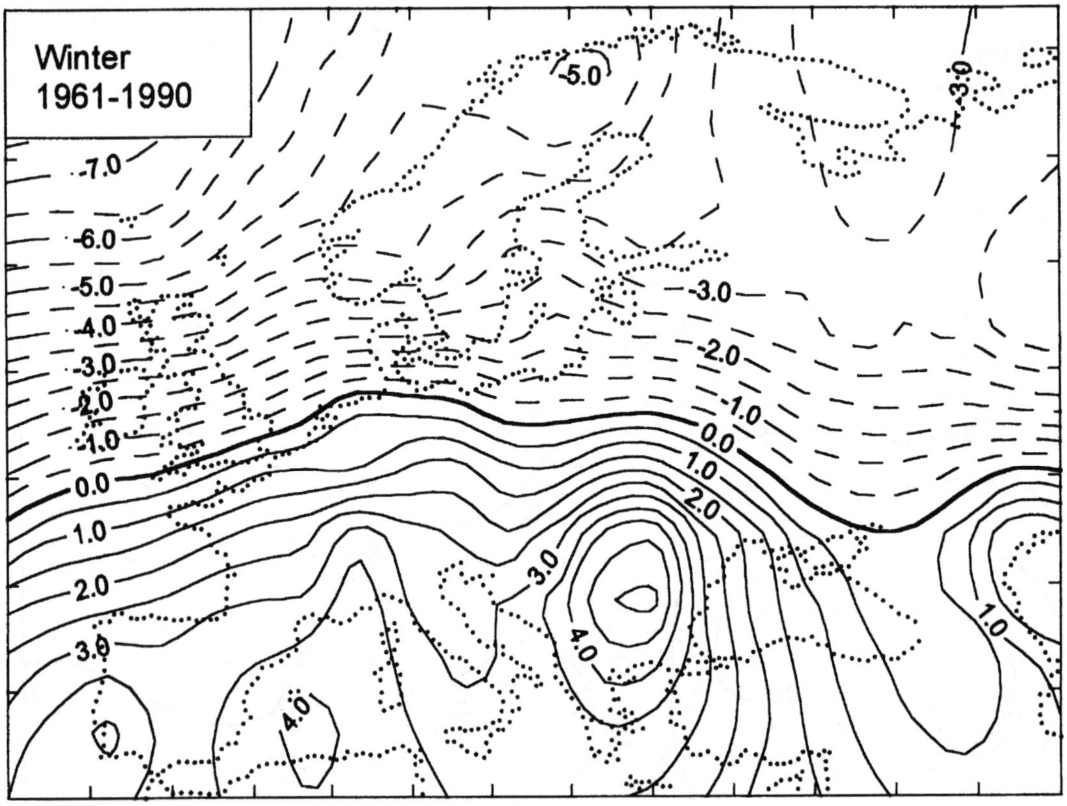

Sea level pressure, trend test

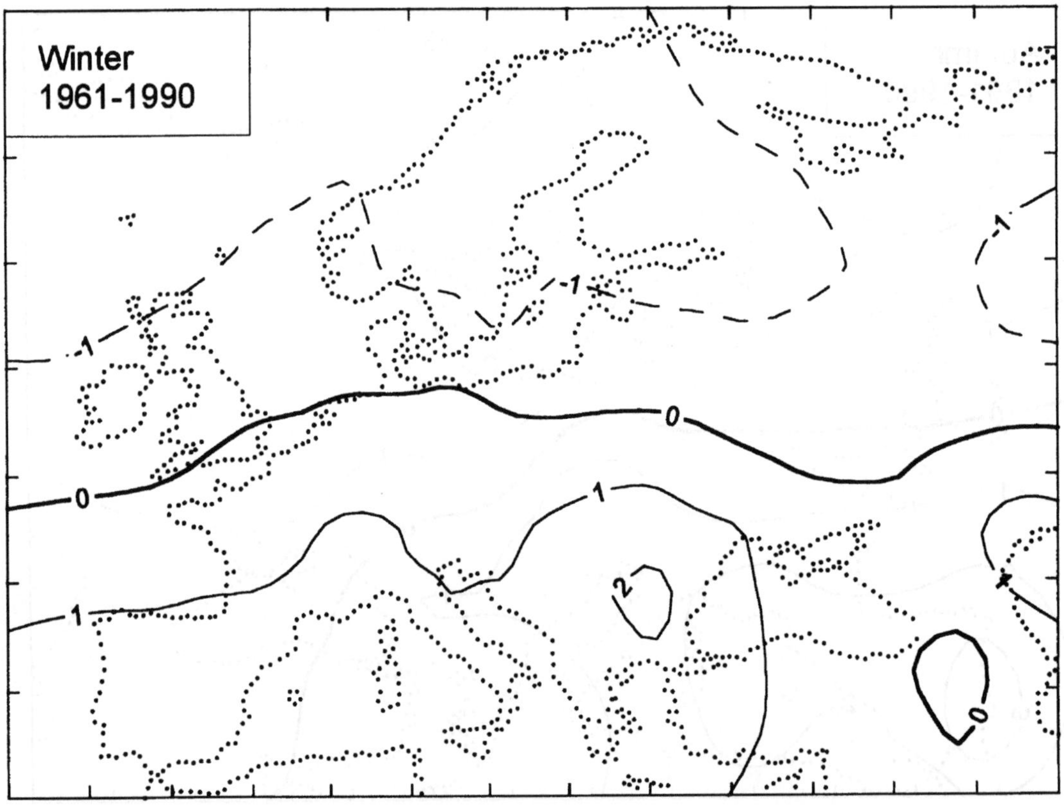

Sea level pressure, linear trend [hpa]

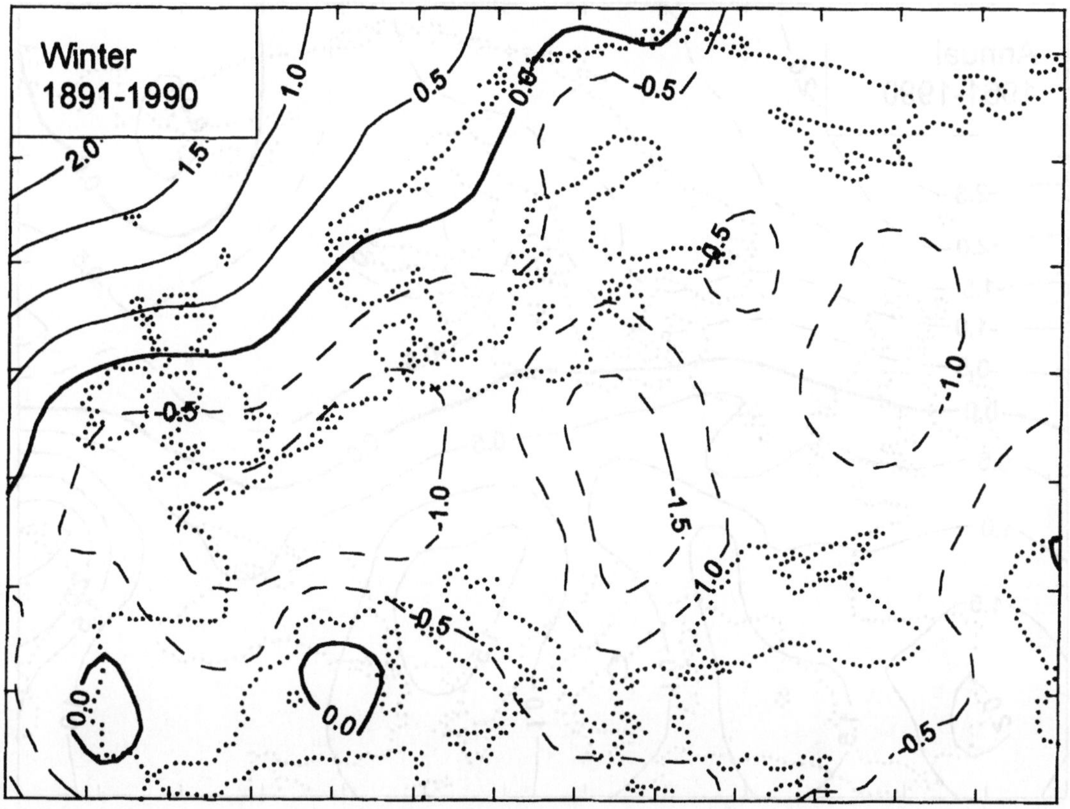

Sea level pressure, trend test

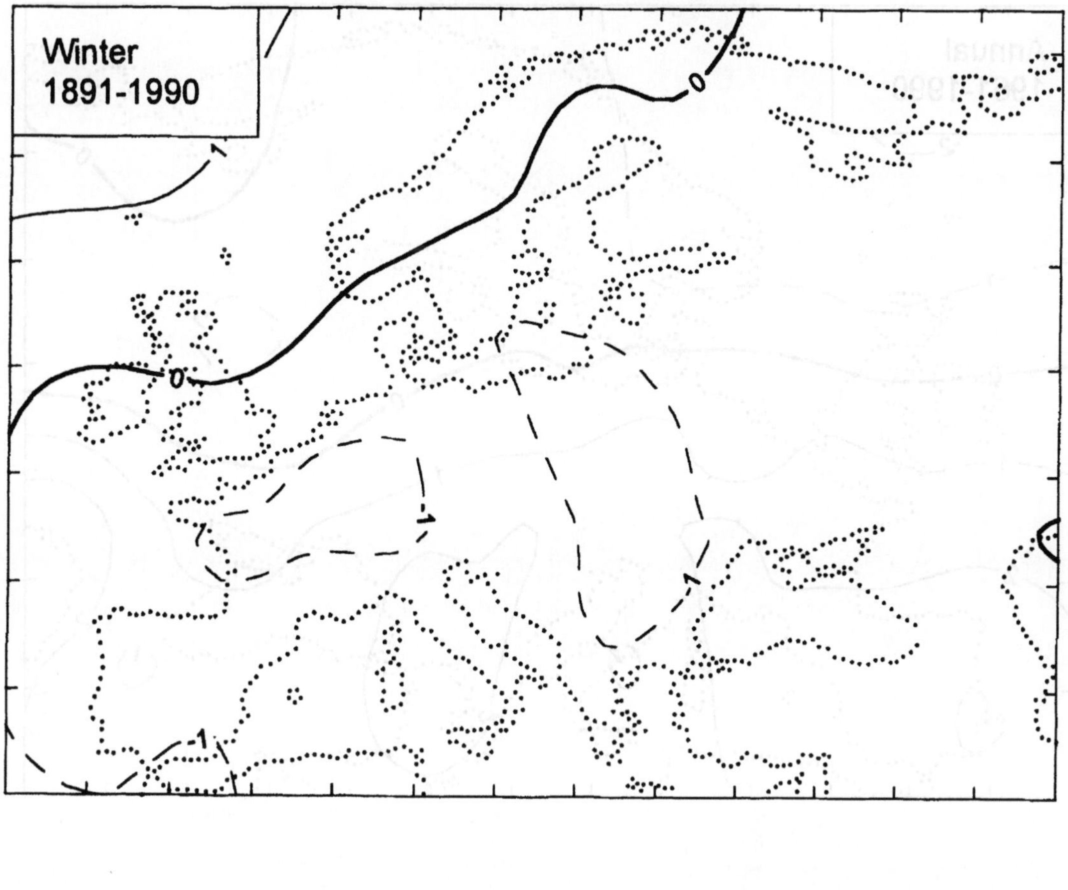

Sea level pressure, linear trend [hpa]

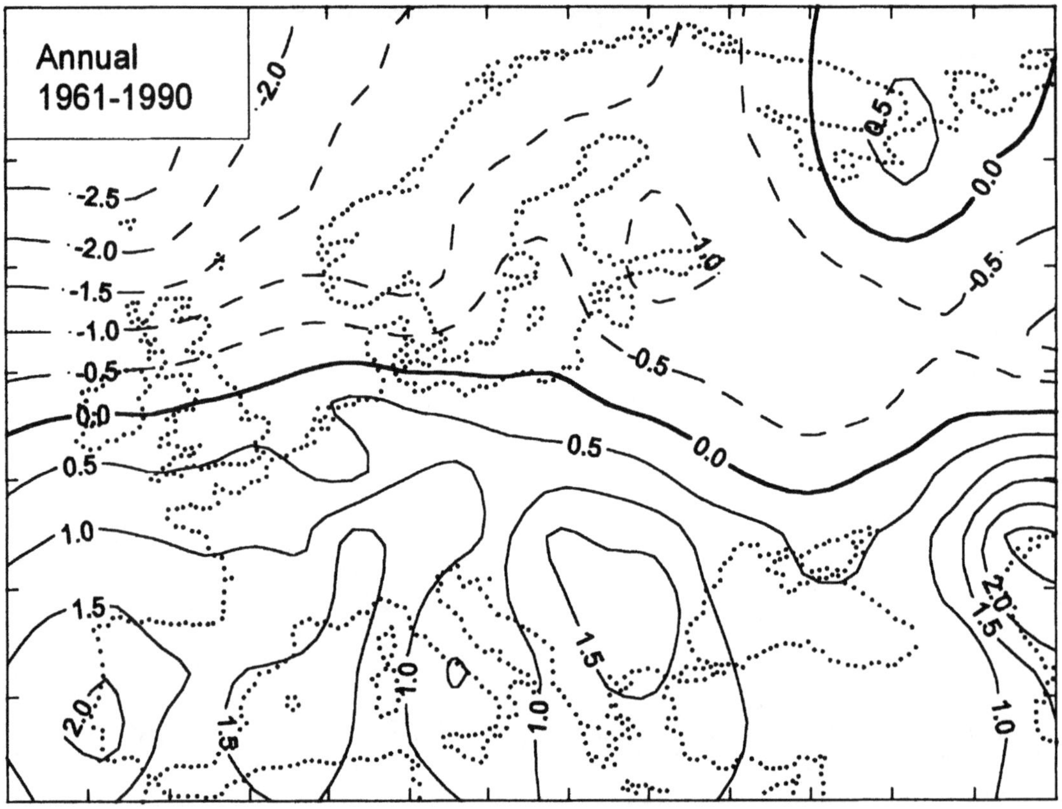

Sea level pressure, trend test

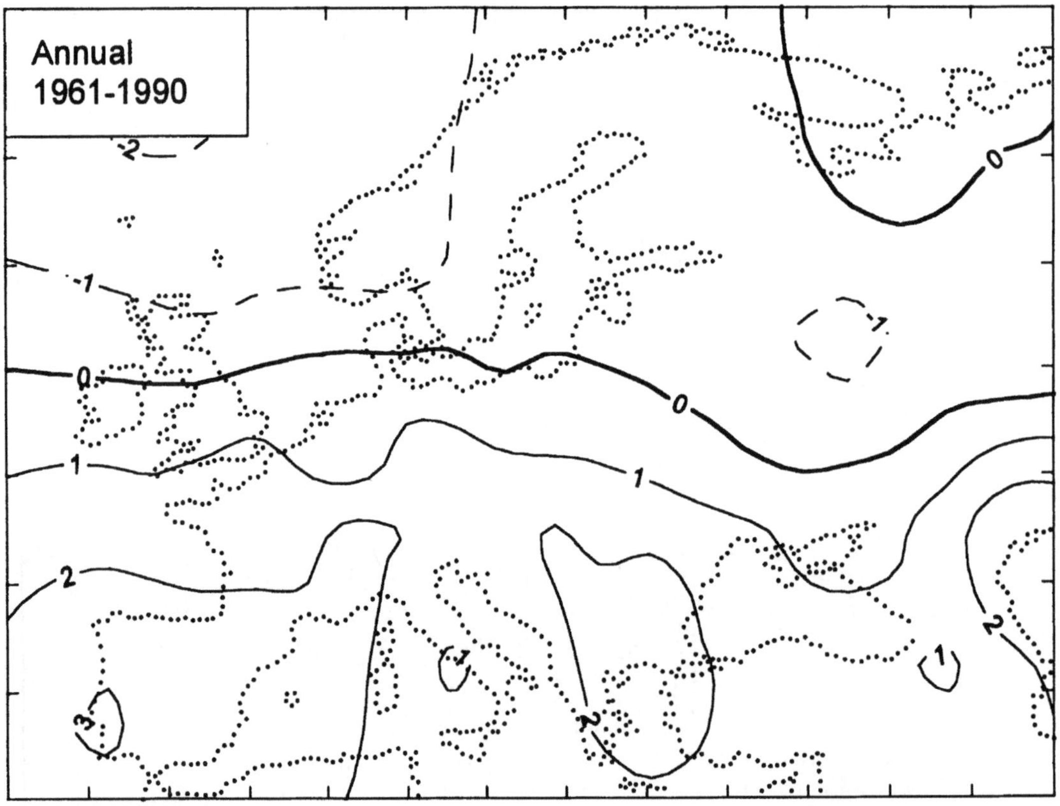

Sea level pressure, linear trend [hpa]

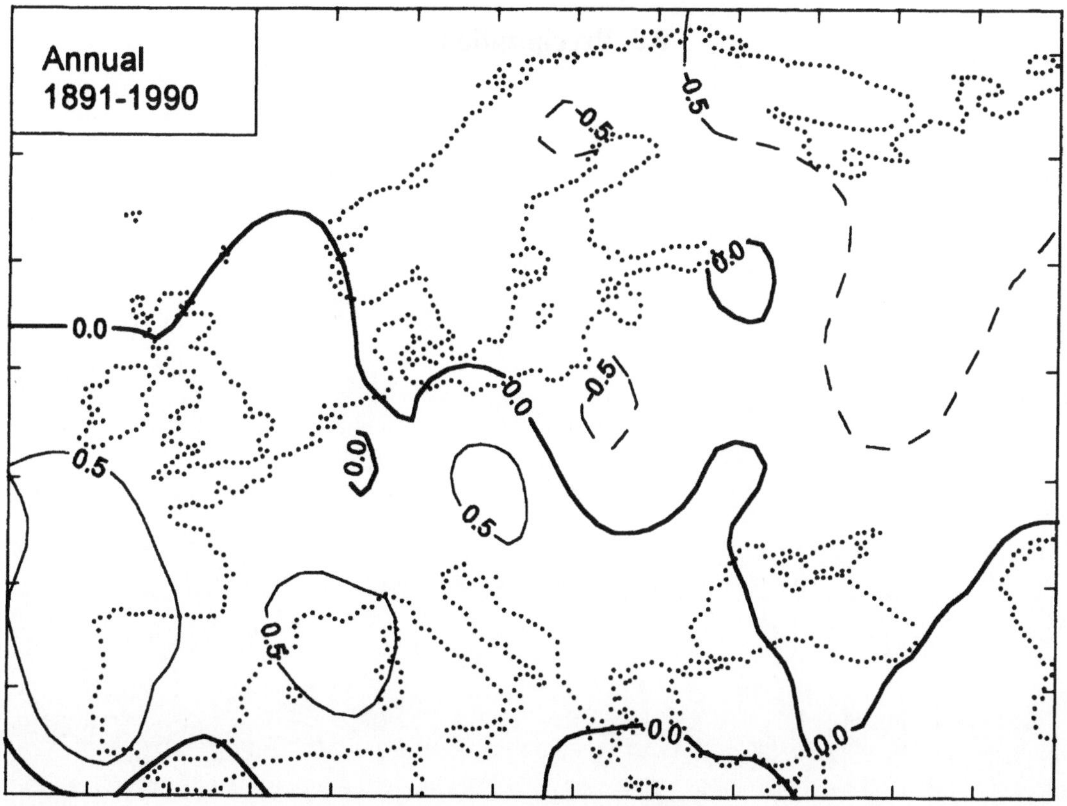

Sea level pressure, trend test

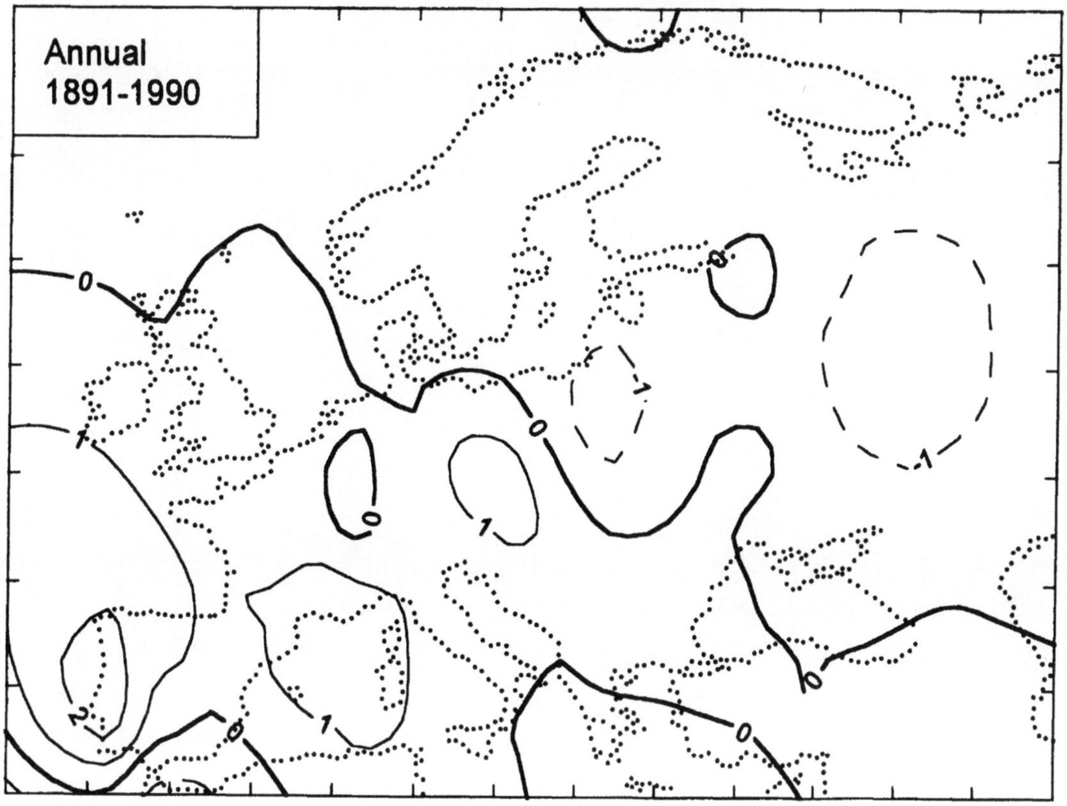

8.3. Precipitation

Precipitation, average [mm]

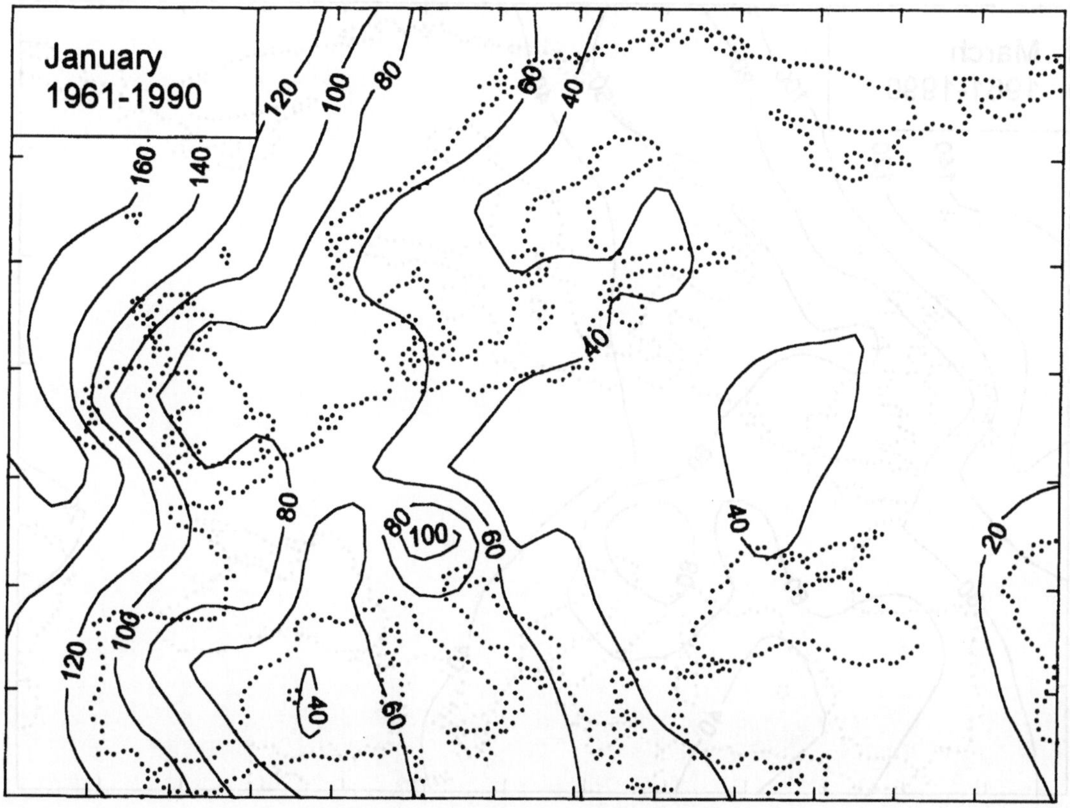

Precipitation, average [mm]

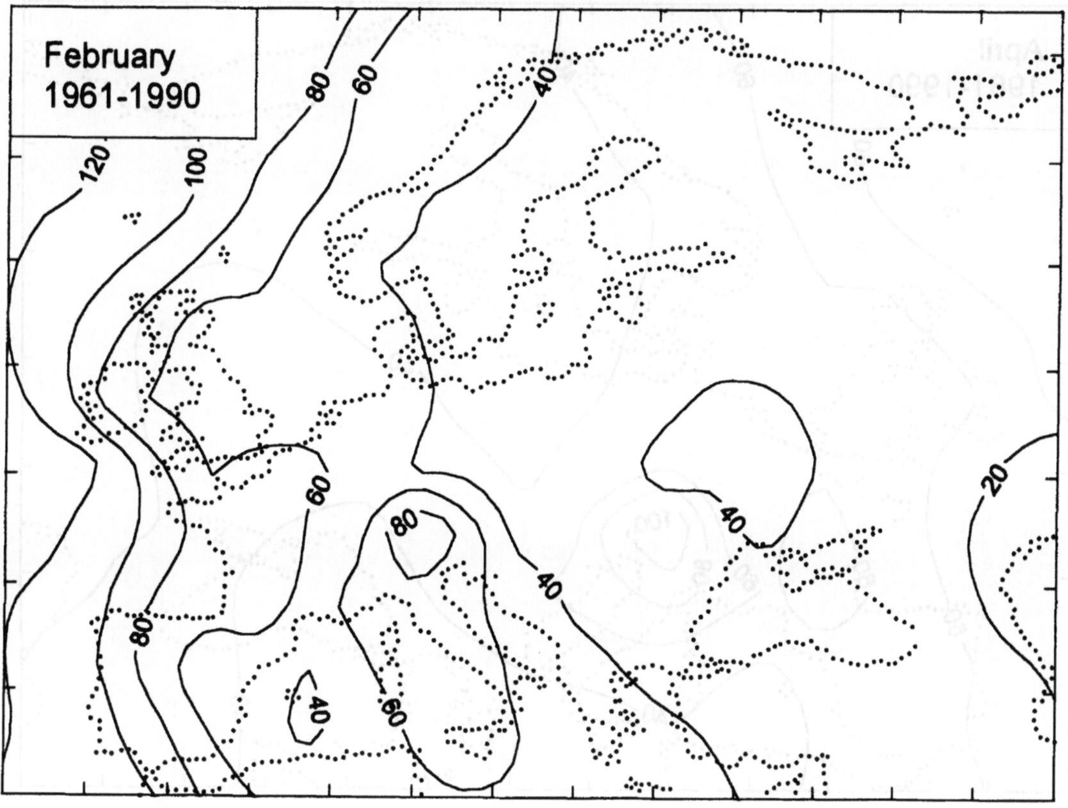

Precipitation, average [mm]

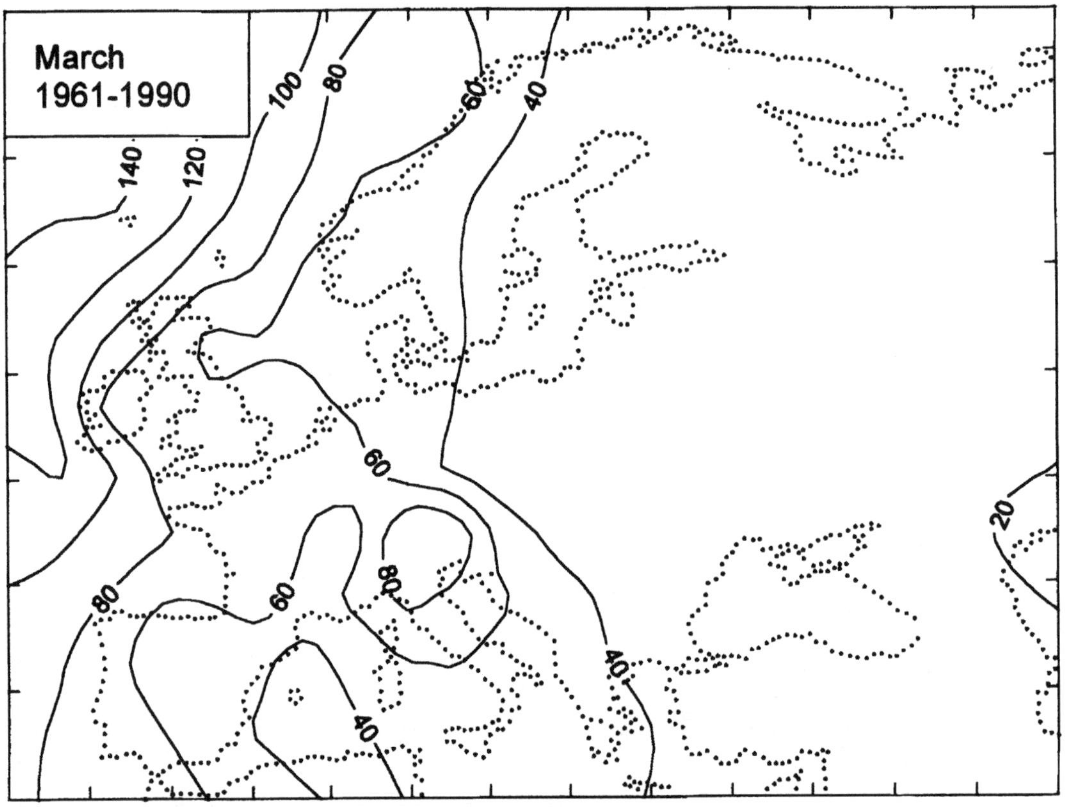

Precipitation, average [mm]

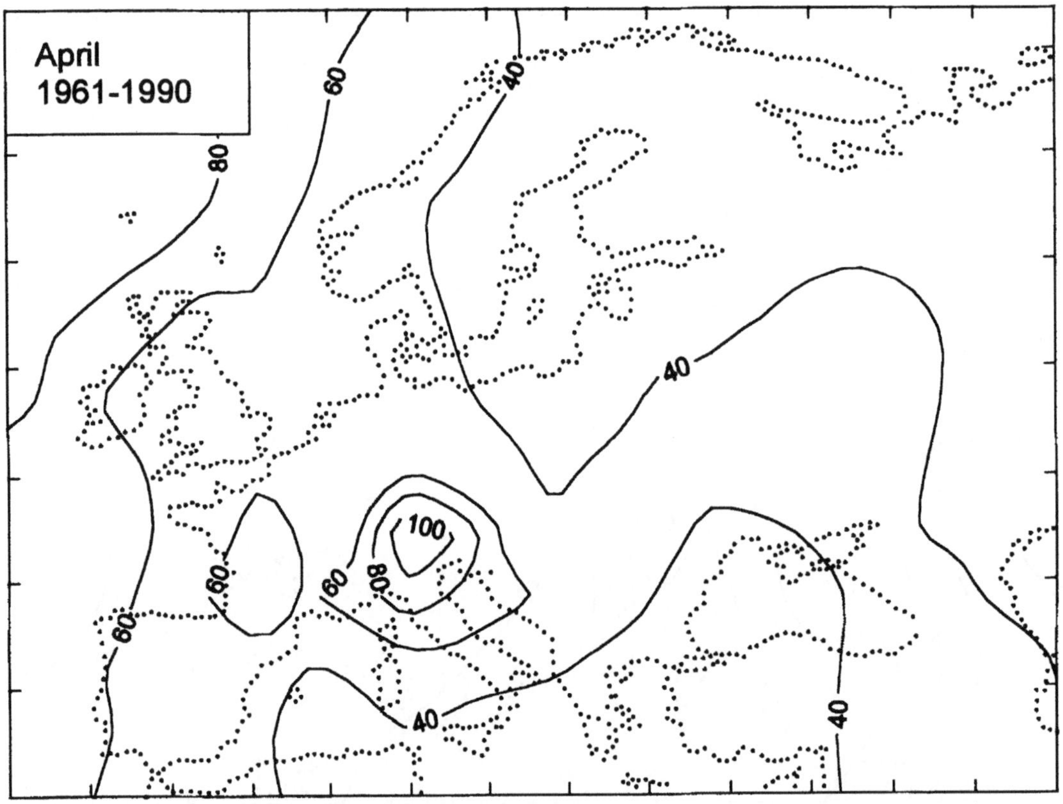

Precipitation, average [mm]

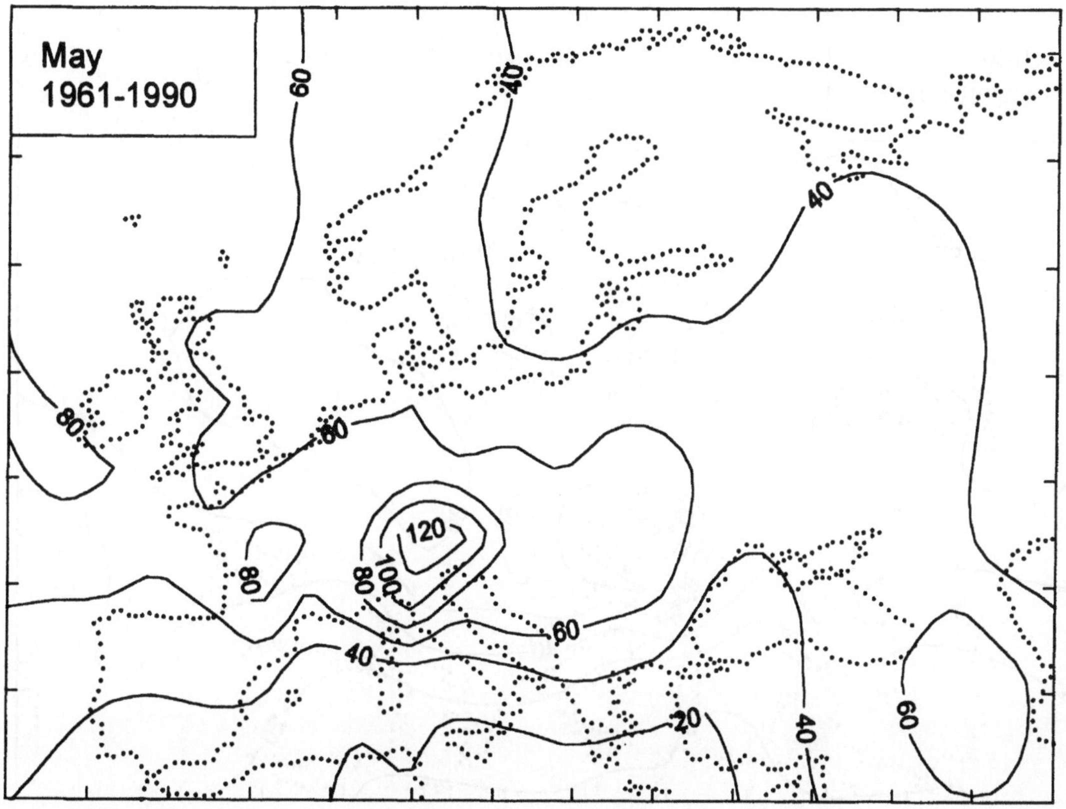

Precipitation, average [mm]

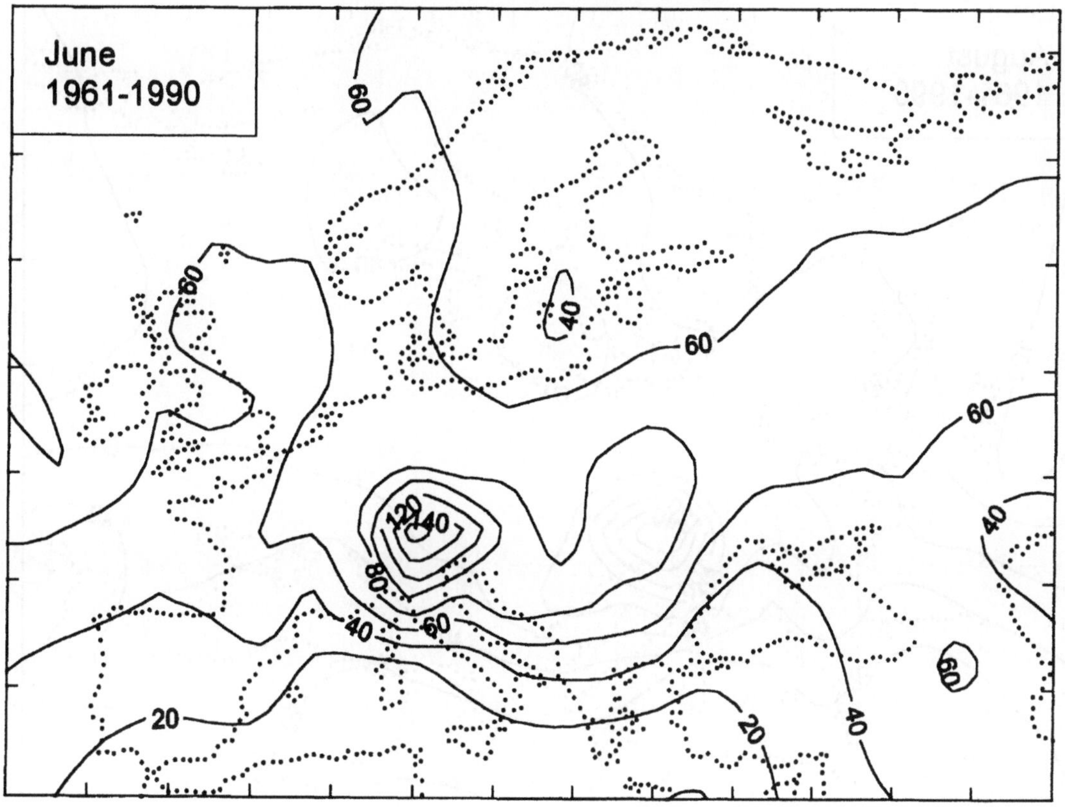

Precipitation, average [mm]

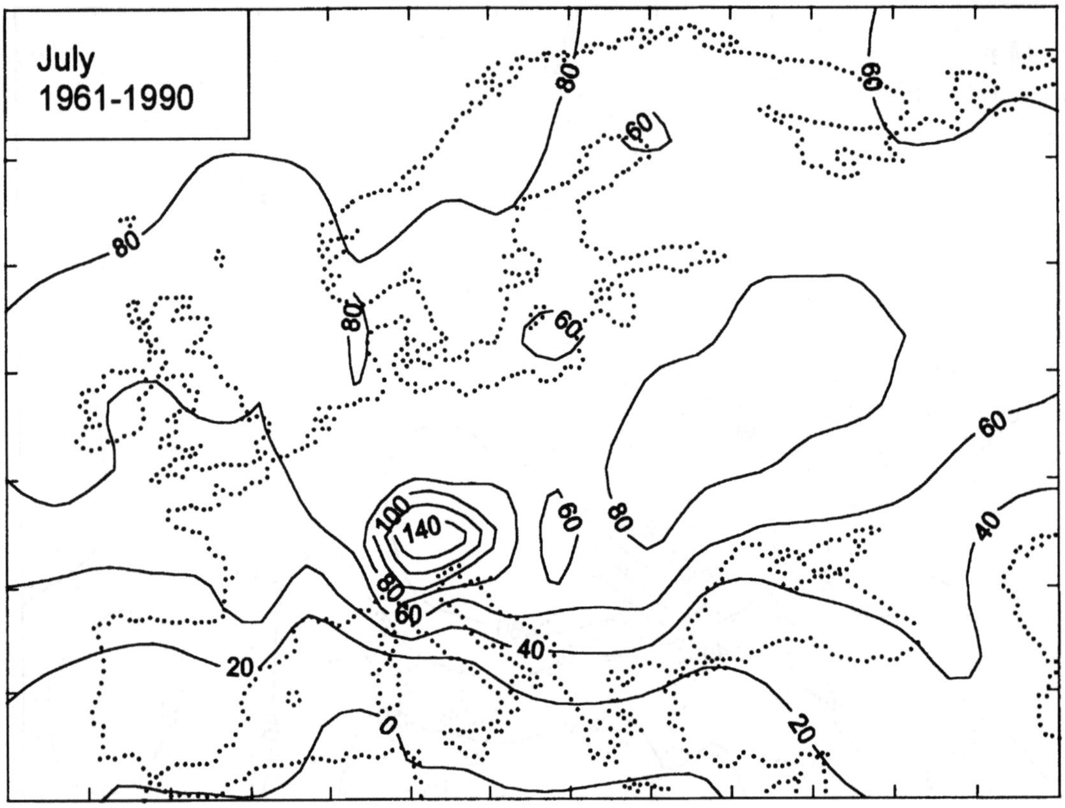

Precipitation, average [mm]

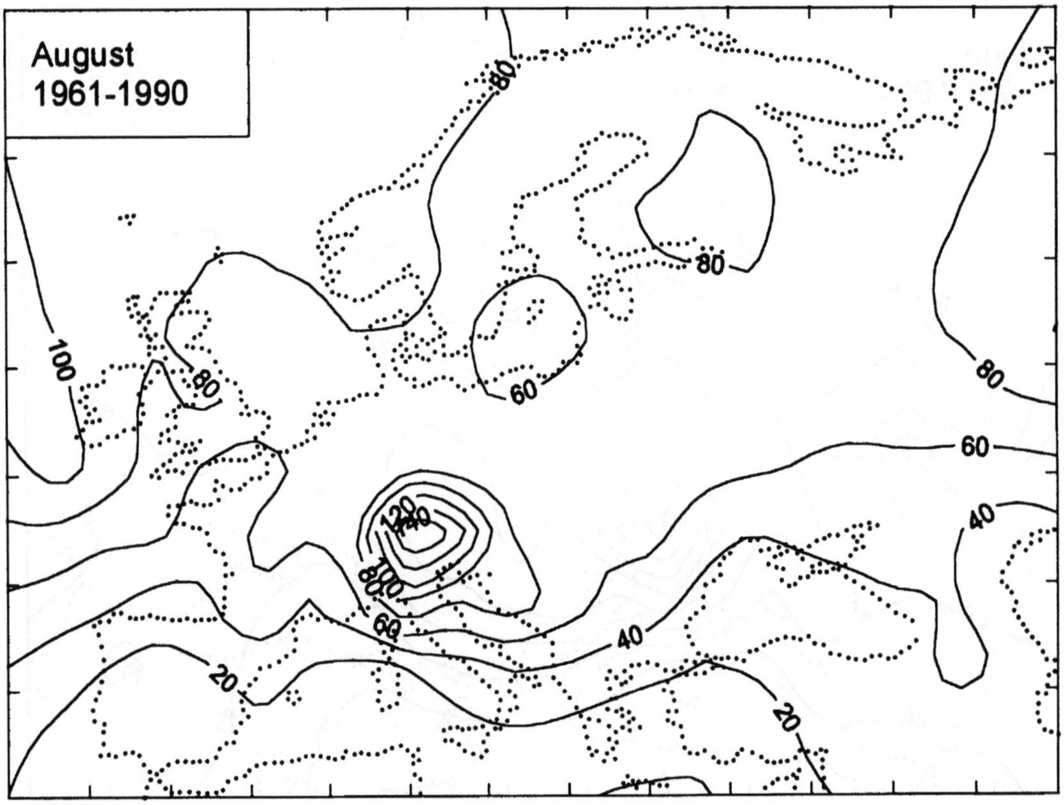

Precipitation, average [mm]

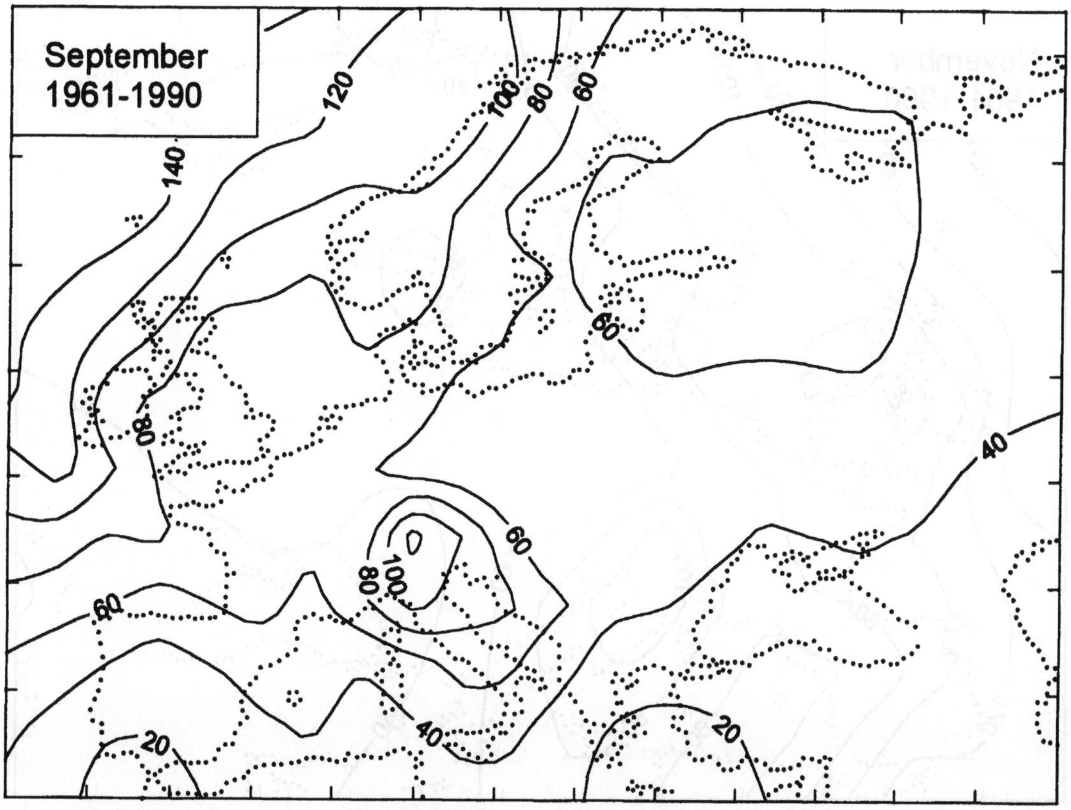

Precipitation, average [mm]

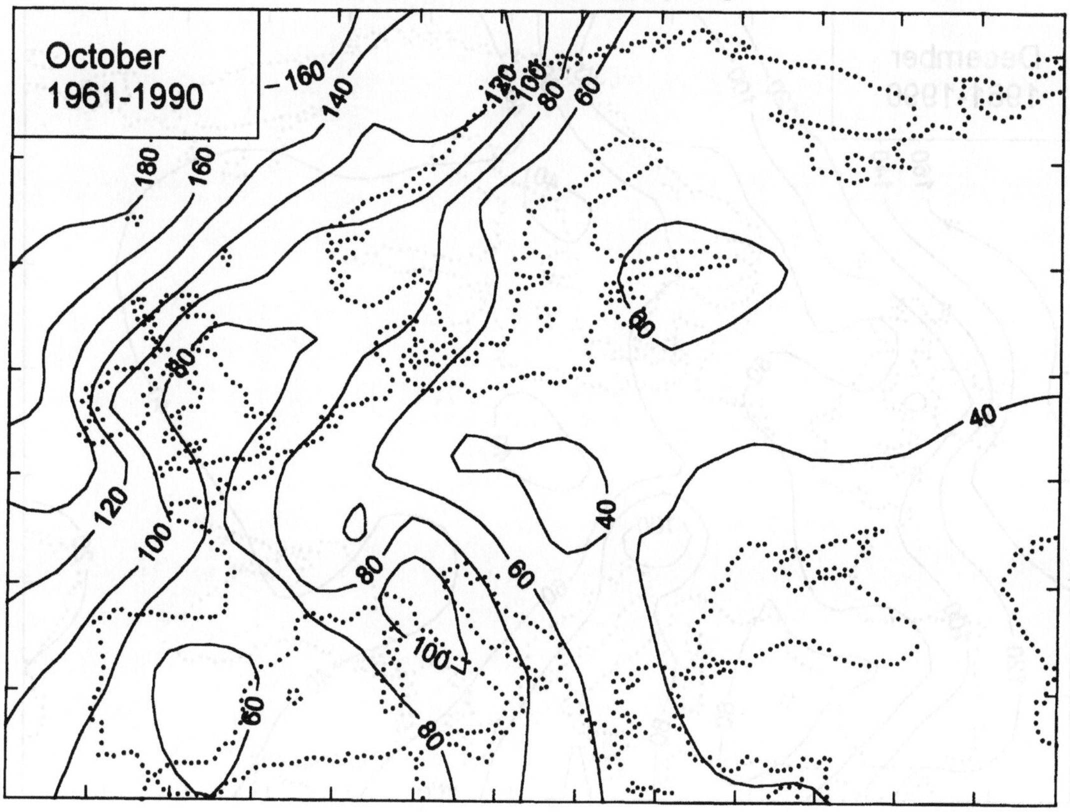

Precipitation, average [mm]

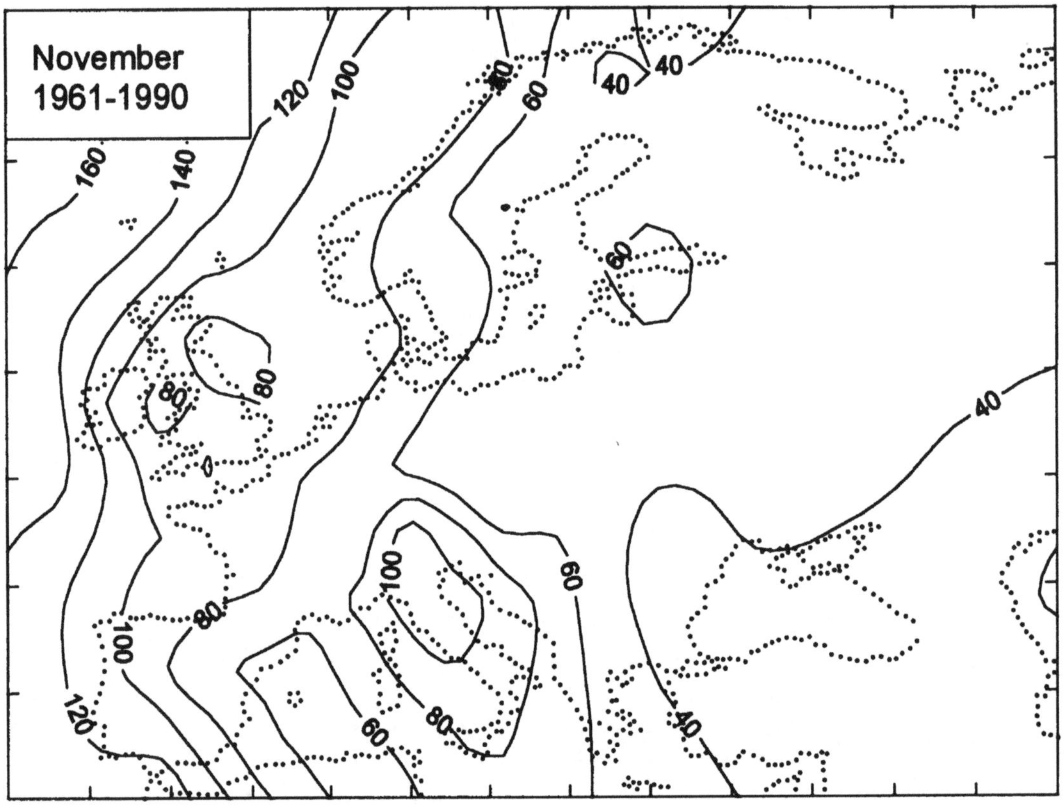

Precipitation, average [mm]

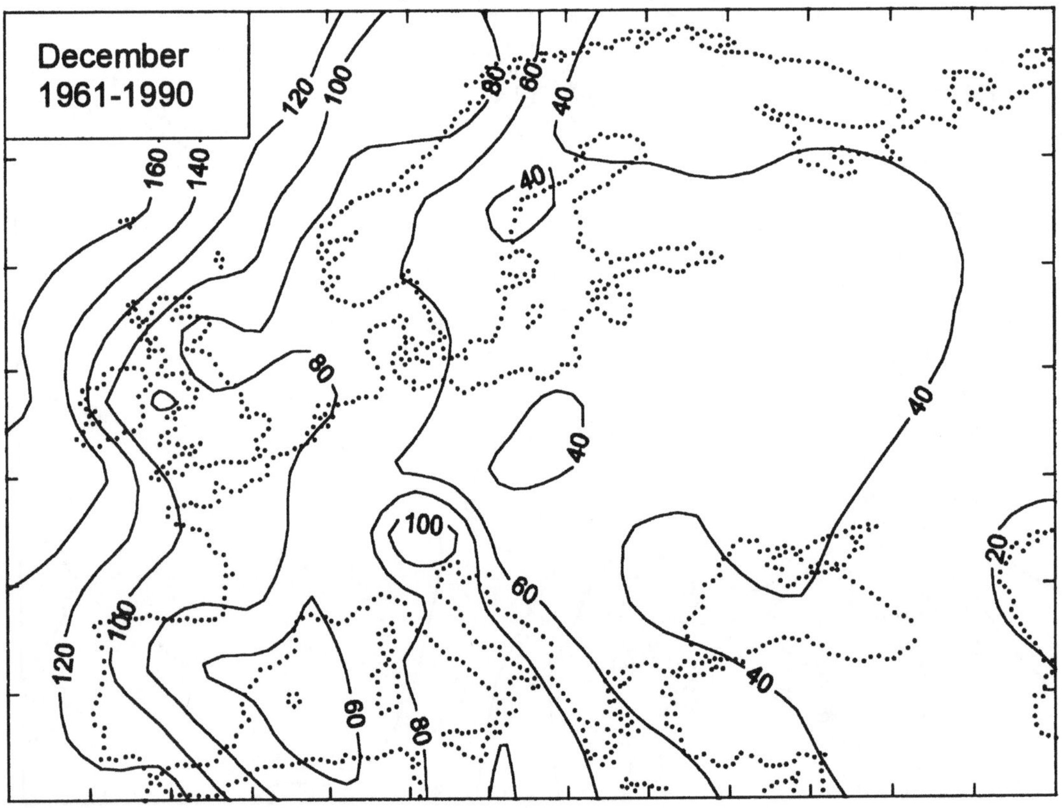

Precipitation, average [mm]

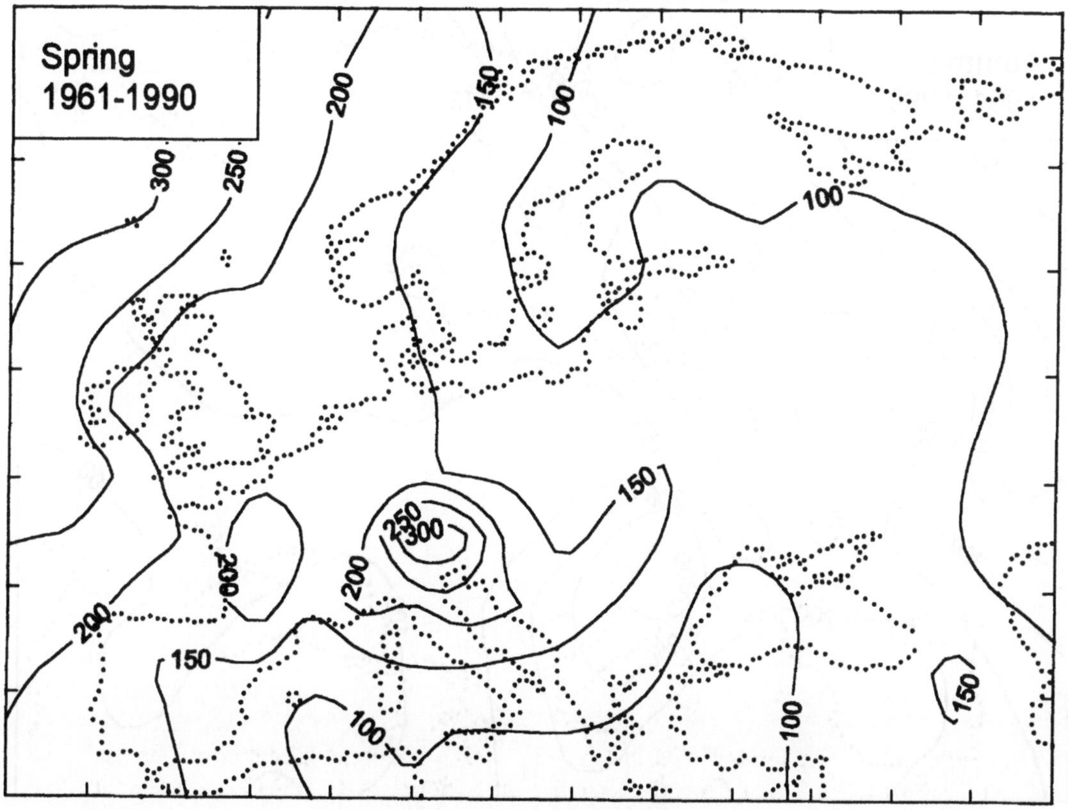

Precipitation, average [mm]

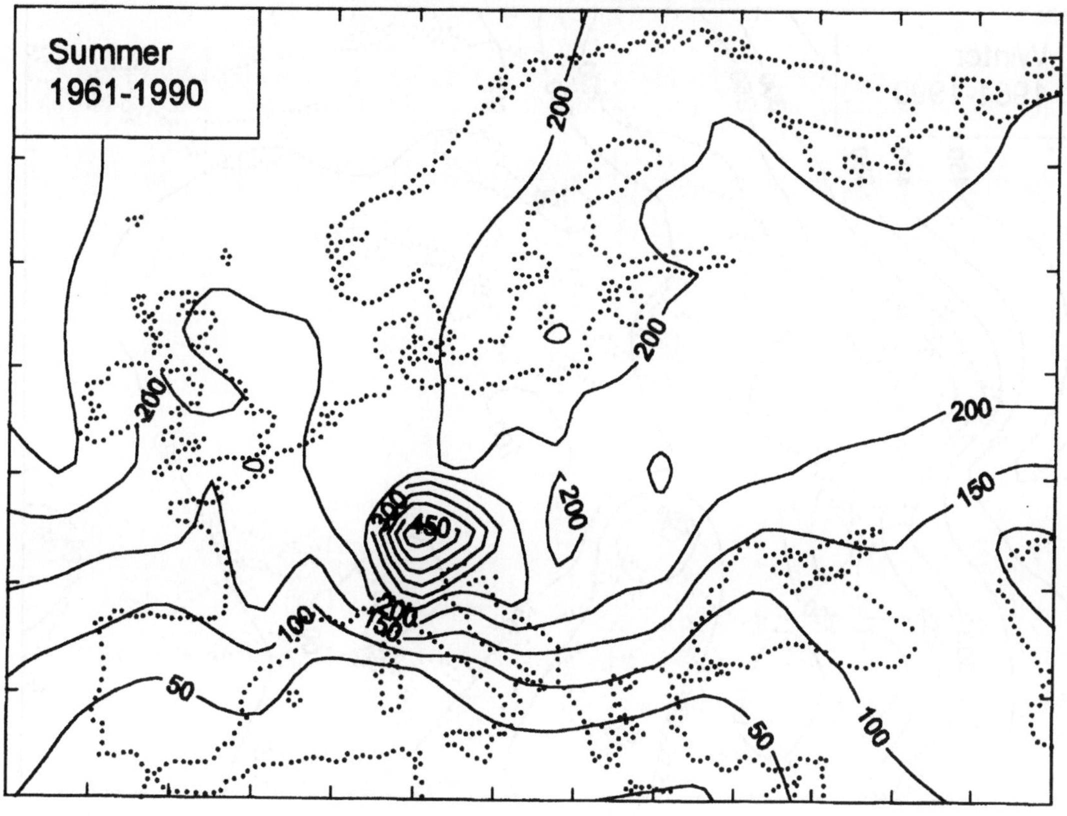

Precipitation, average [mm]

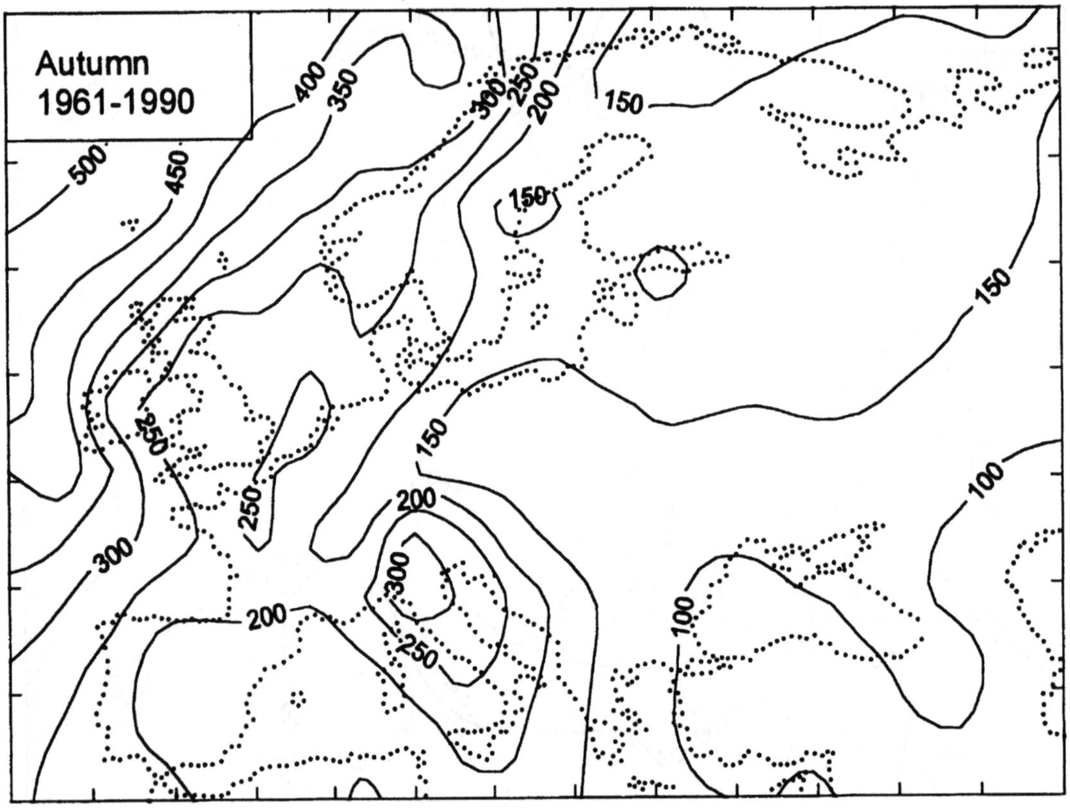

Precipitation, average [mm]

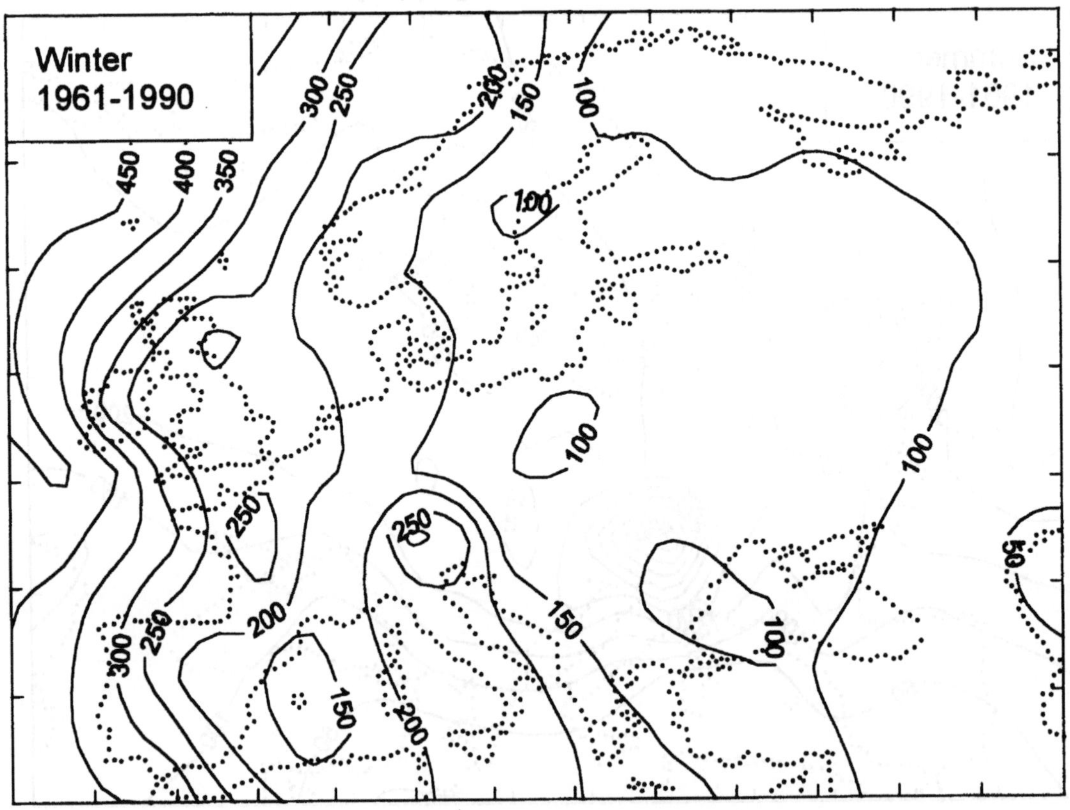

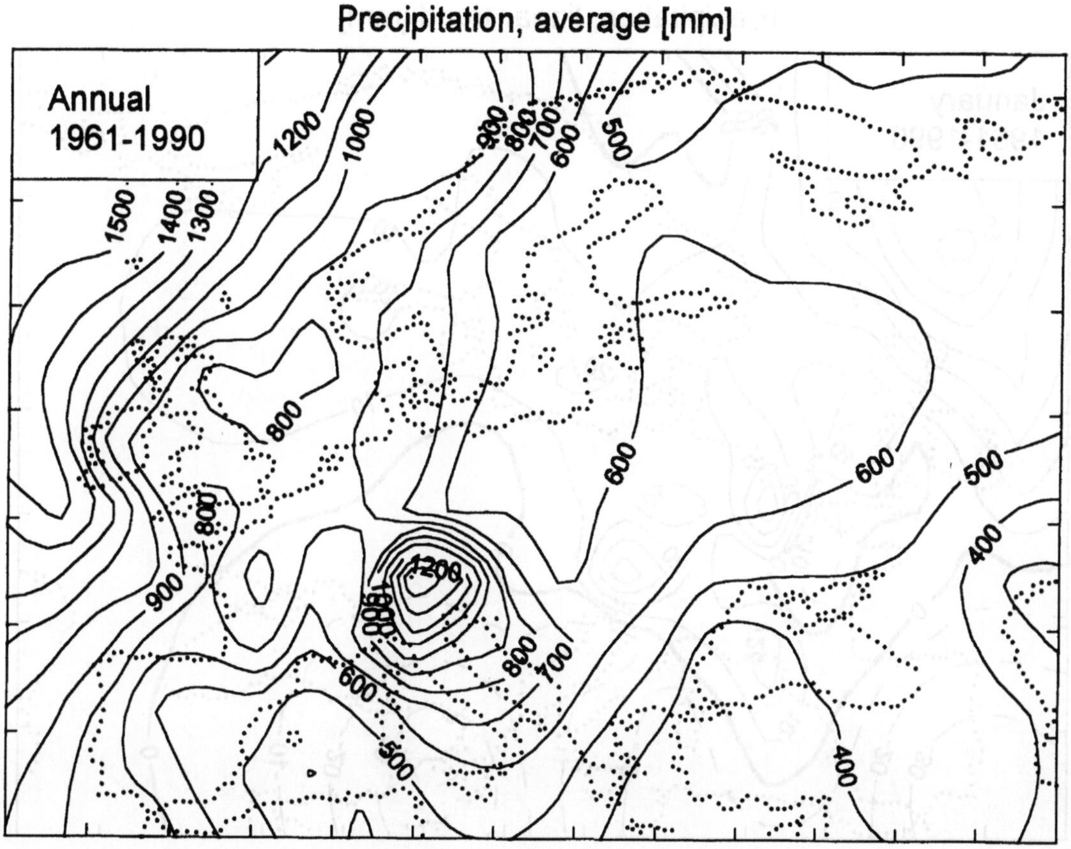
Precipitation, average [mm]
Annual
1961-1990
1500
1400
1300
1200
1000
900
800
700
600
500
400

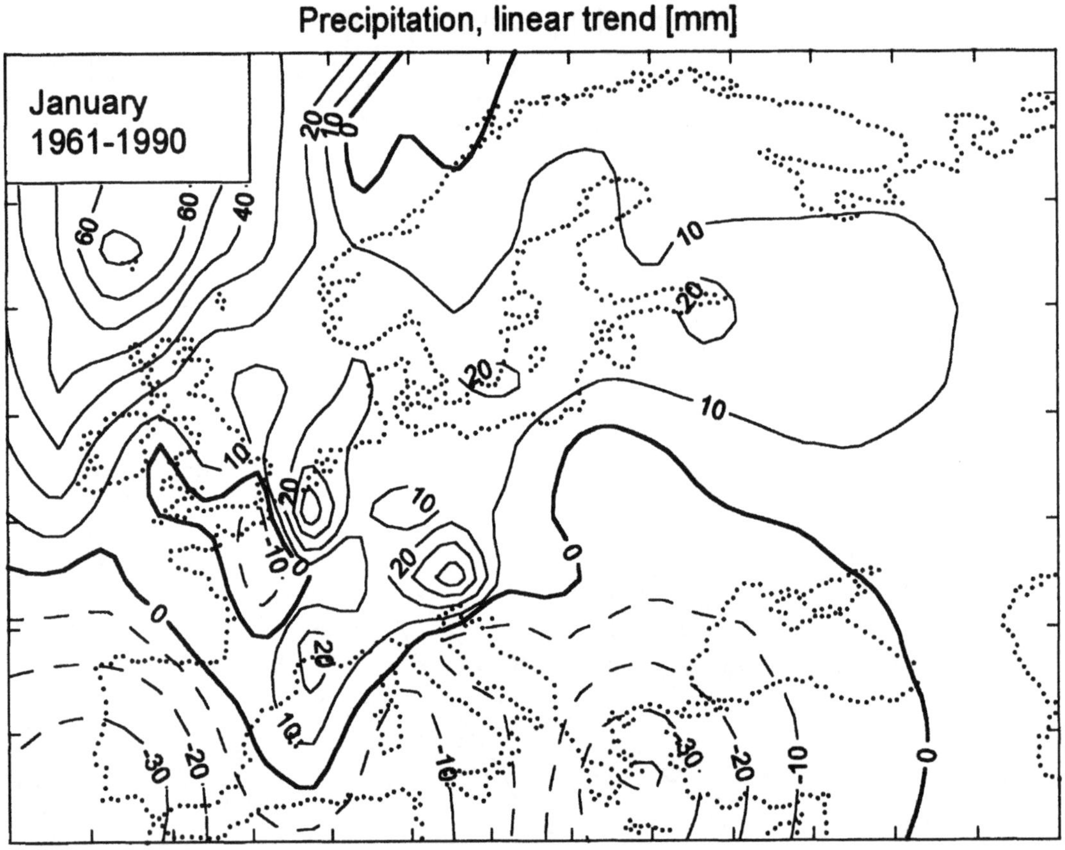
Precipitation, linear trend [mm]
January
1961-1990

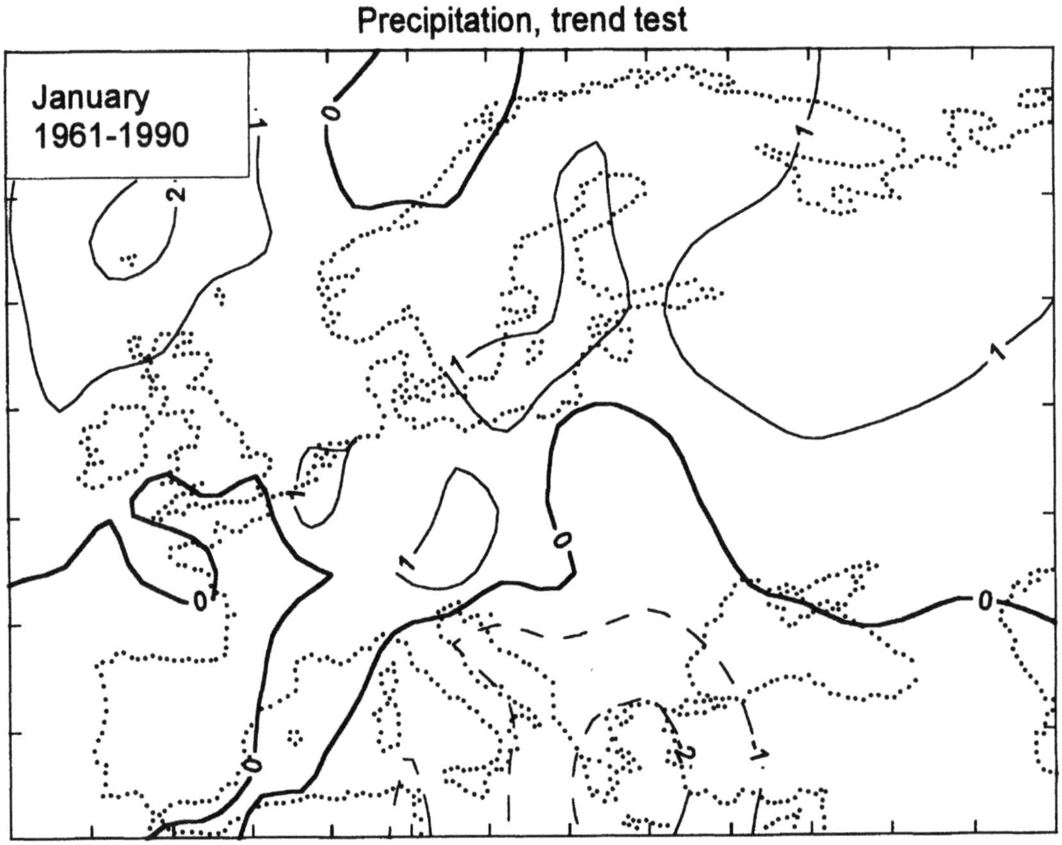
Precipitation, trend test
January
1961-1990

Precipitation, linear trend [mm]

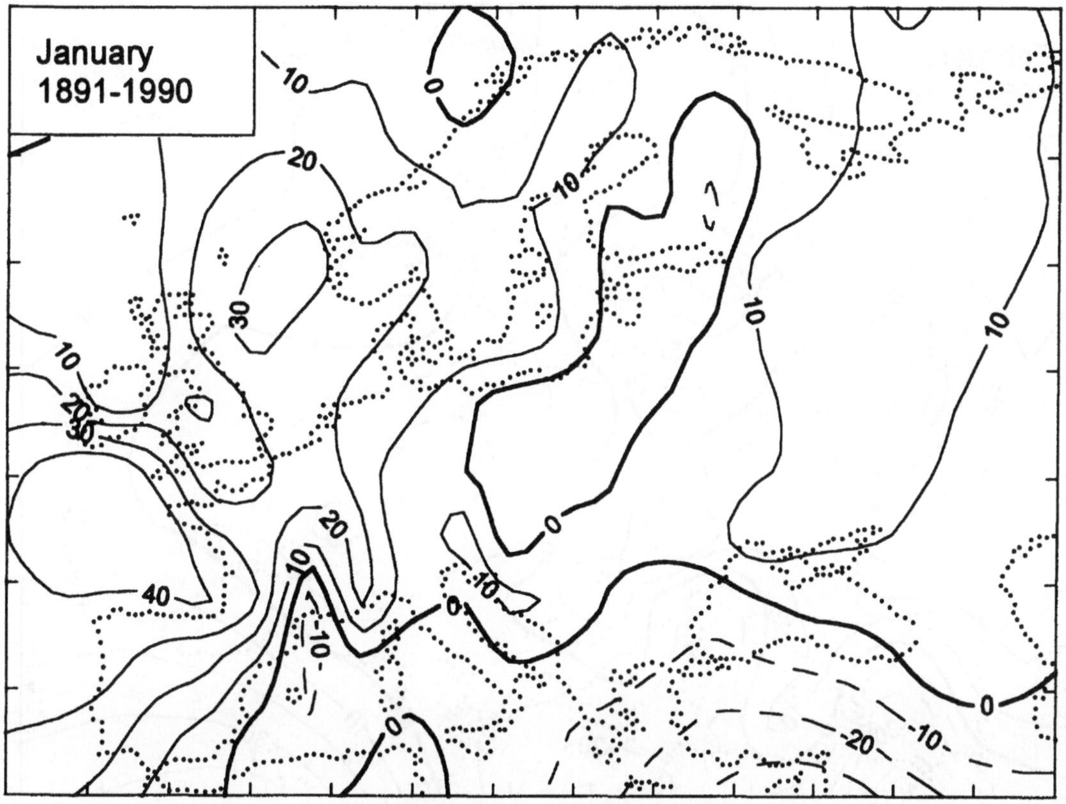

Precipitation, trend test

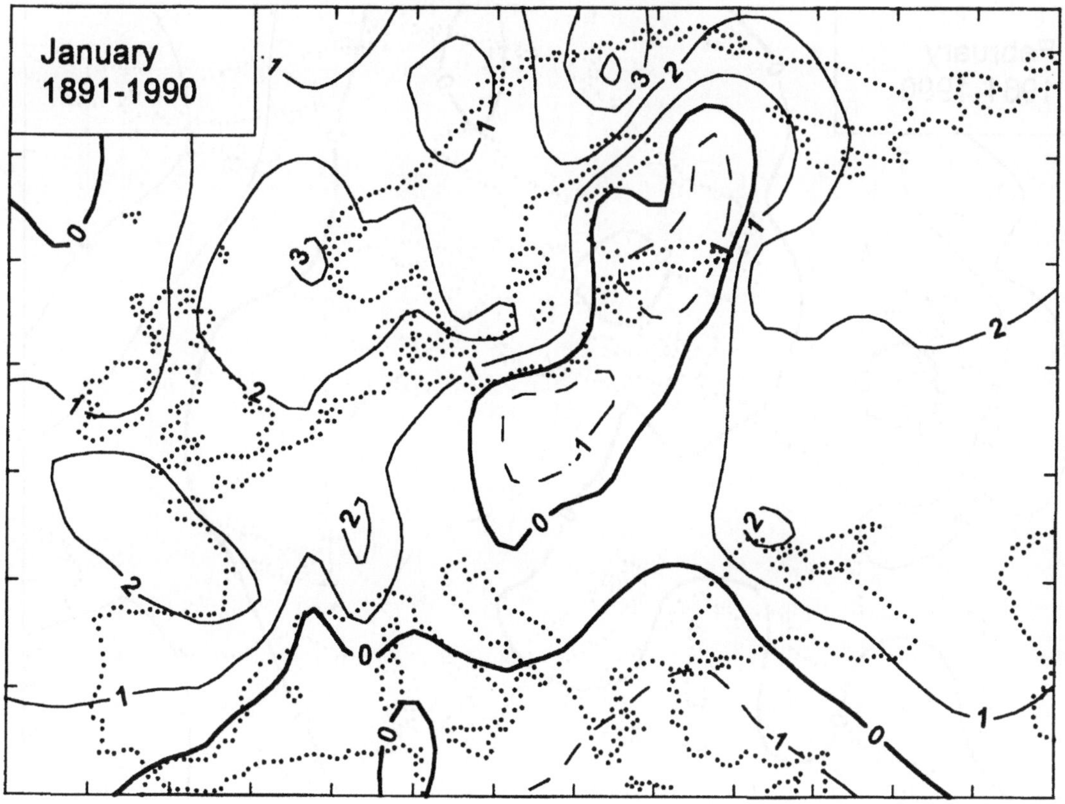

Precipitation, linear trend [mm]

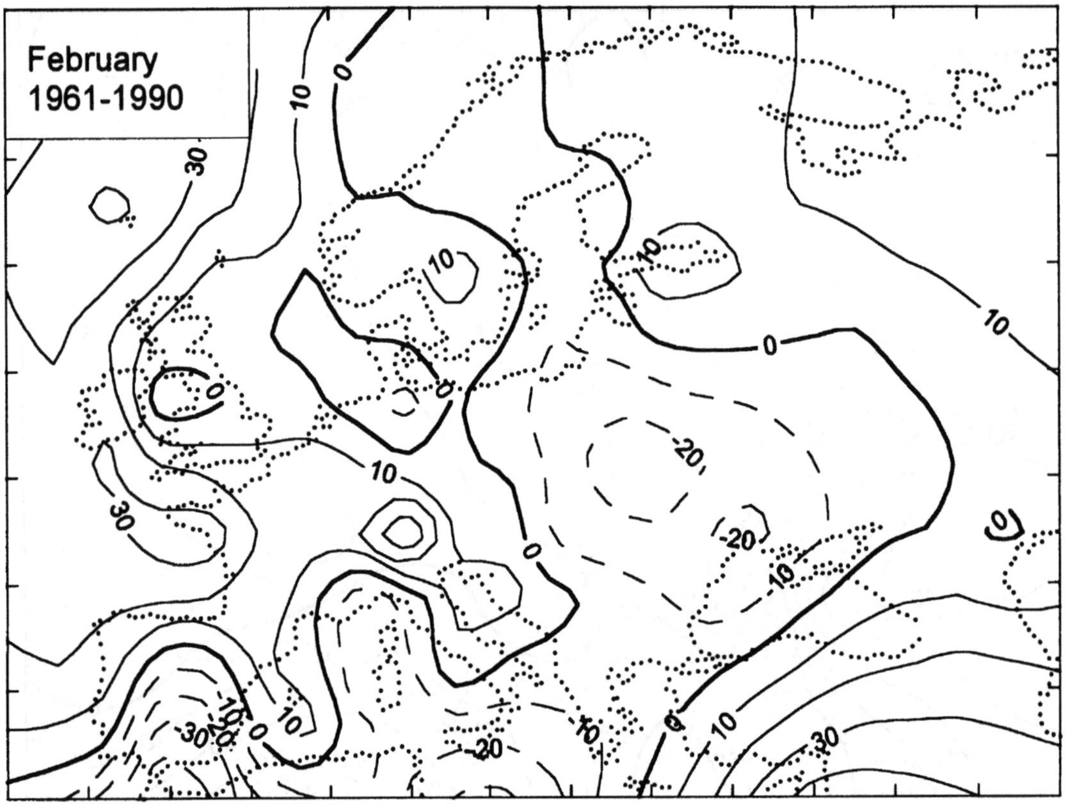

Precipitation, trend test

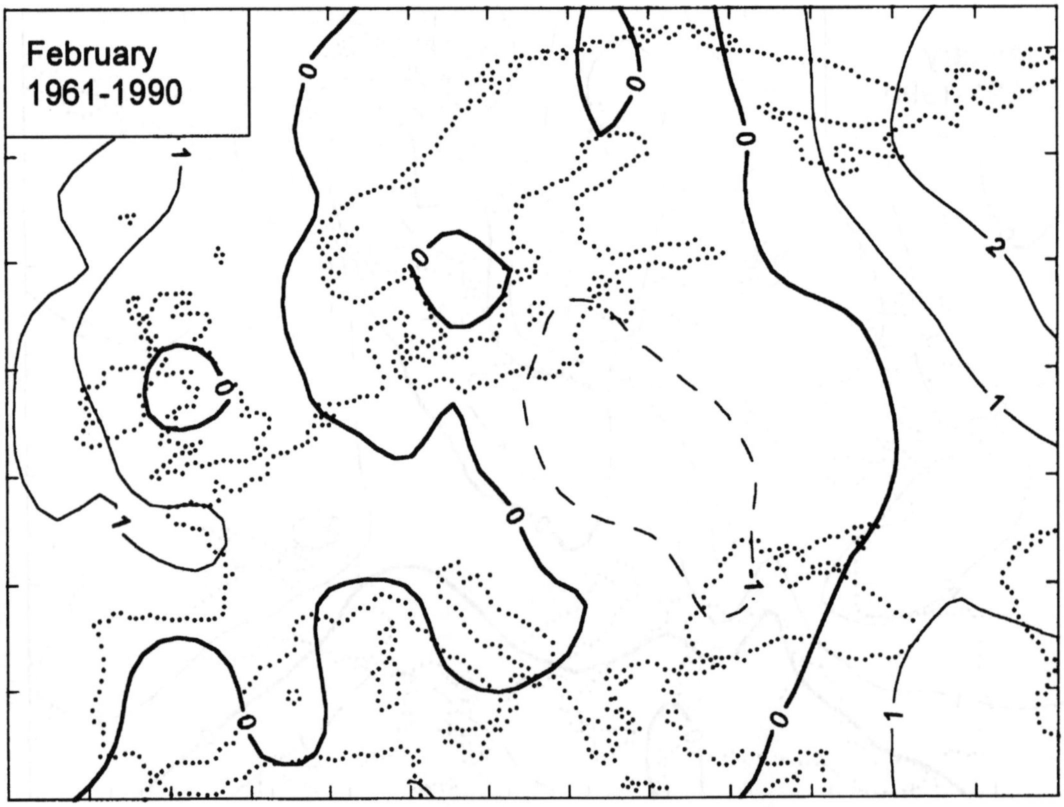

Precipitation, linear trend [mm]

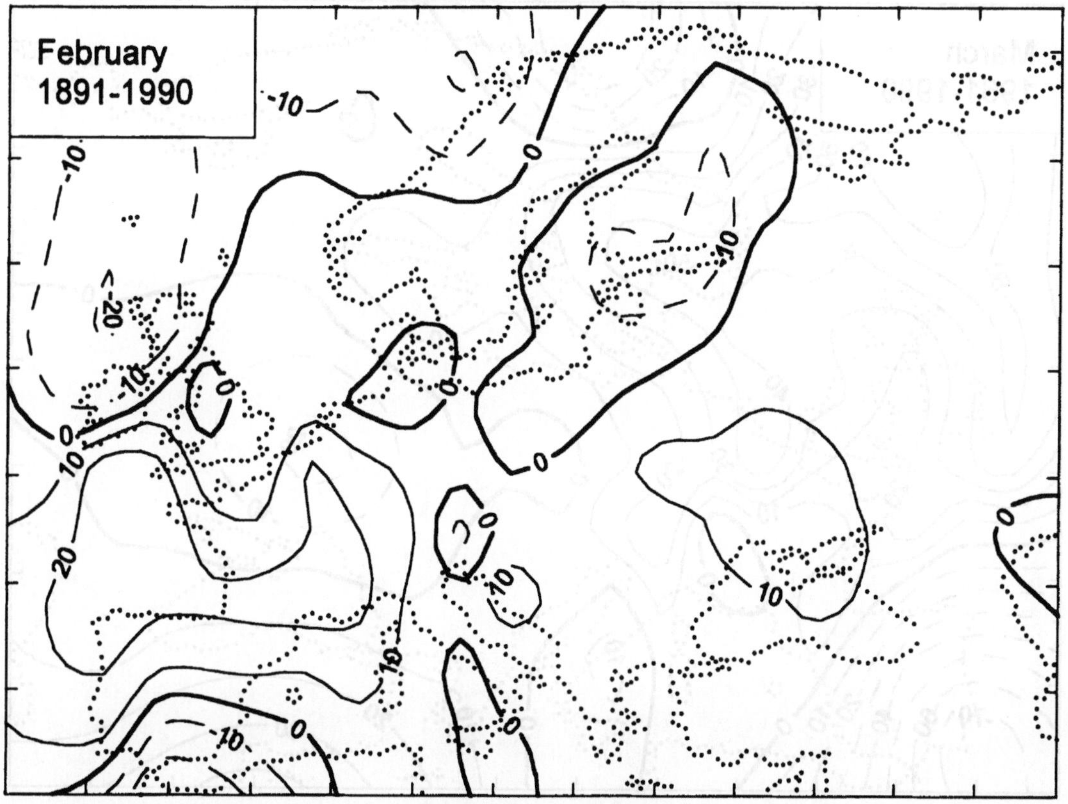

Precipitation, trend test

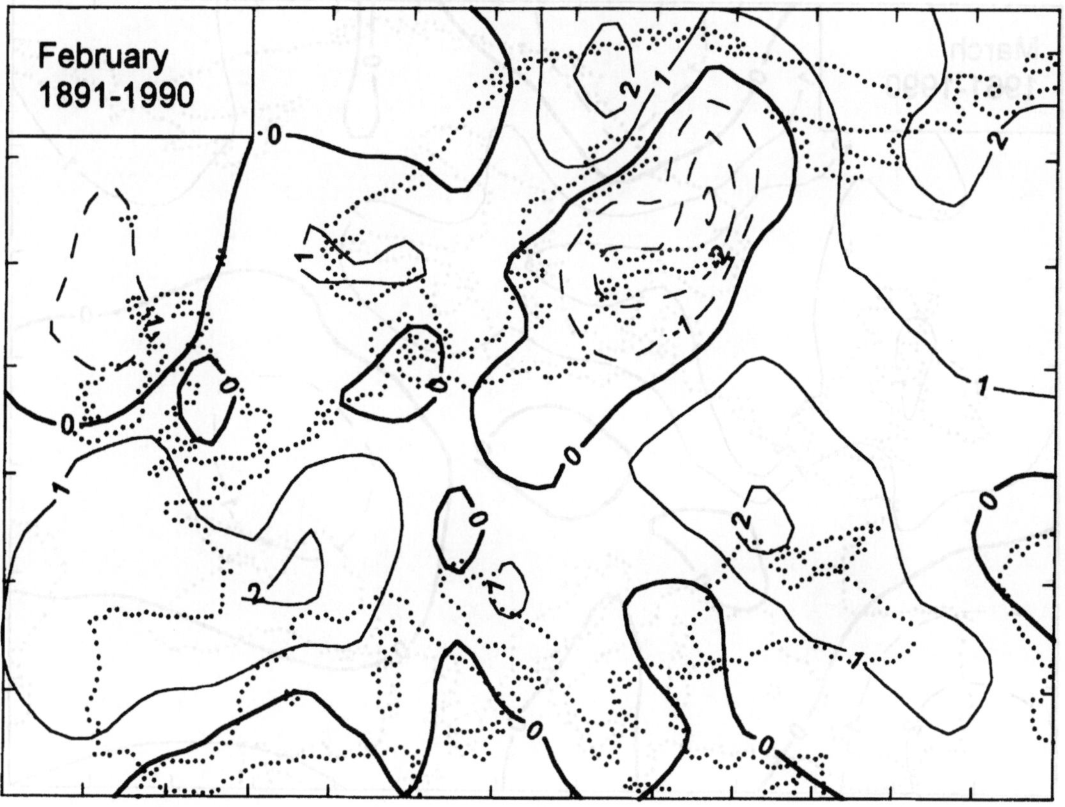

Precipitation, linear trend [mm]

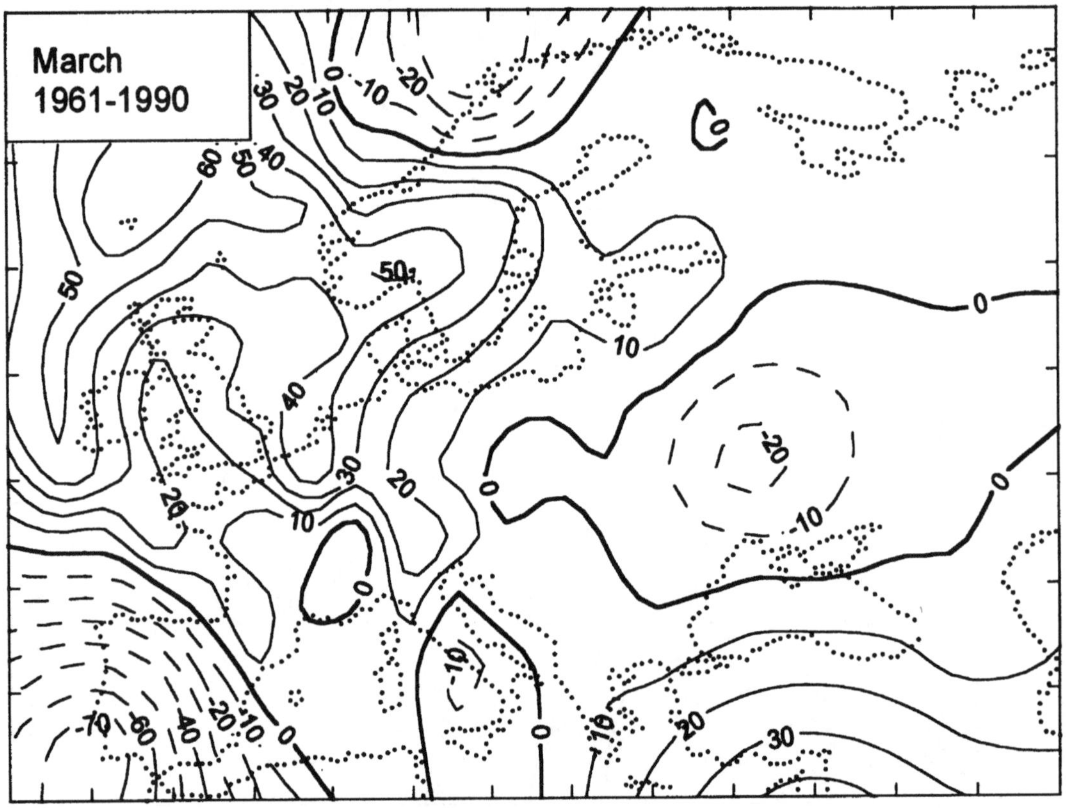

Precipitation, trend test

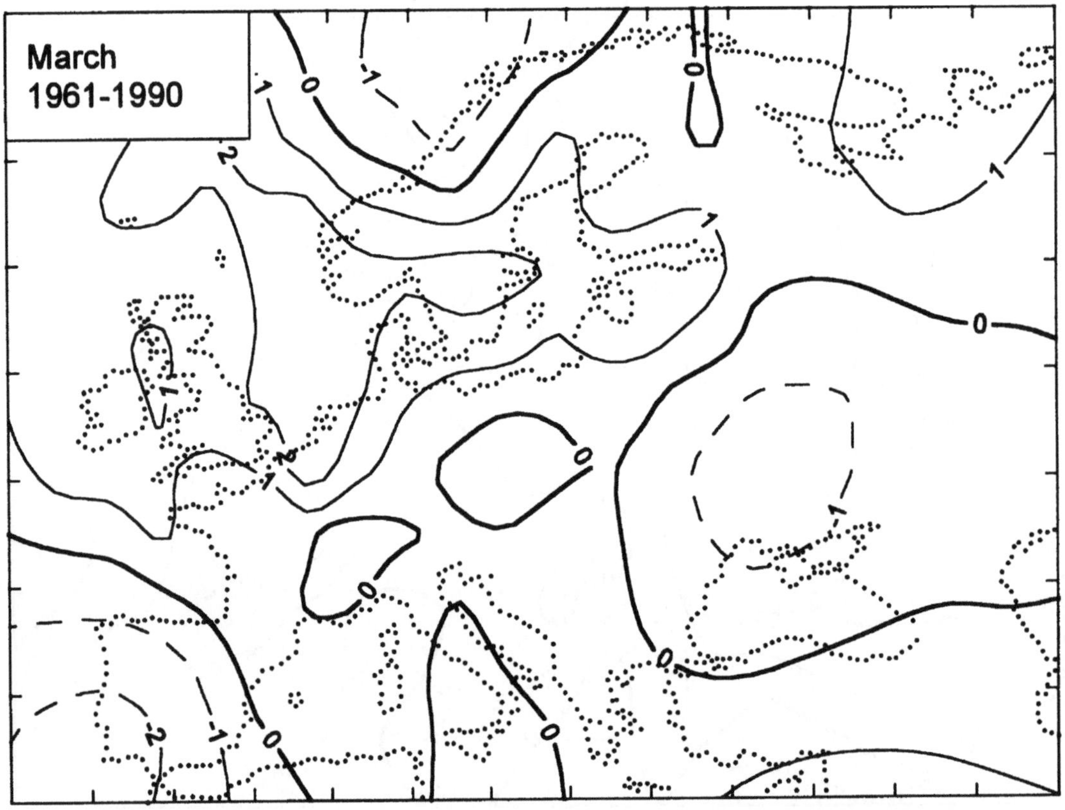

Precipitation, linear trend [mm]

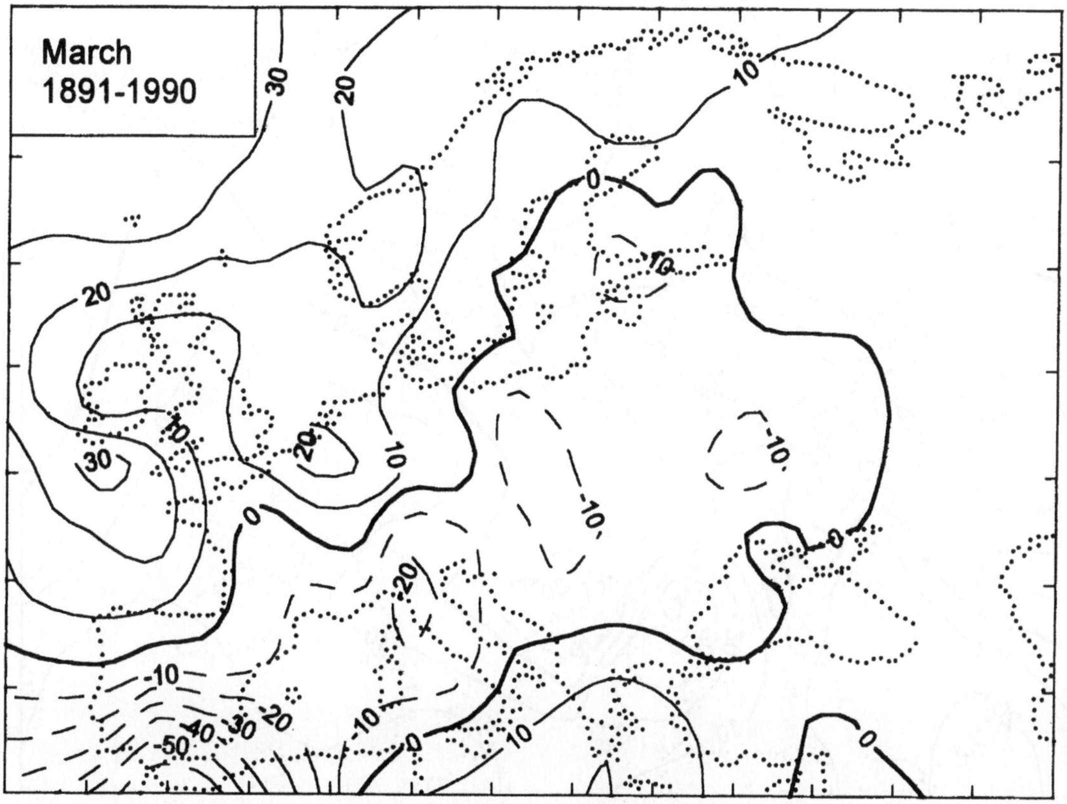

Precipitation, trend test

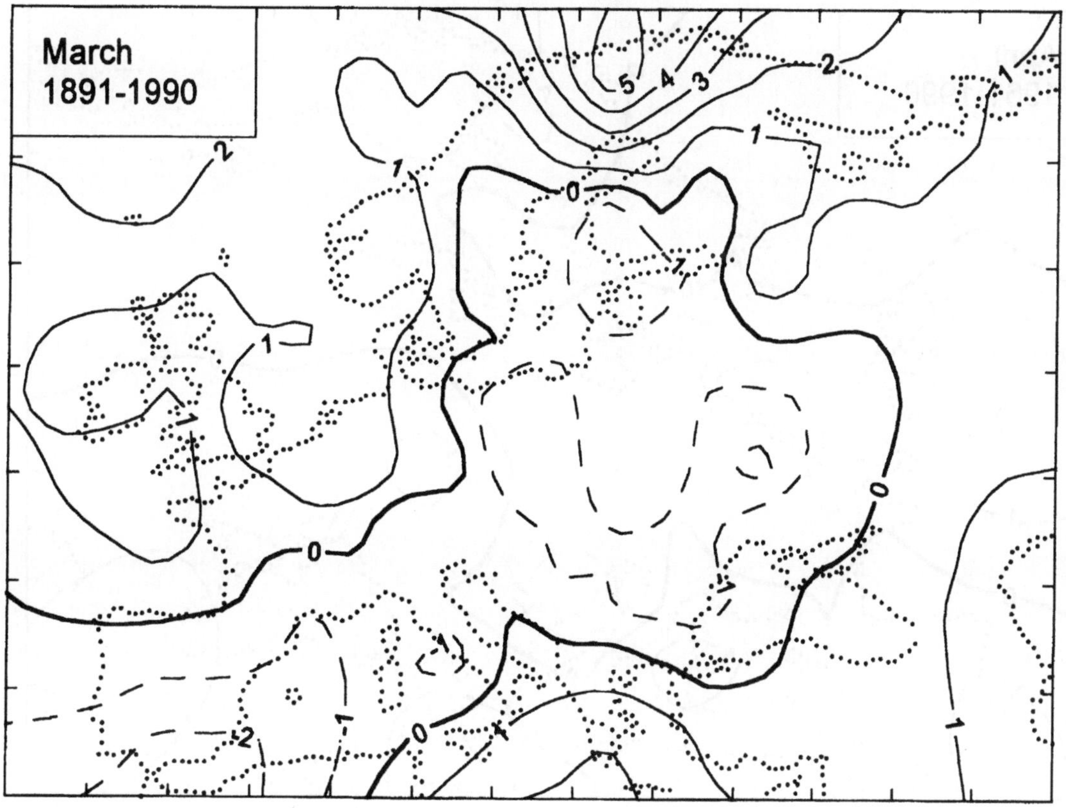

Precipitation, linear trend [mm]

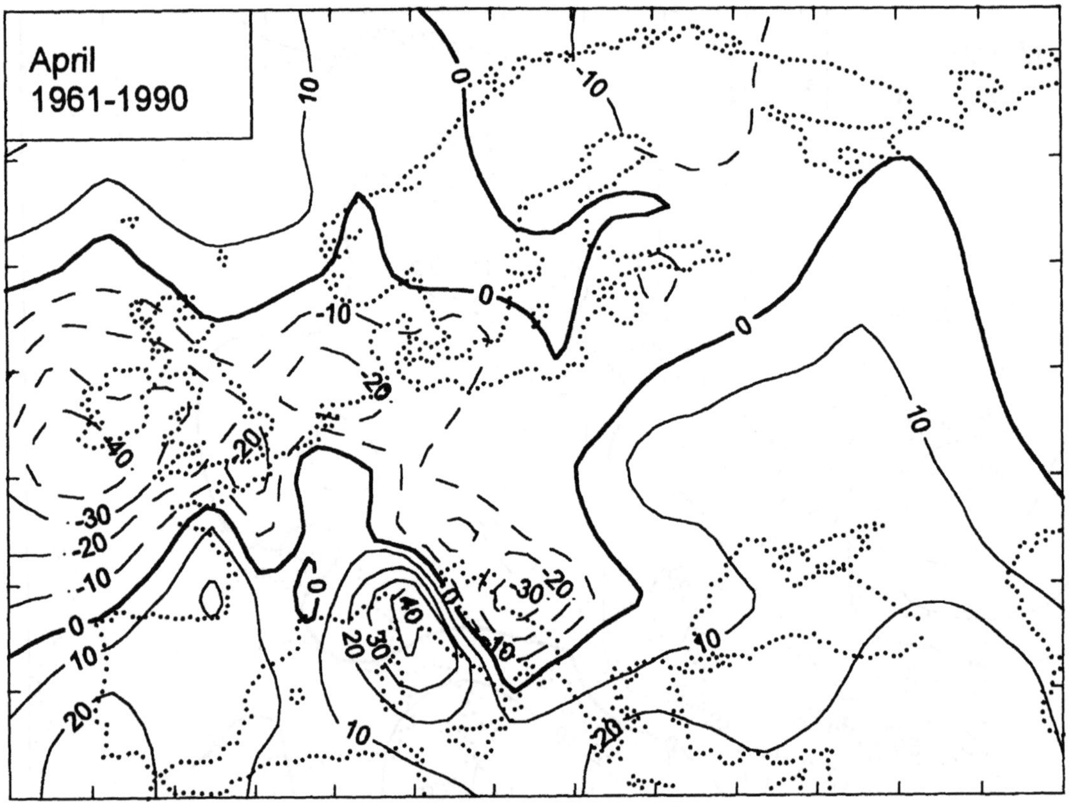

Precipitation, trend test

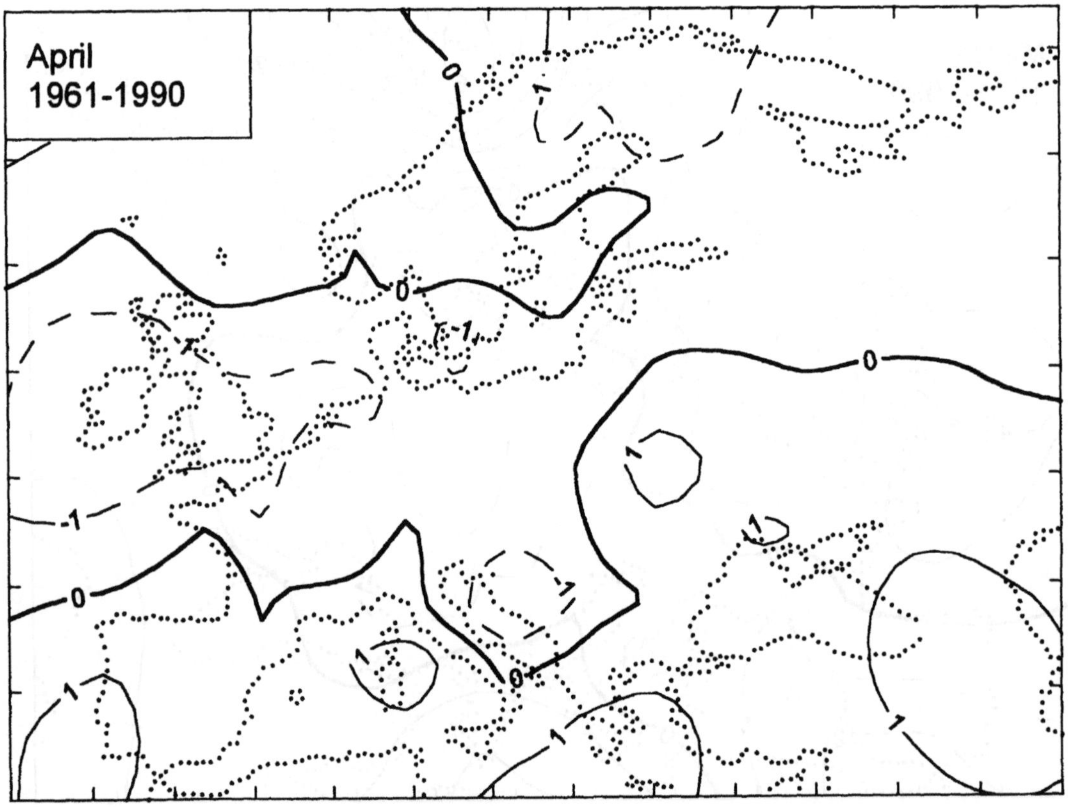

Precipitation, linear trend [mm]

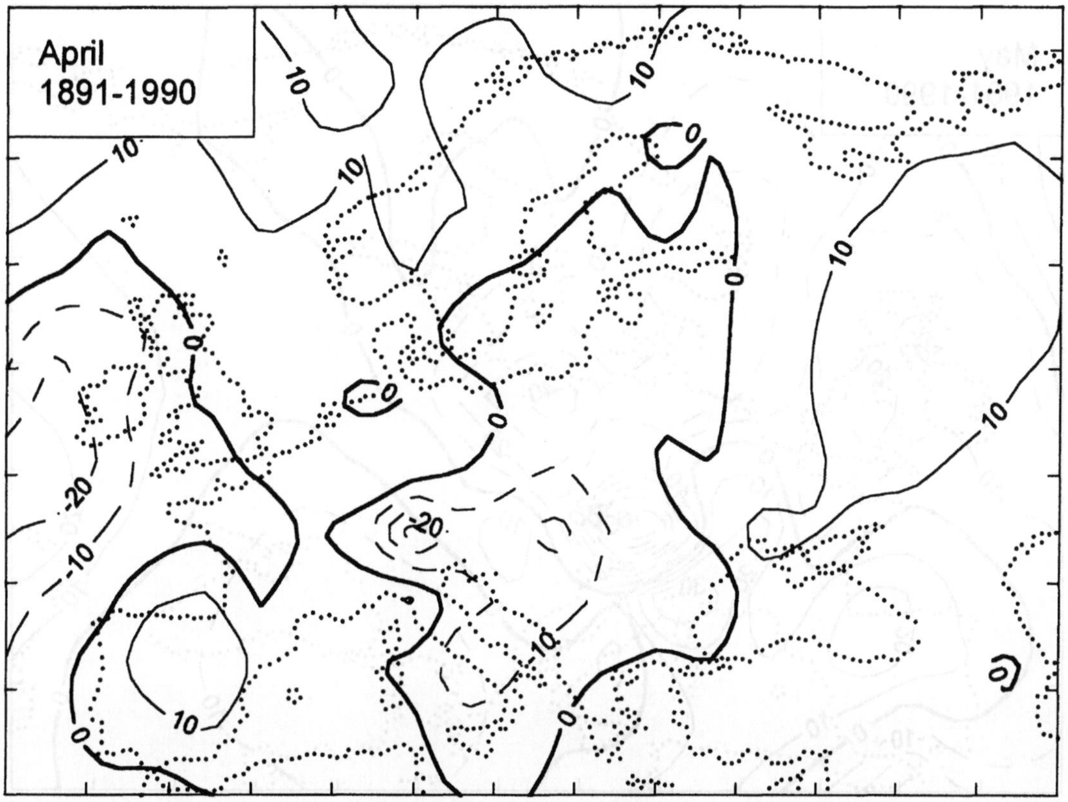

Precipitation, trend test

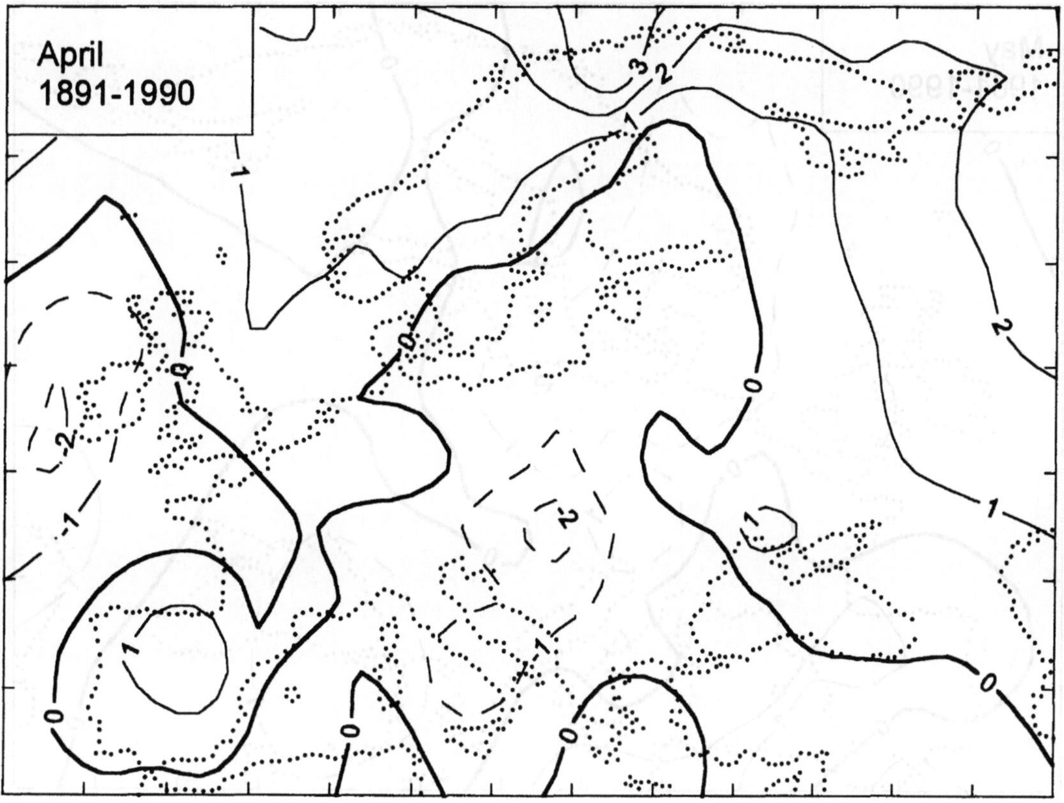

Precipitation, linear trend [mm]

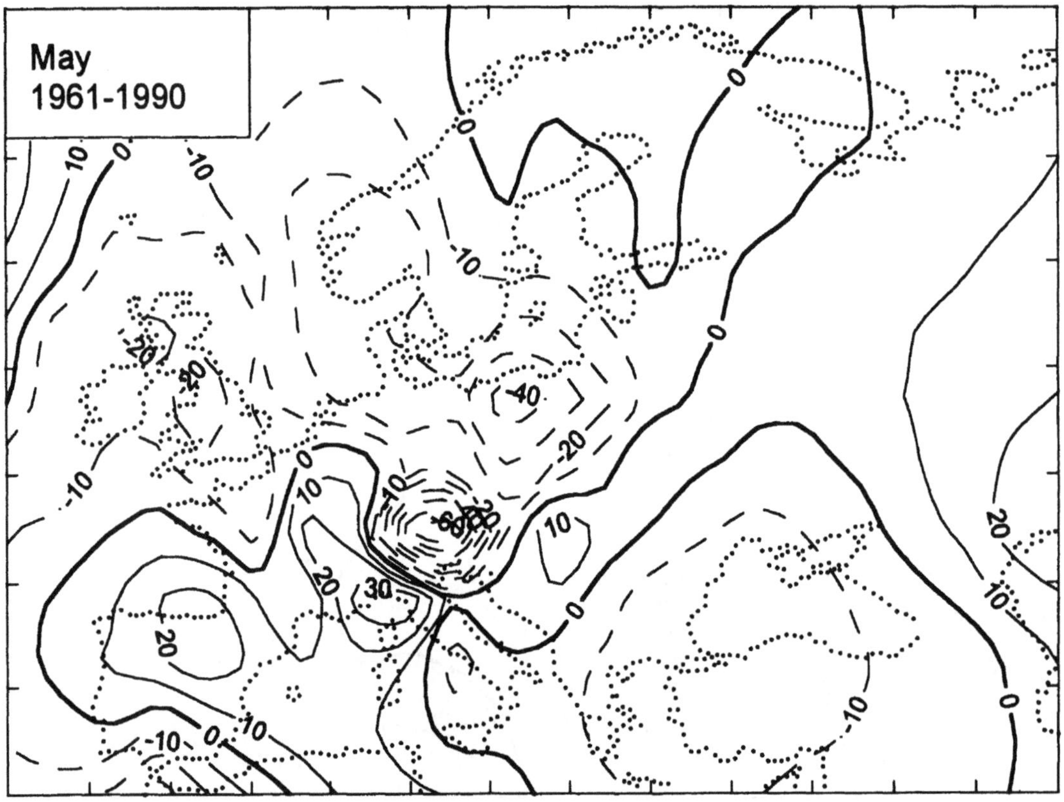

Precipitation, trend test

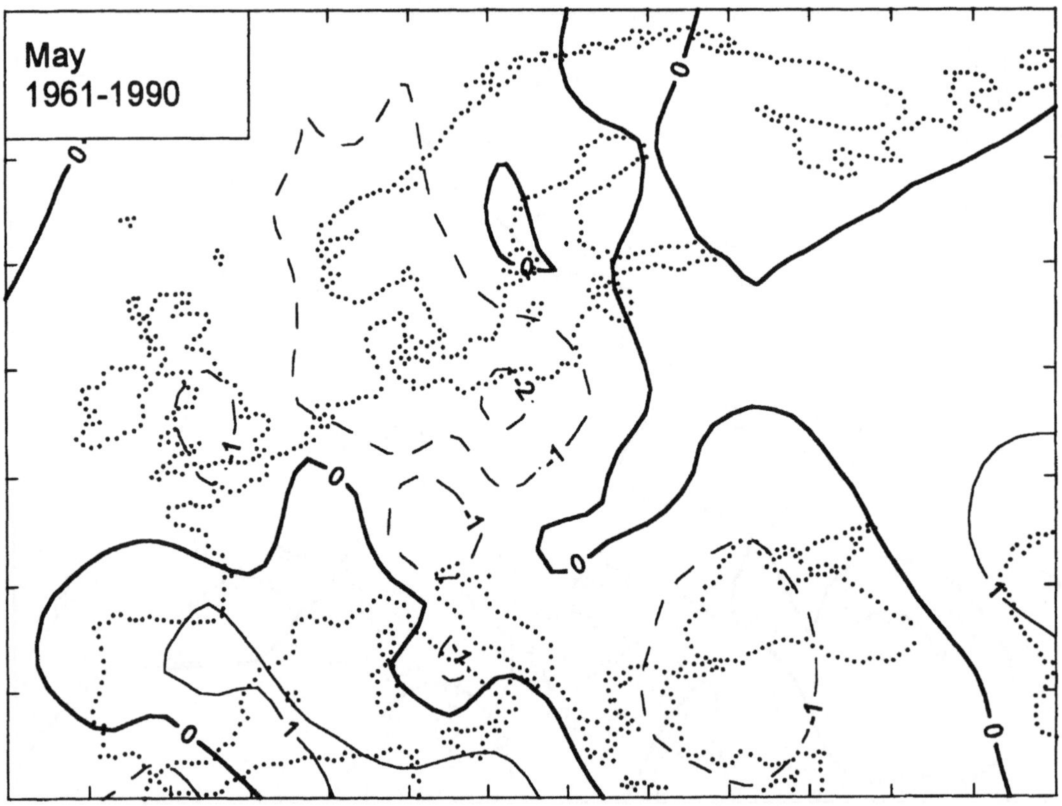

Precipitation, linear trend [mm]

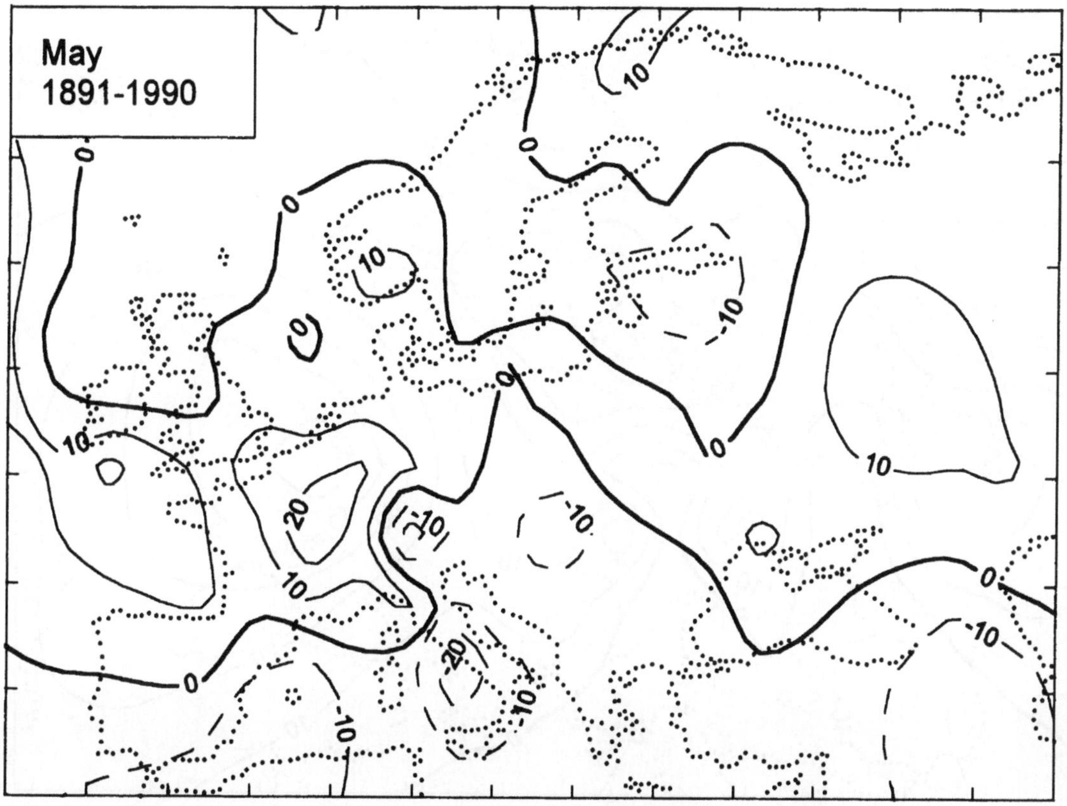

Precipitation, trend test

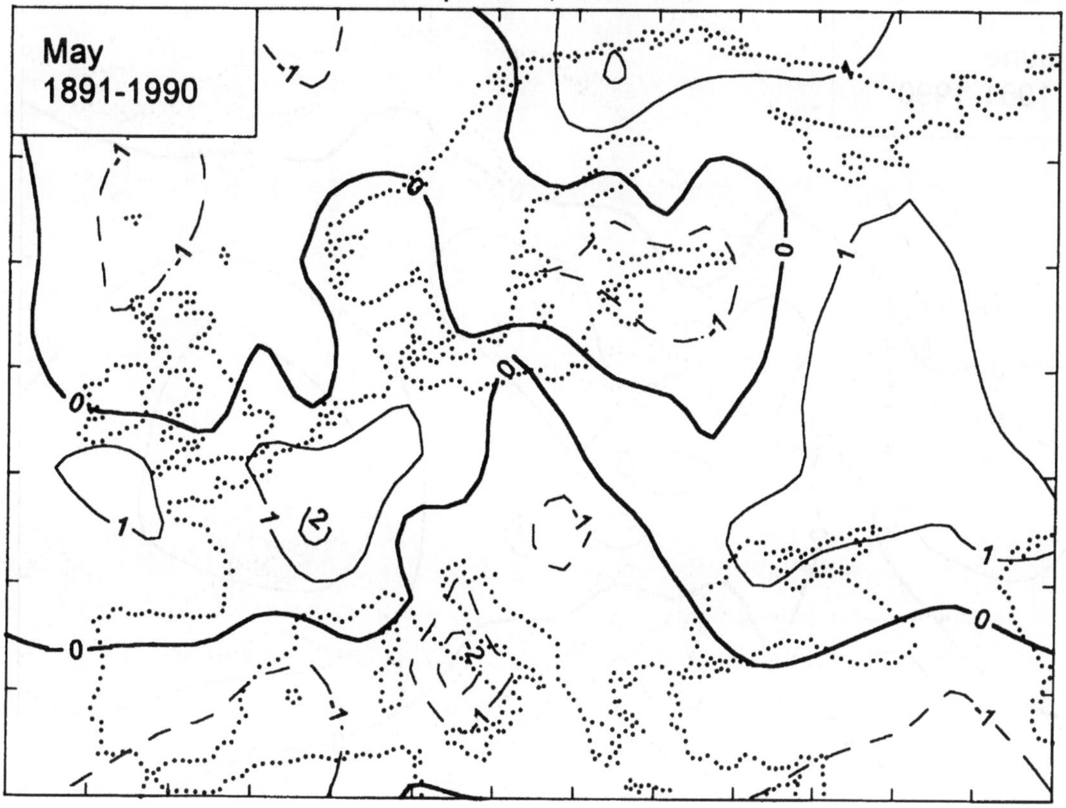

Precipitation, linear trend [mm]

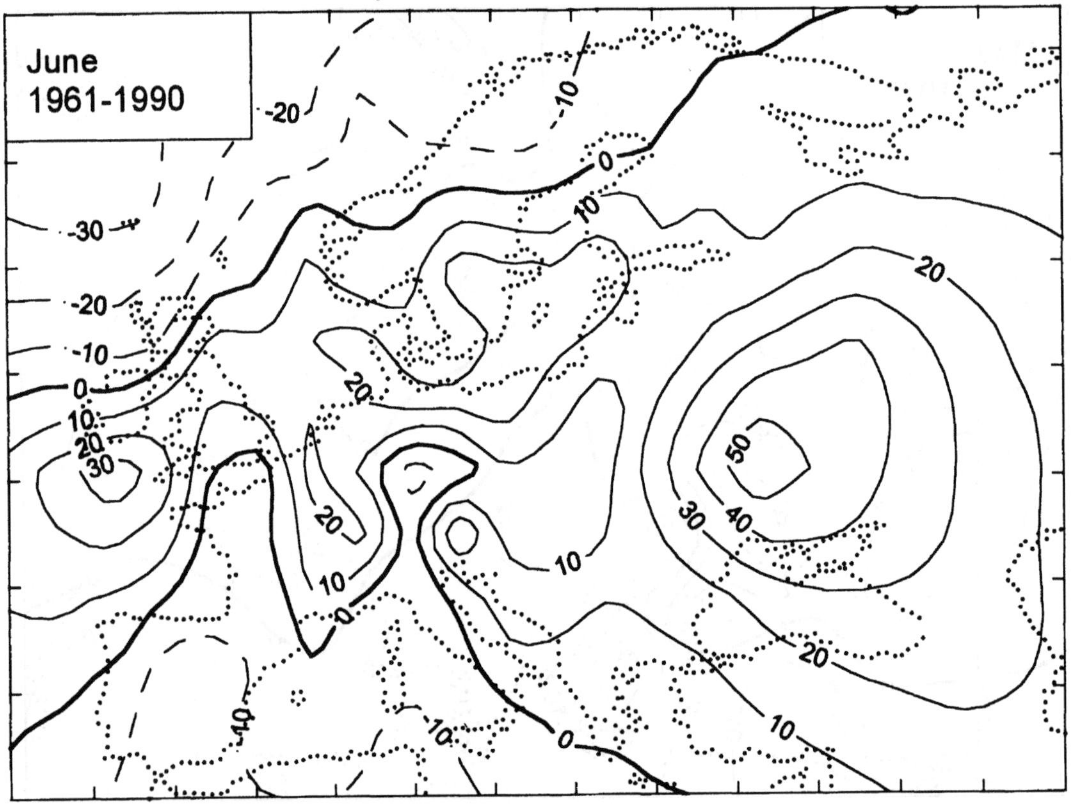

Precipitation, trend test

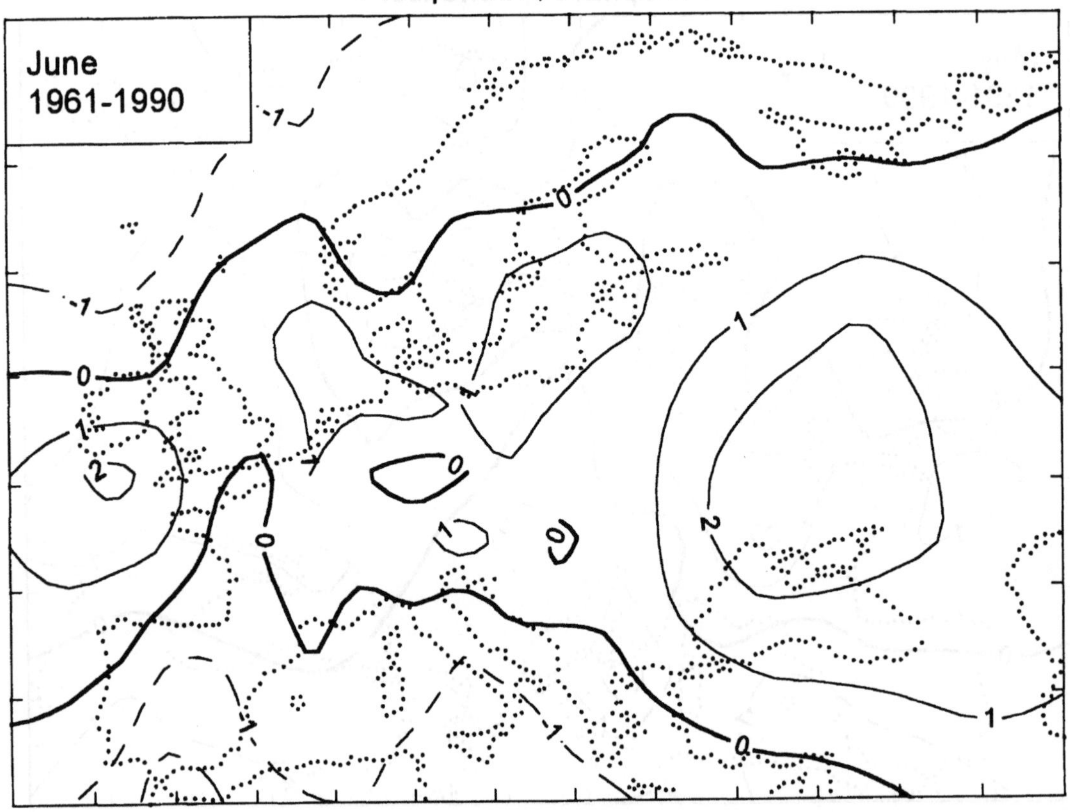

Precipitation, linear trend [mm]

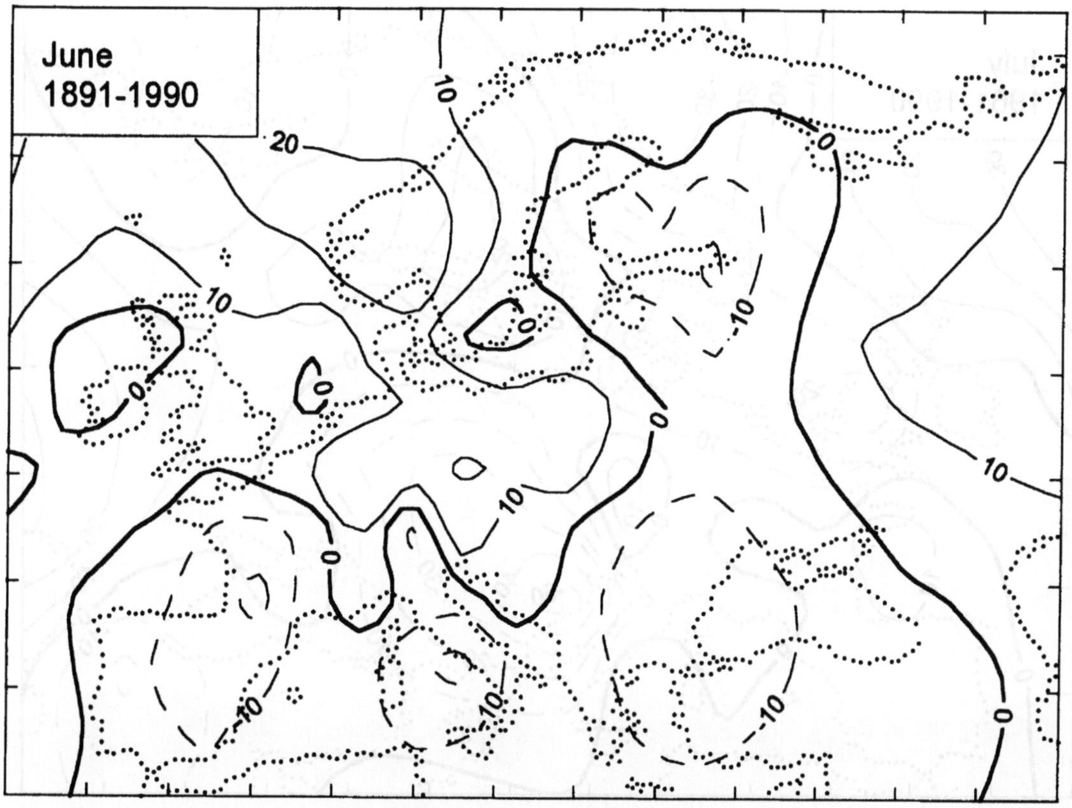

Precipitation, trend test

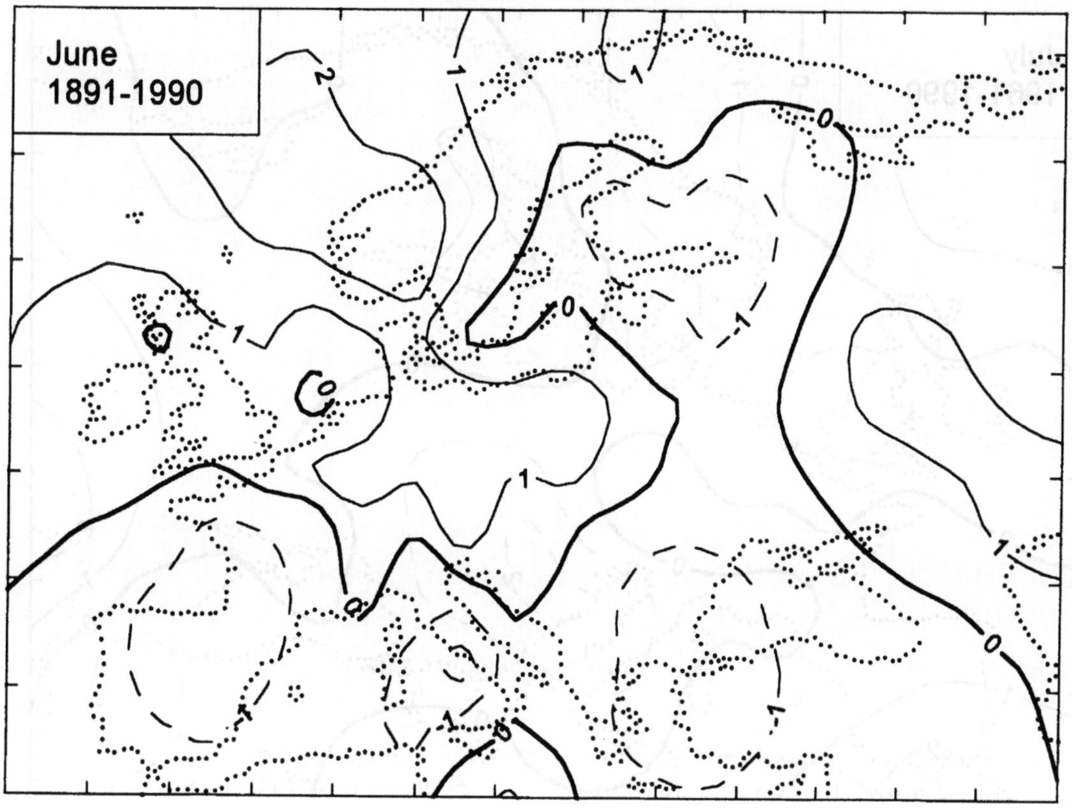

Precipitation, linear trend [mm]

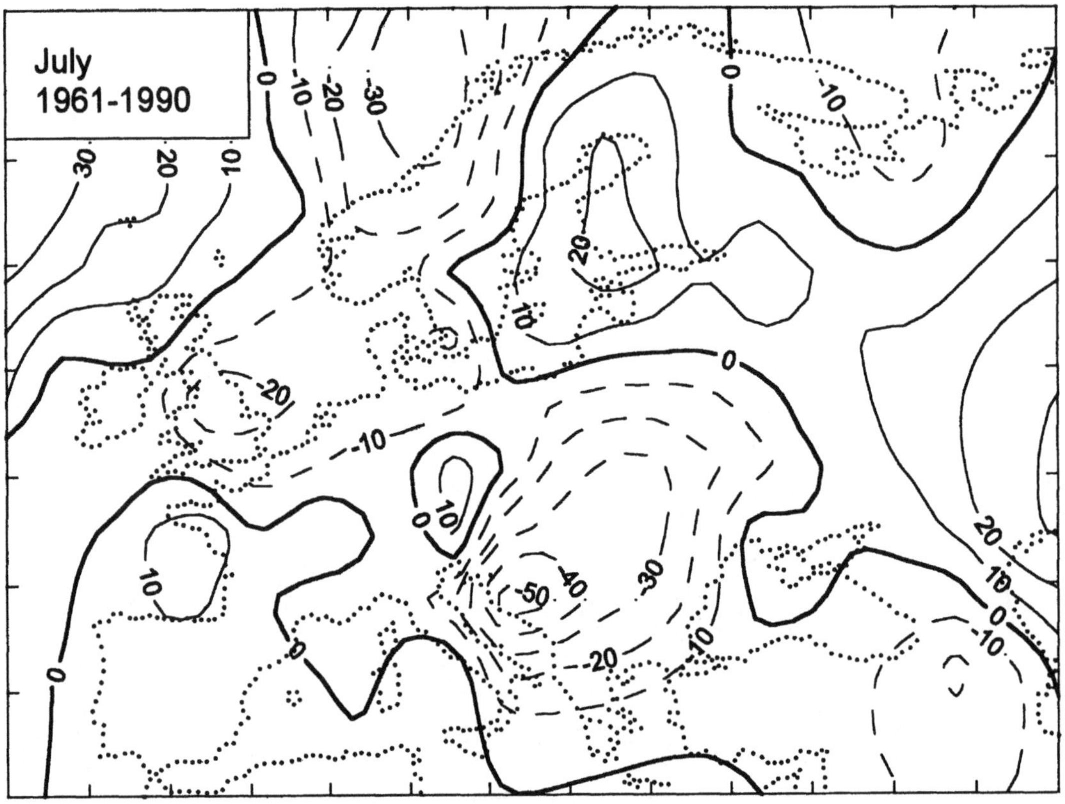

Precipitation, trend test

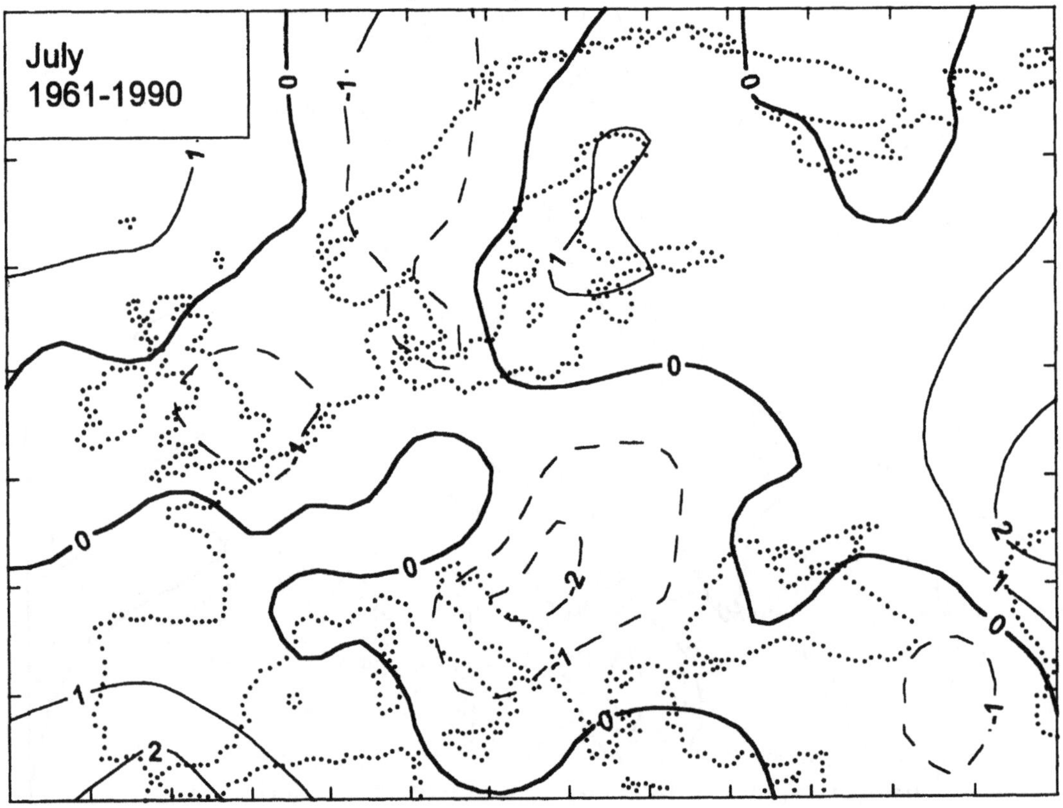

Precipitation, linear trend [mm]

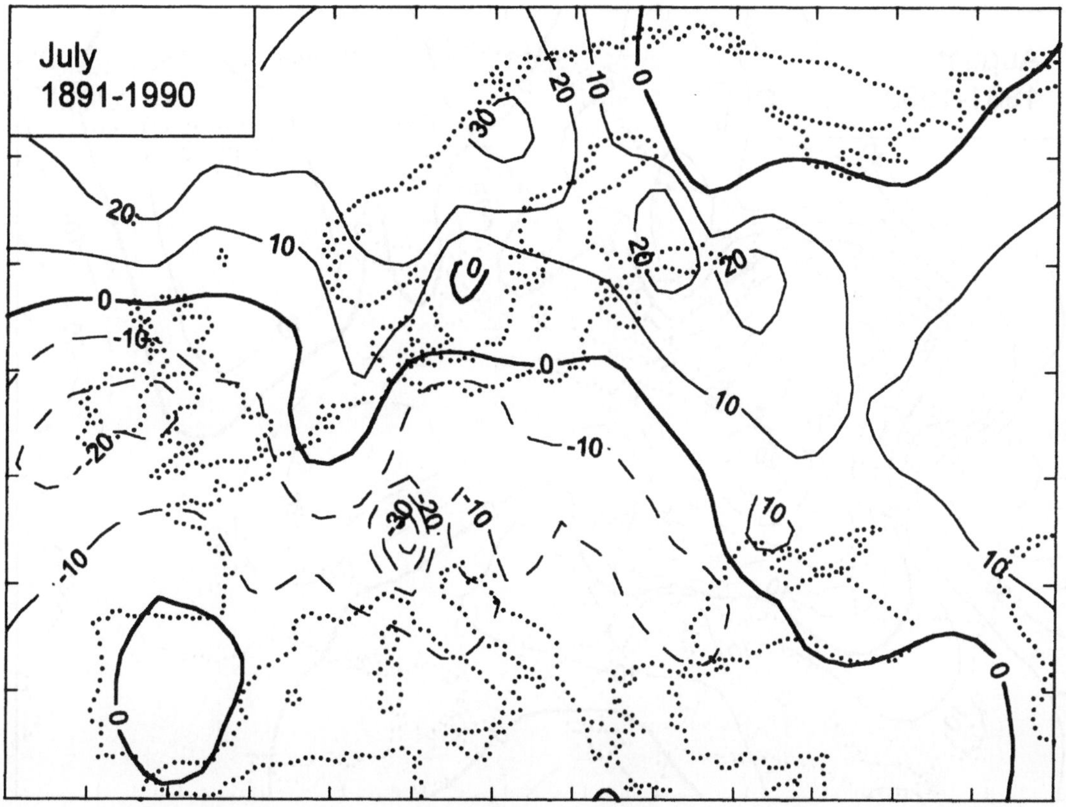

Precipitation, trend test

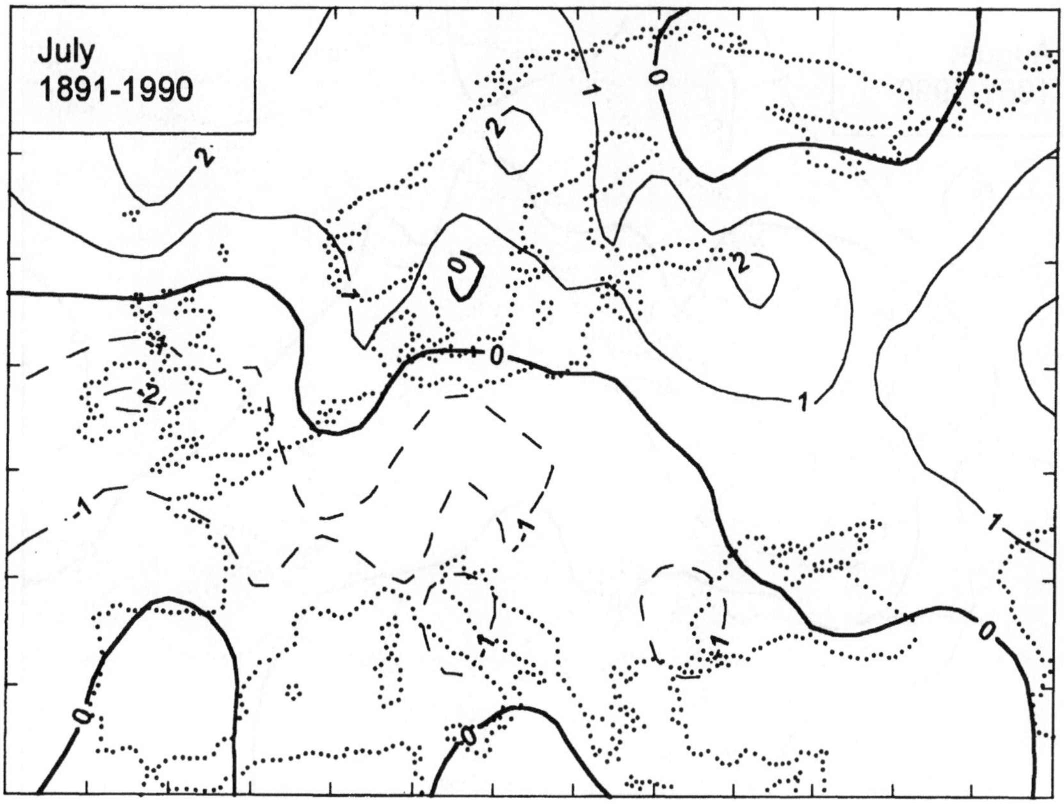

Precipitation, linear trend [mm]

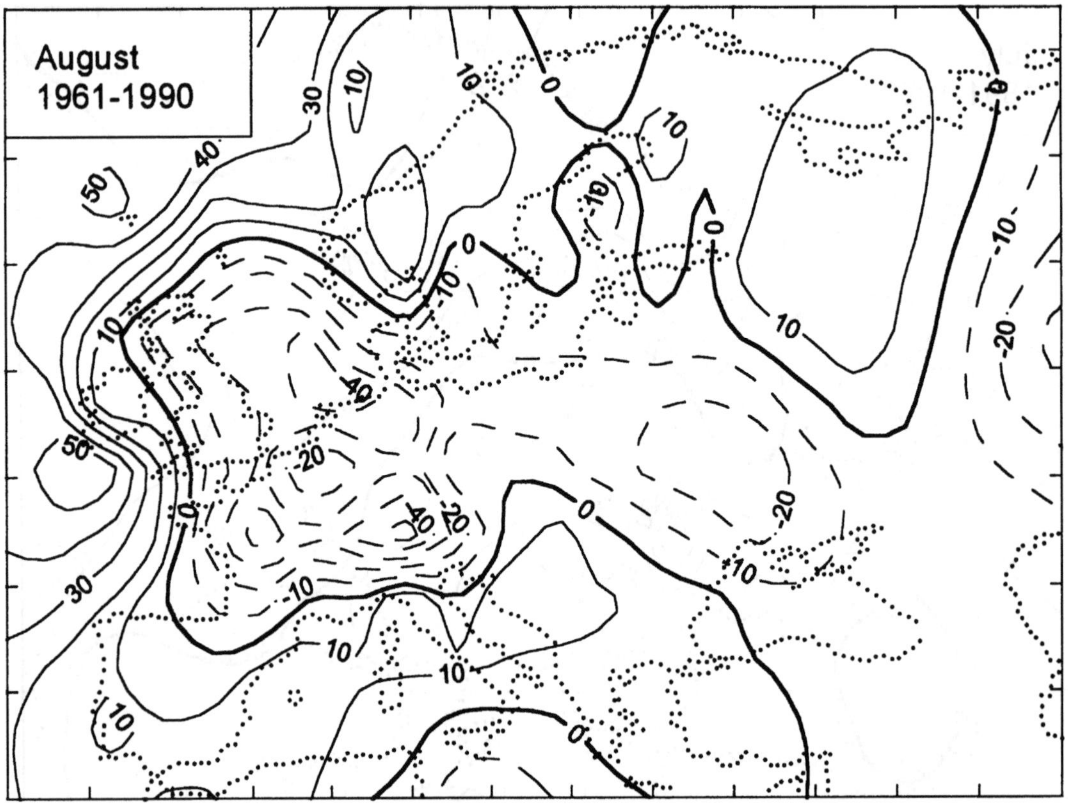

Precipitation, trend test

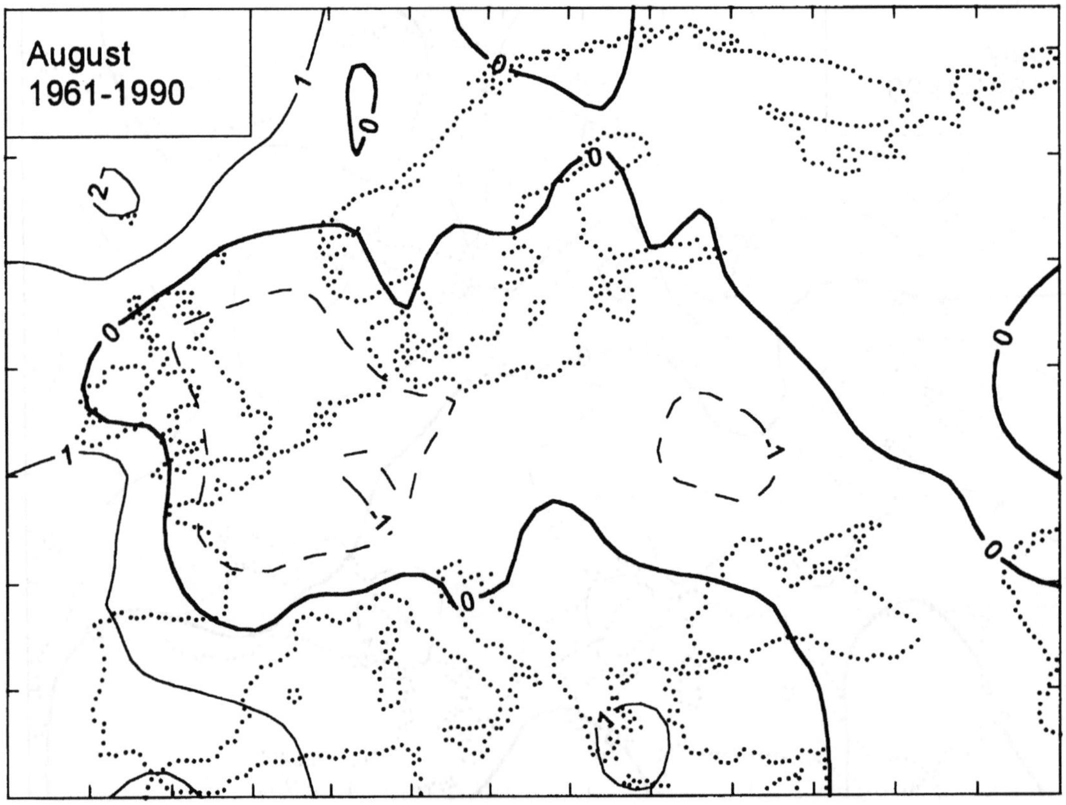

Precipitation, linear trend [mm]

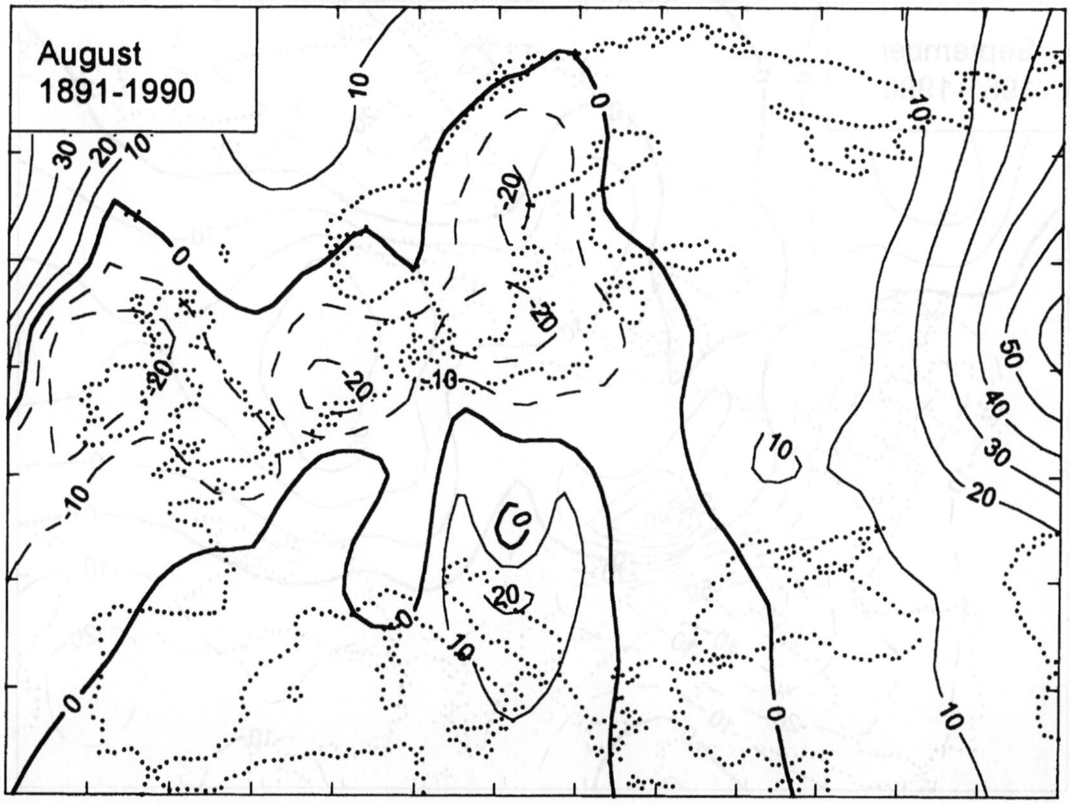

Precipitation, trend test

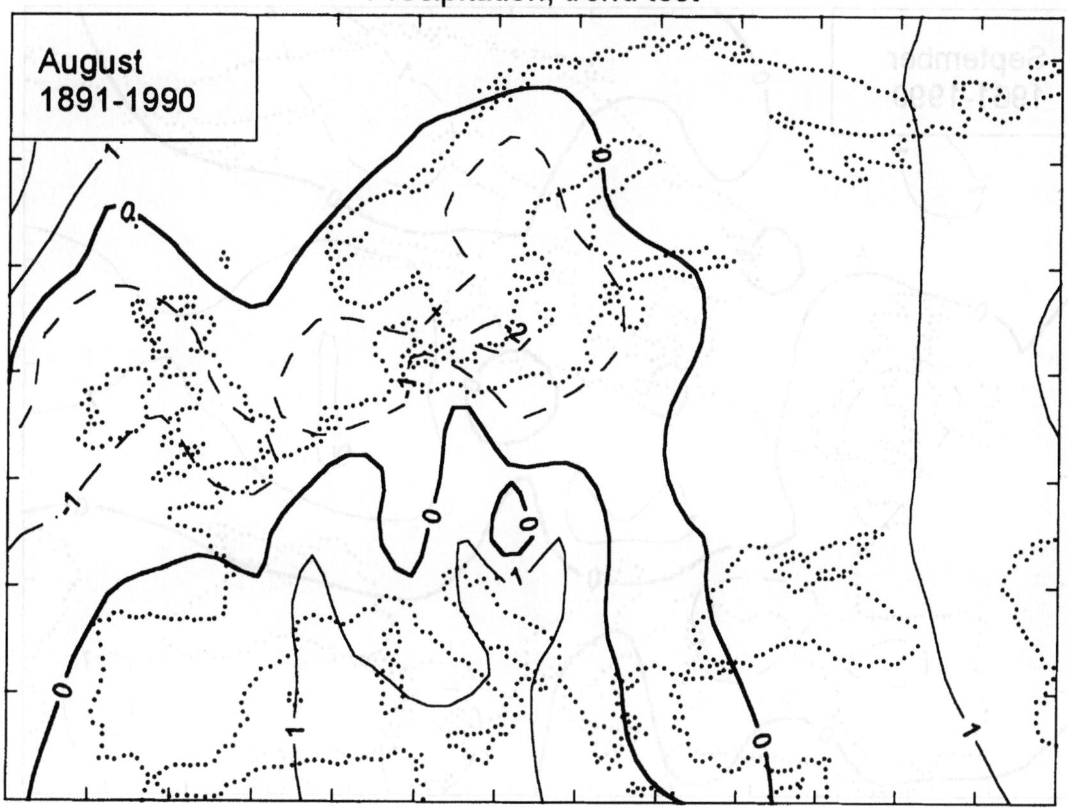

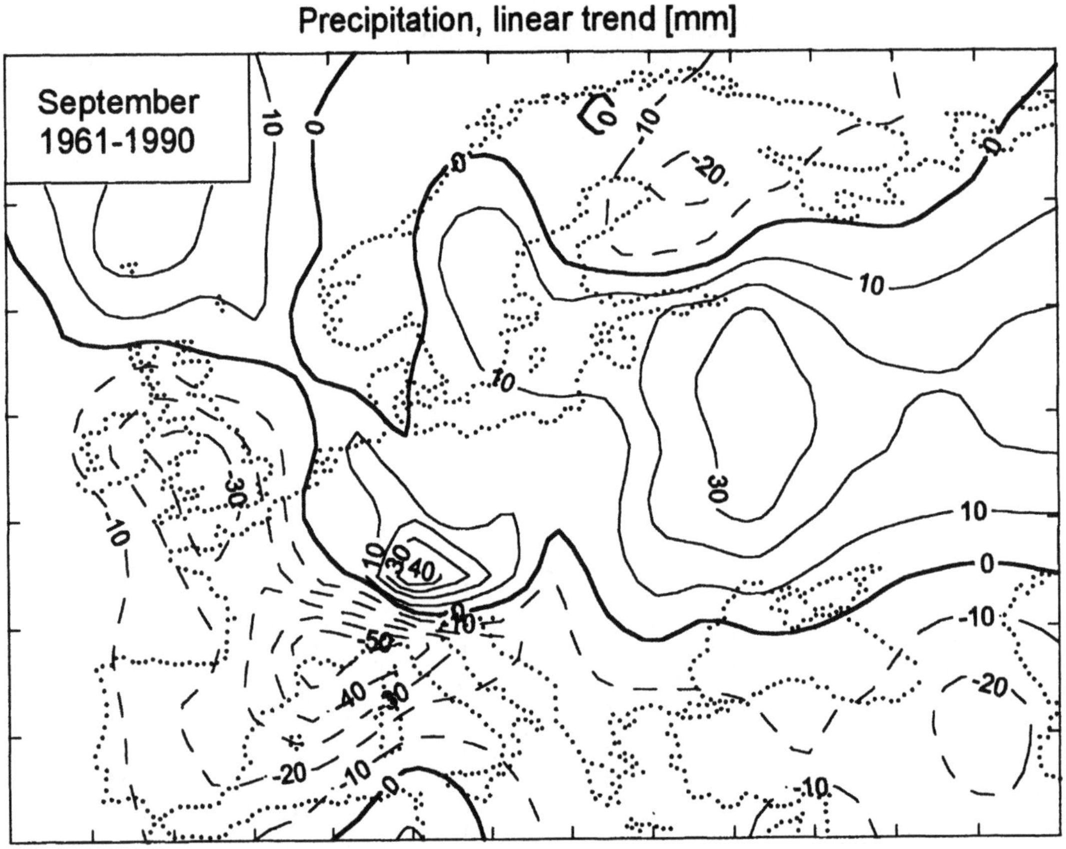
Precipitation, linear trend [mm]
September
1961-1990
10
0
-10
-20
30
40
-50
-40
-30

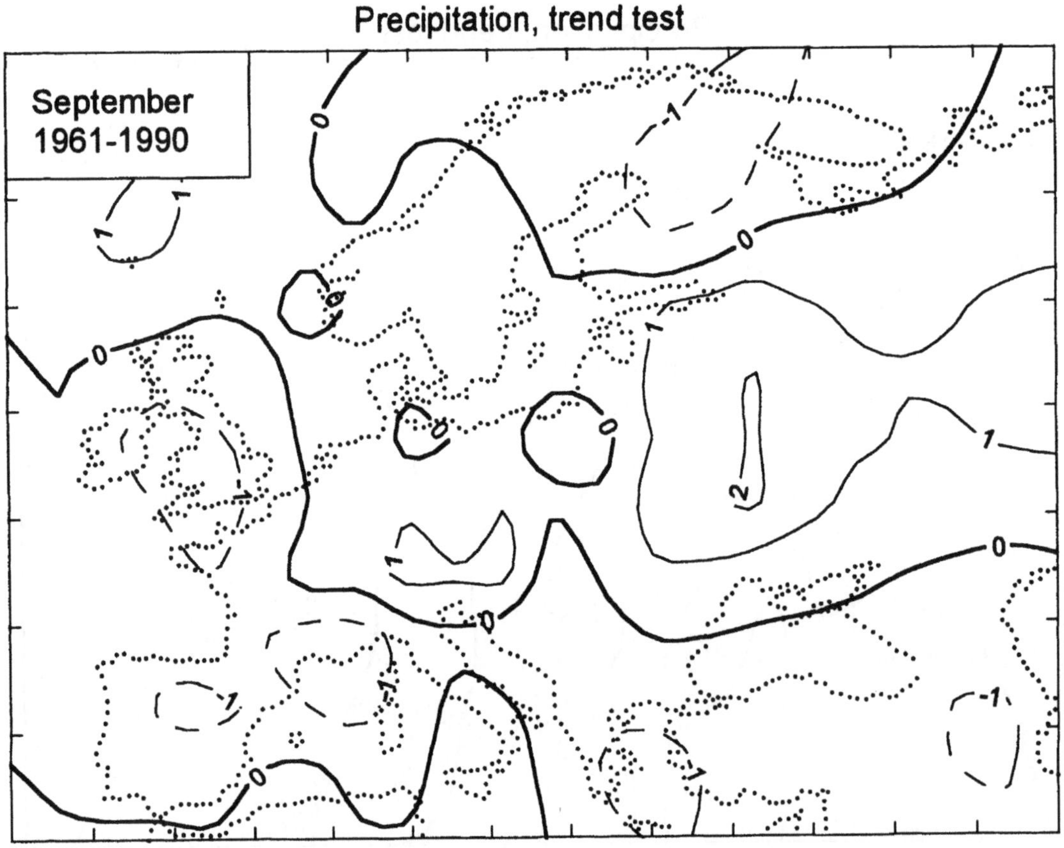
Precipitation, trend test
September
1961-1990
0
1
-1
2

Precipitation, linear trend [mm]

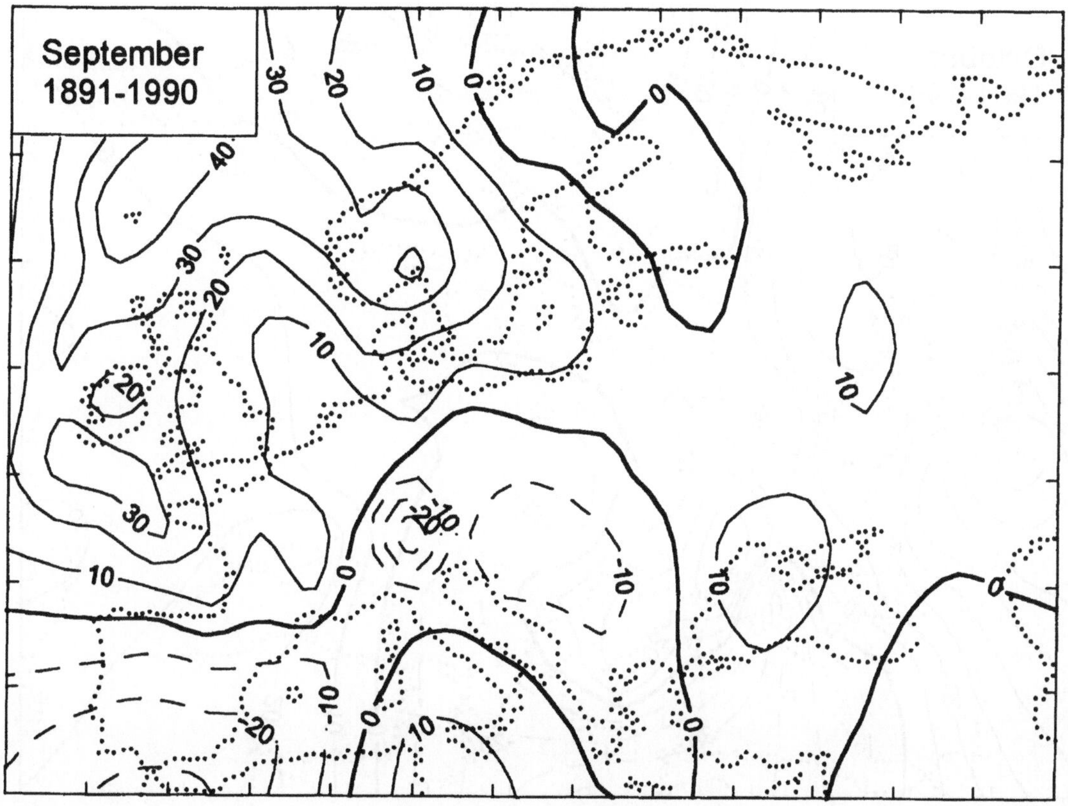

Precipitation, trend test

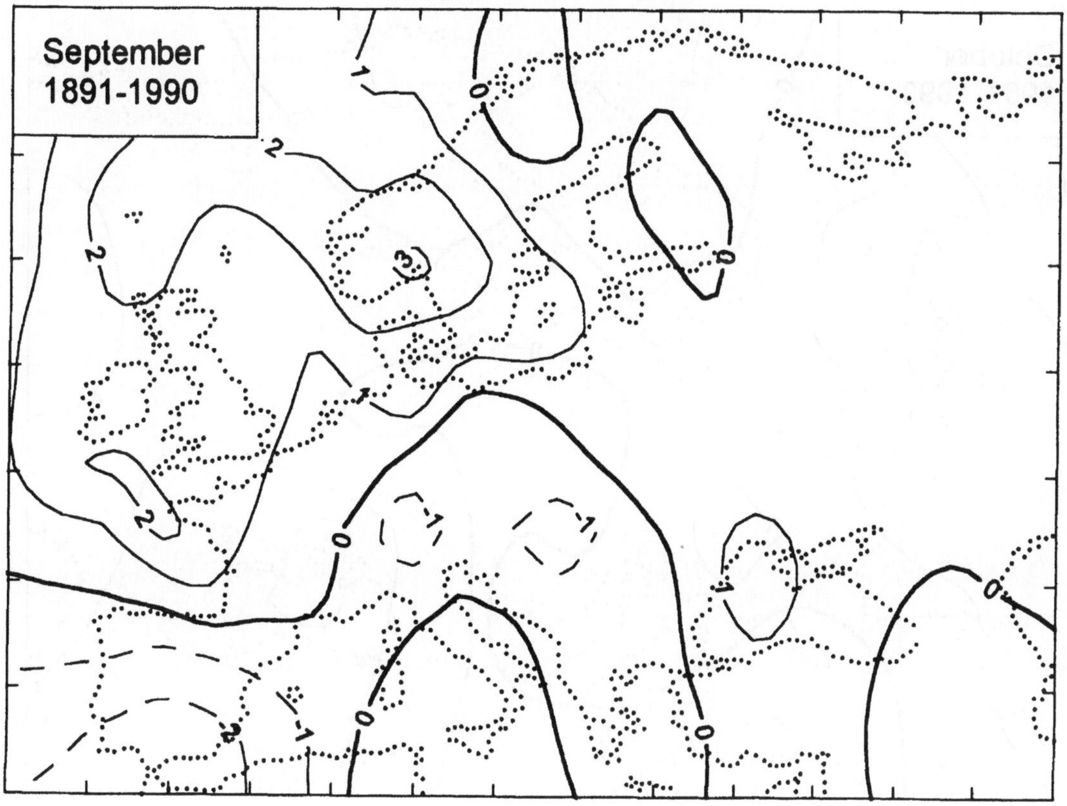

Precipitation, linear trend [mm]

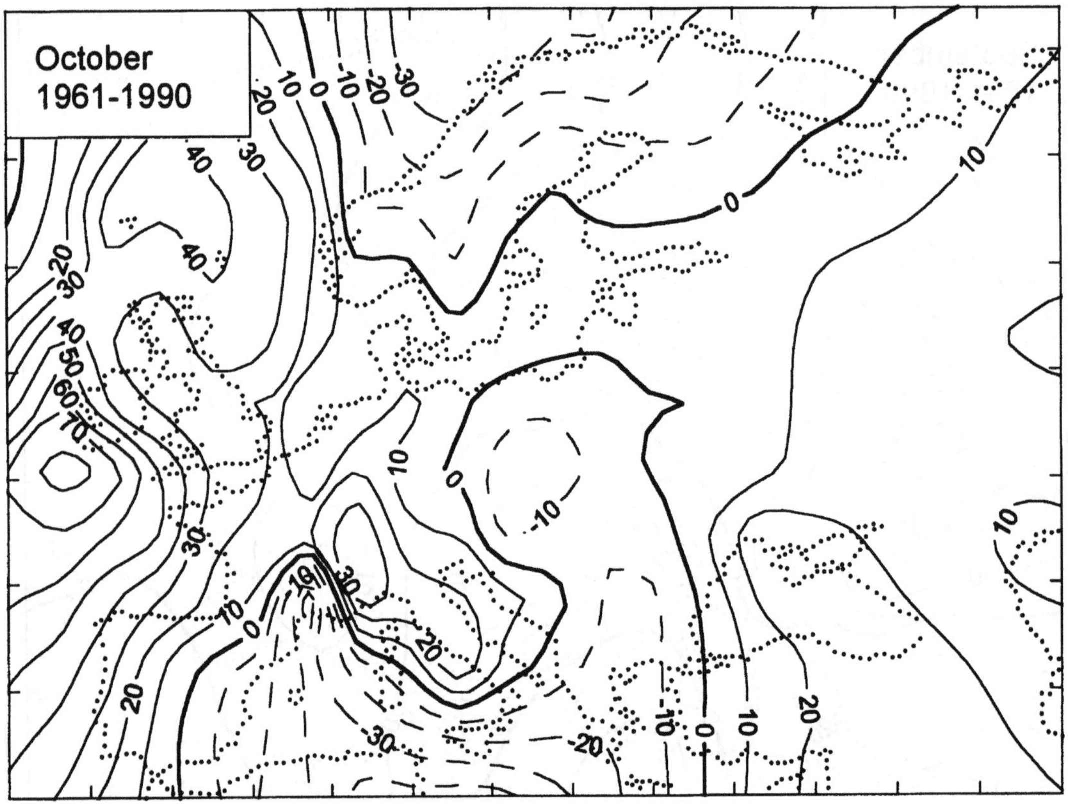

Precipitation, trend test

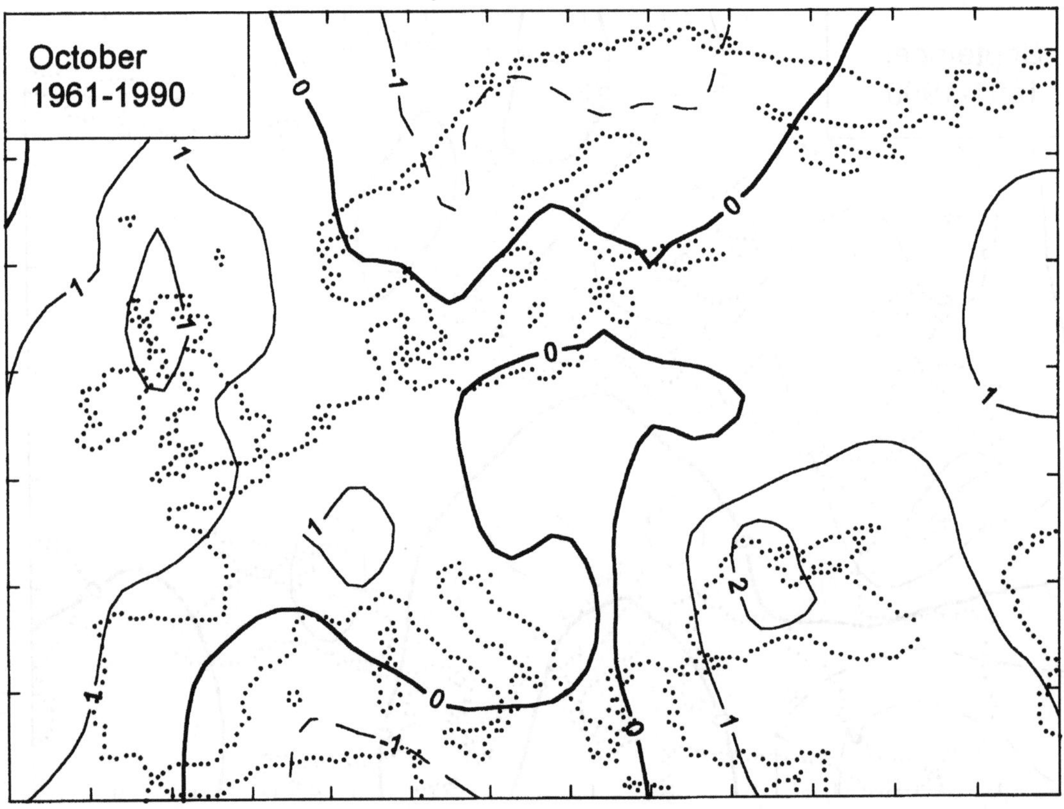

Precipitation, linear trend [mm]

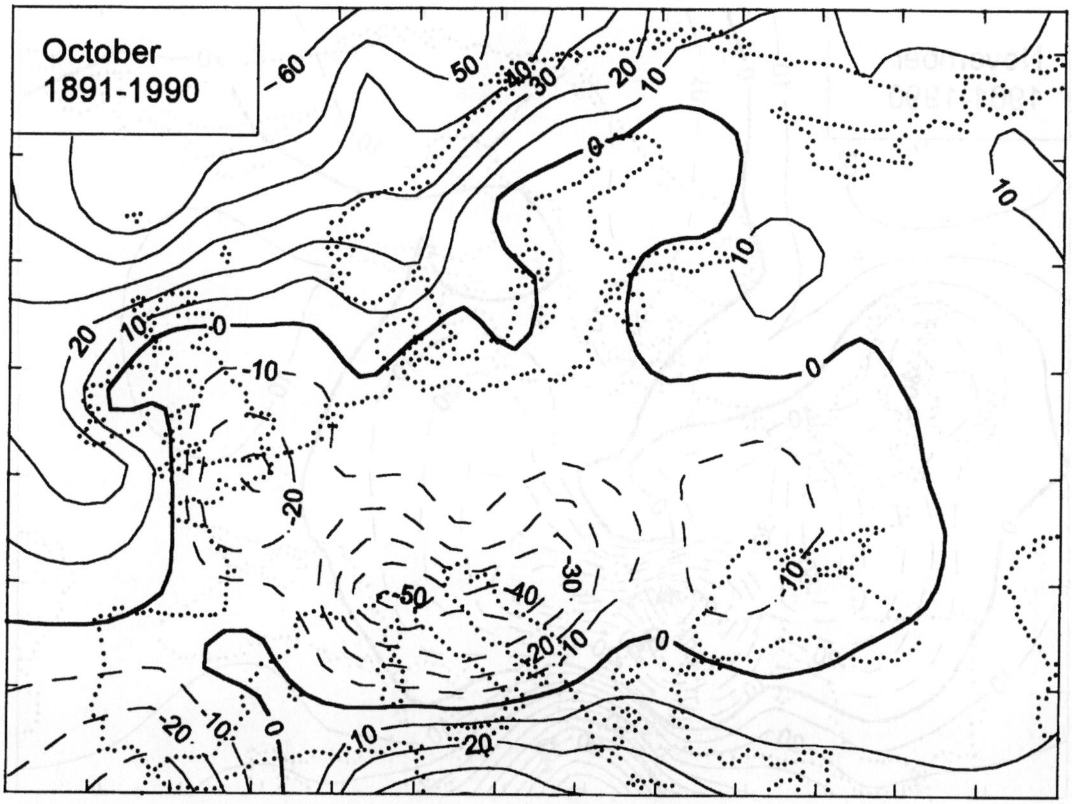

Precipitation, trend test

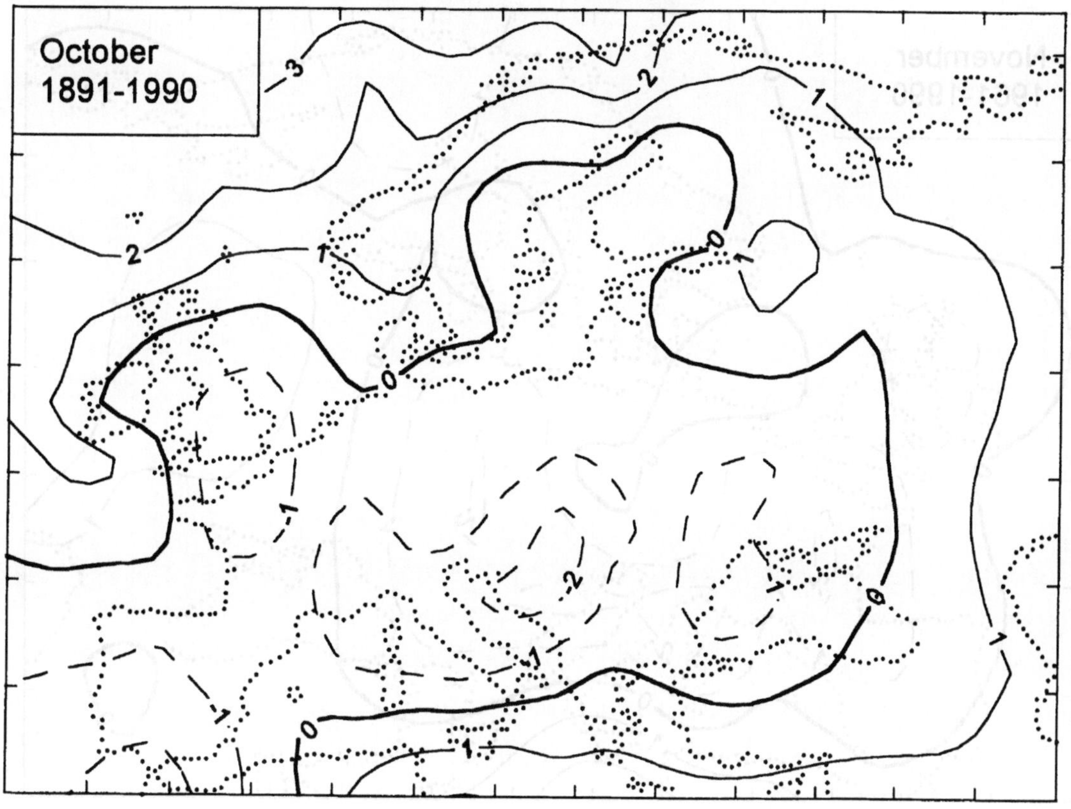

Precipitation, linear trend [mm]

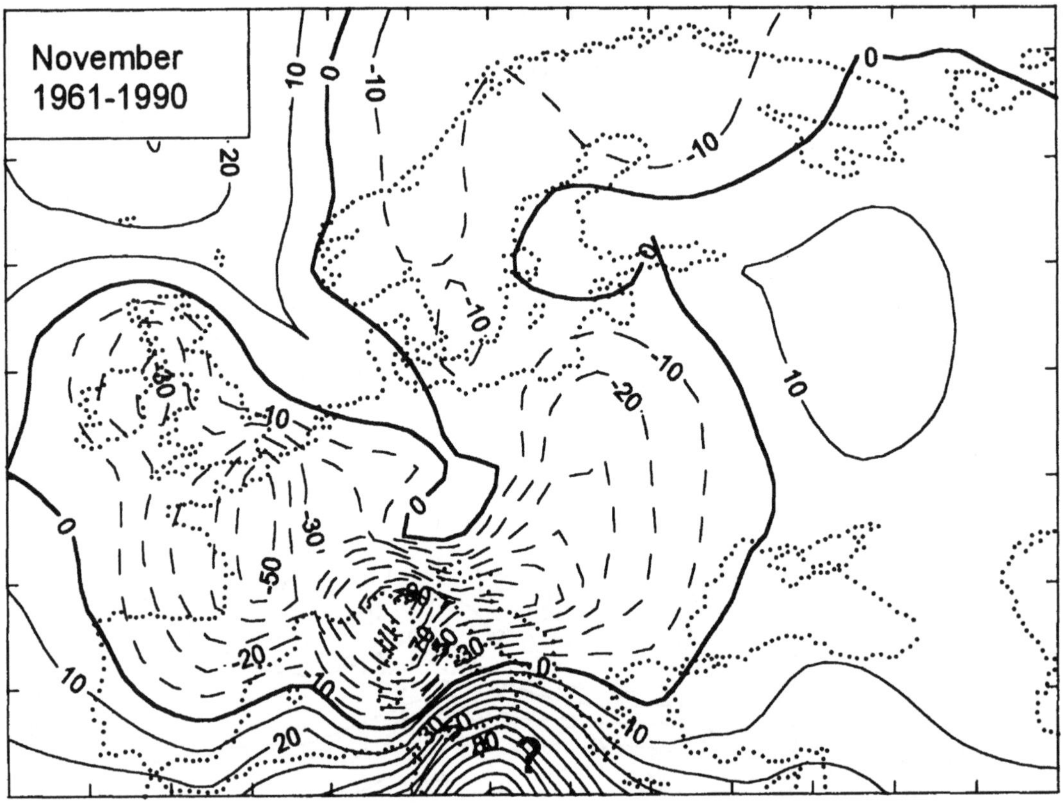

Precipitation, trend test

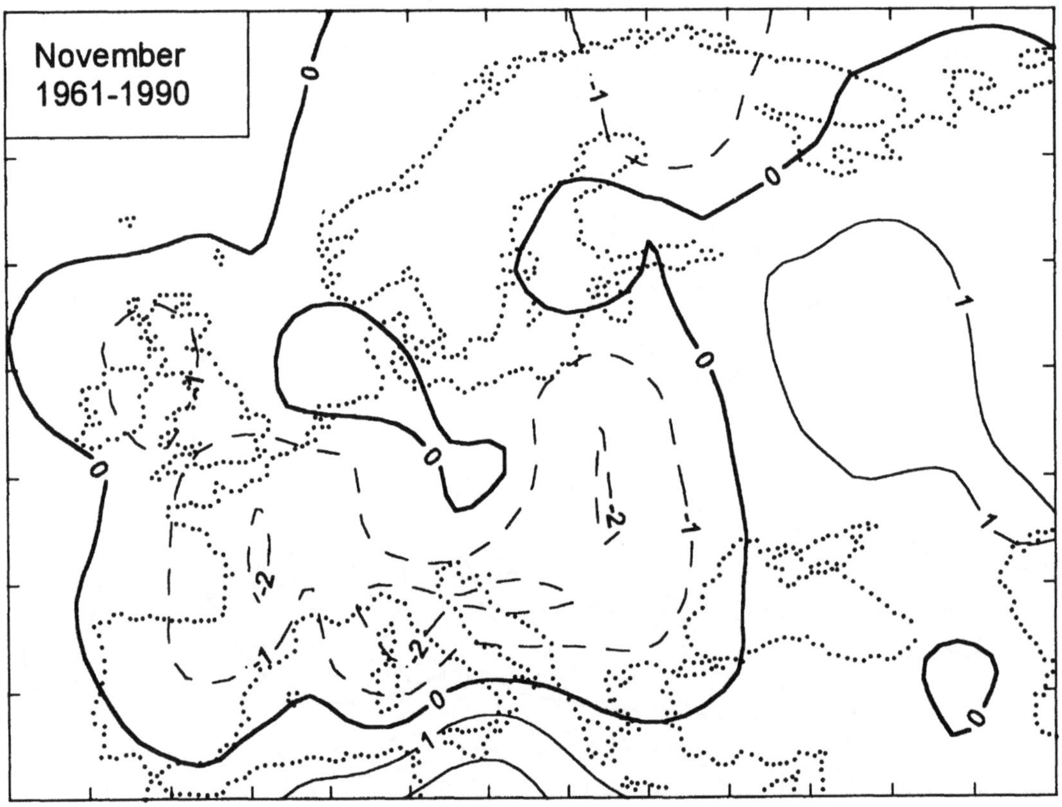

Precipitation, linear trend [mm]

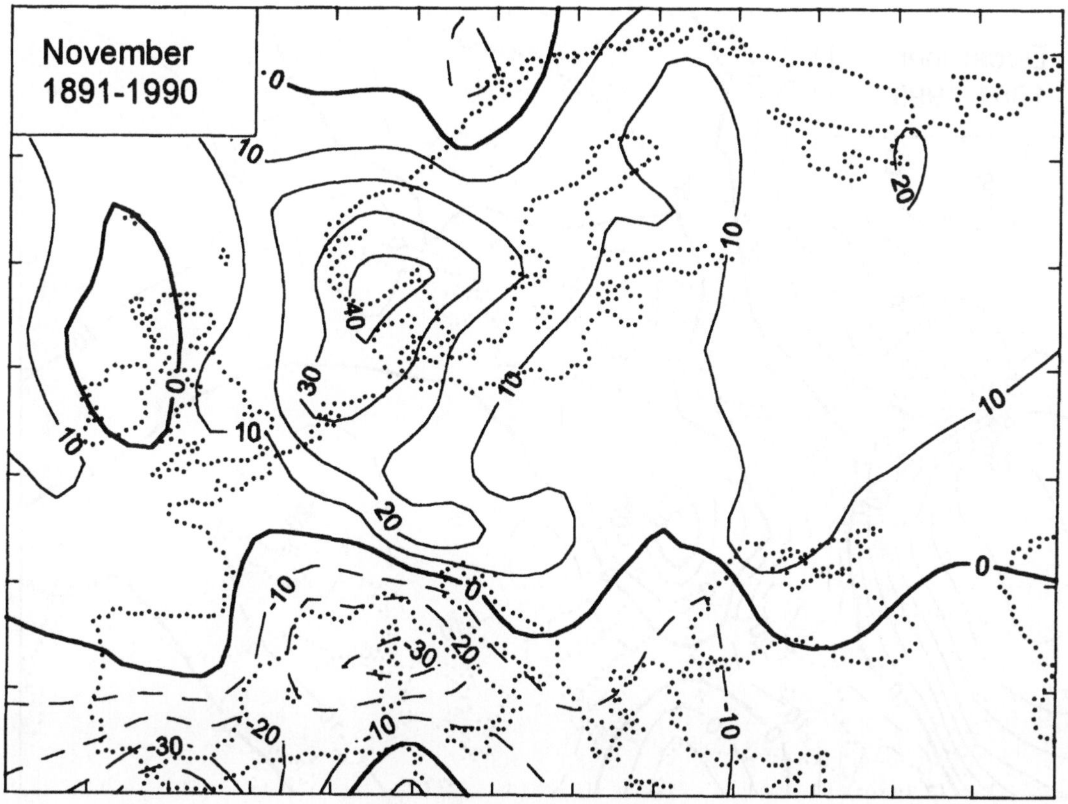

Precipitation, trend test

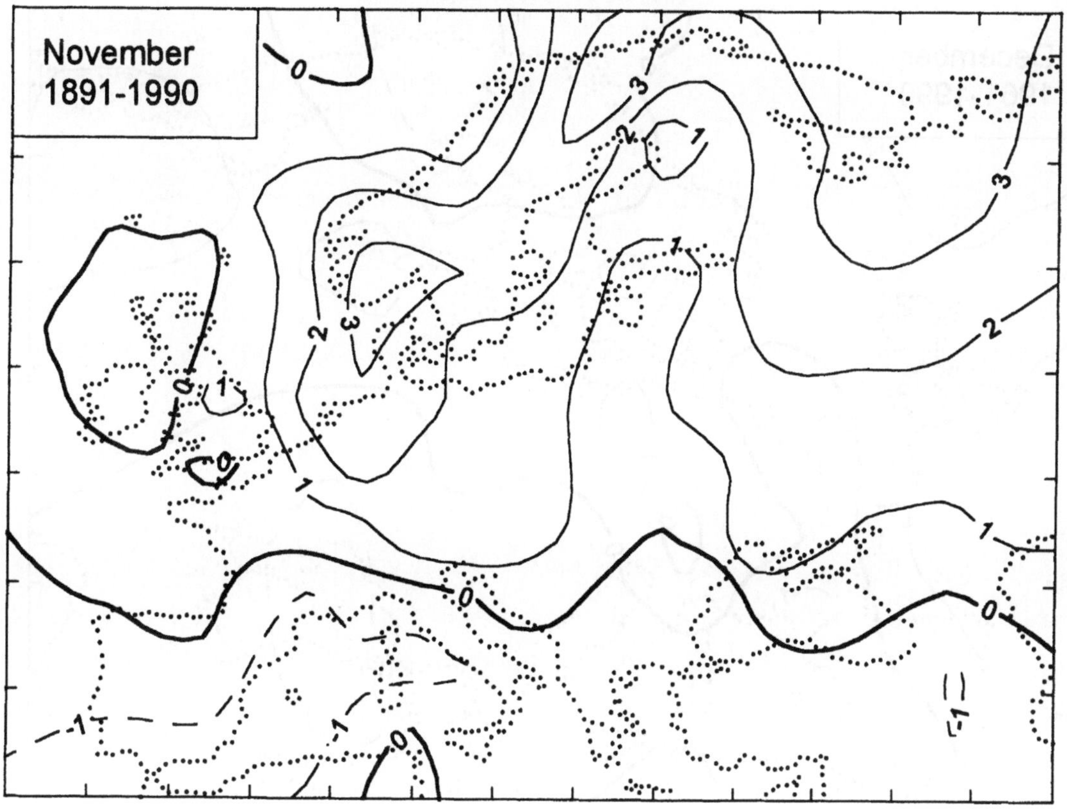

Precipitation, linear trend [mm]

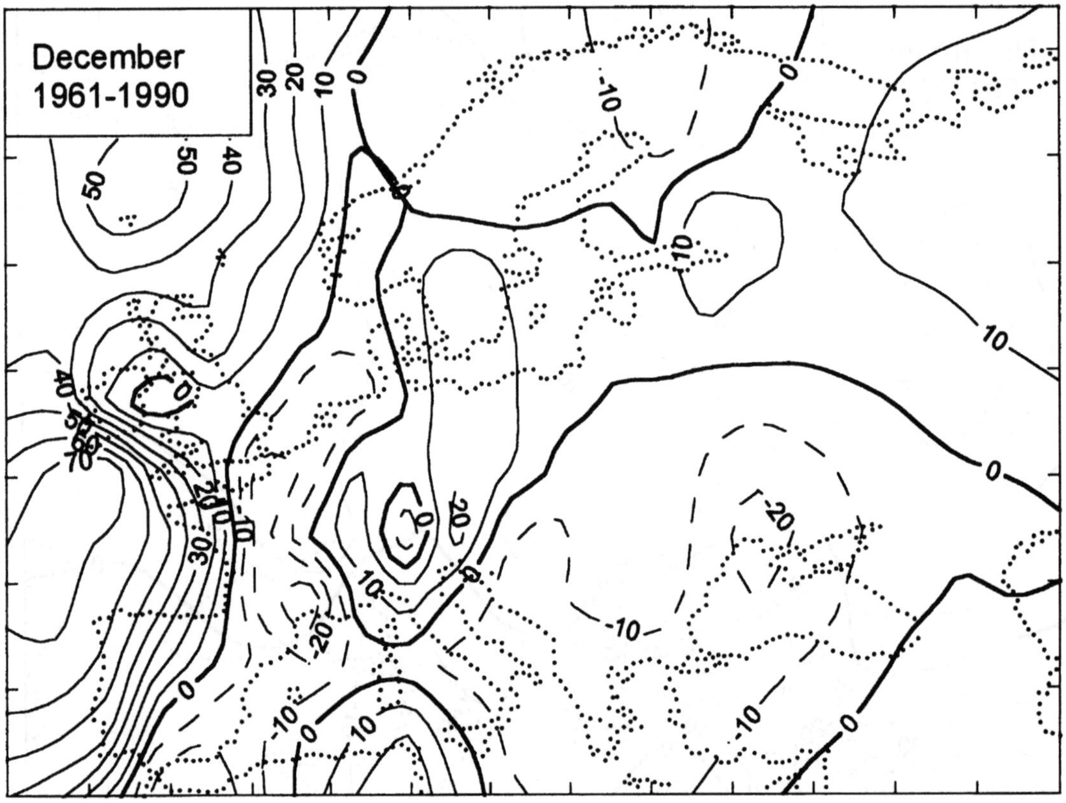

Precipitation, trend test

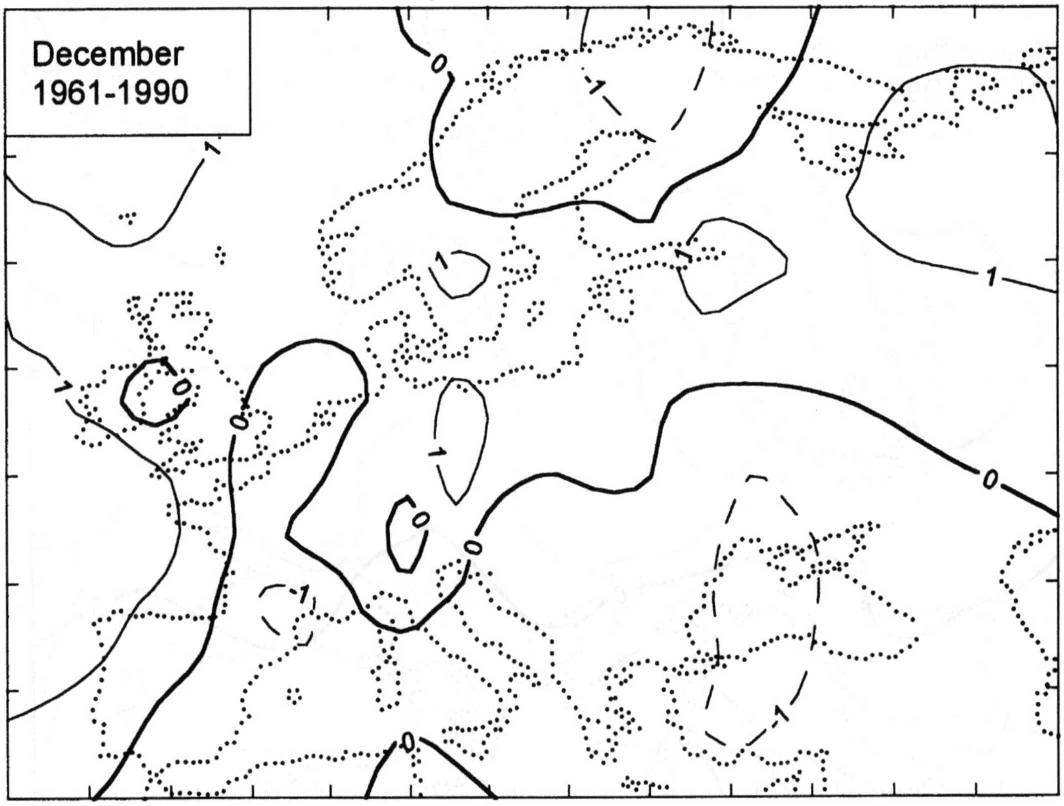

Precipitation, linear trend [mm]

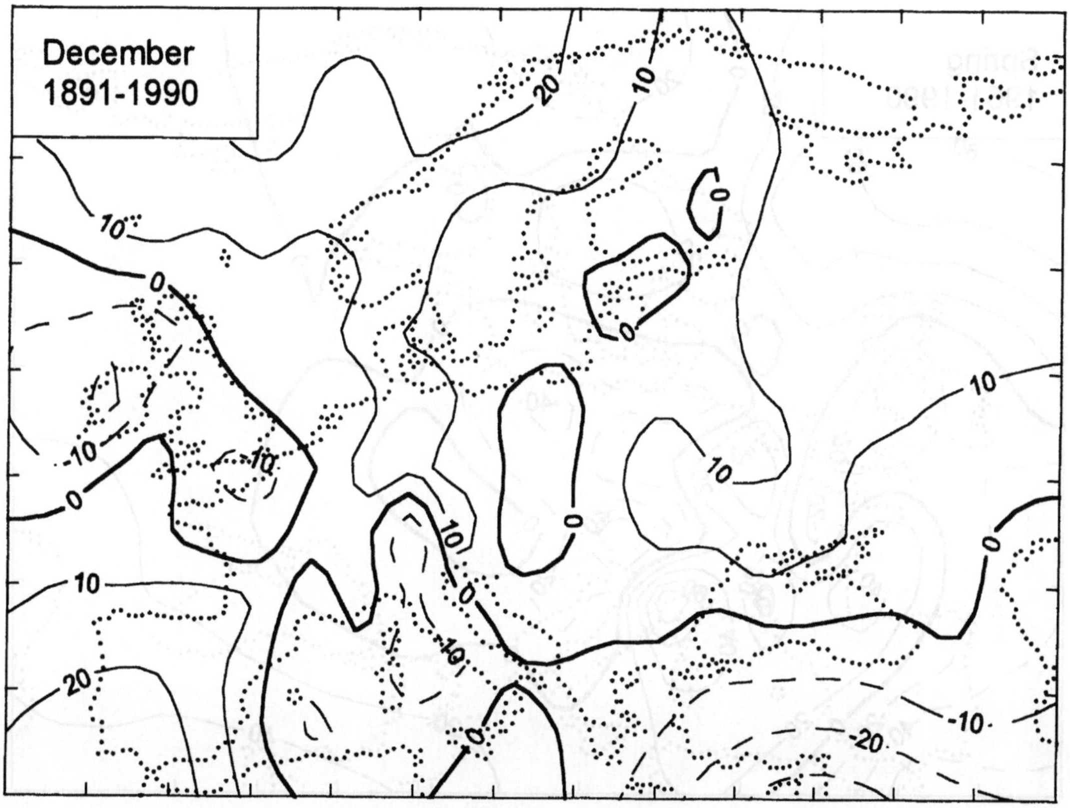

Precipitation, trend test

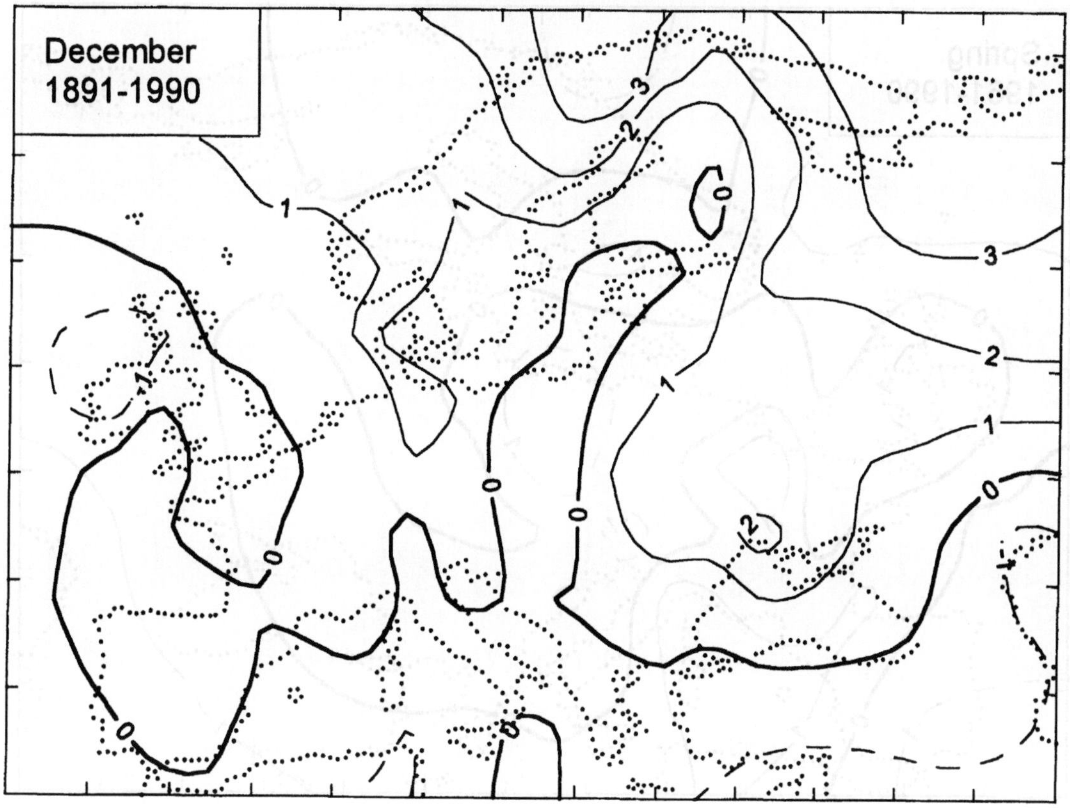

Precipitation, linear trend [mm]

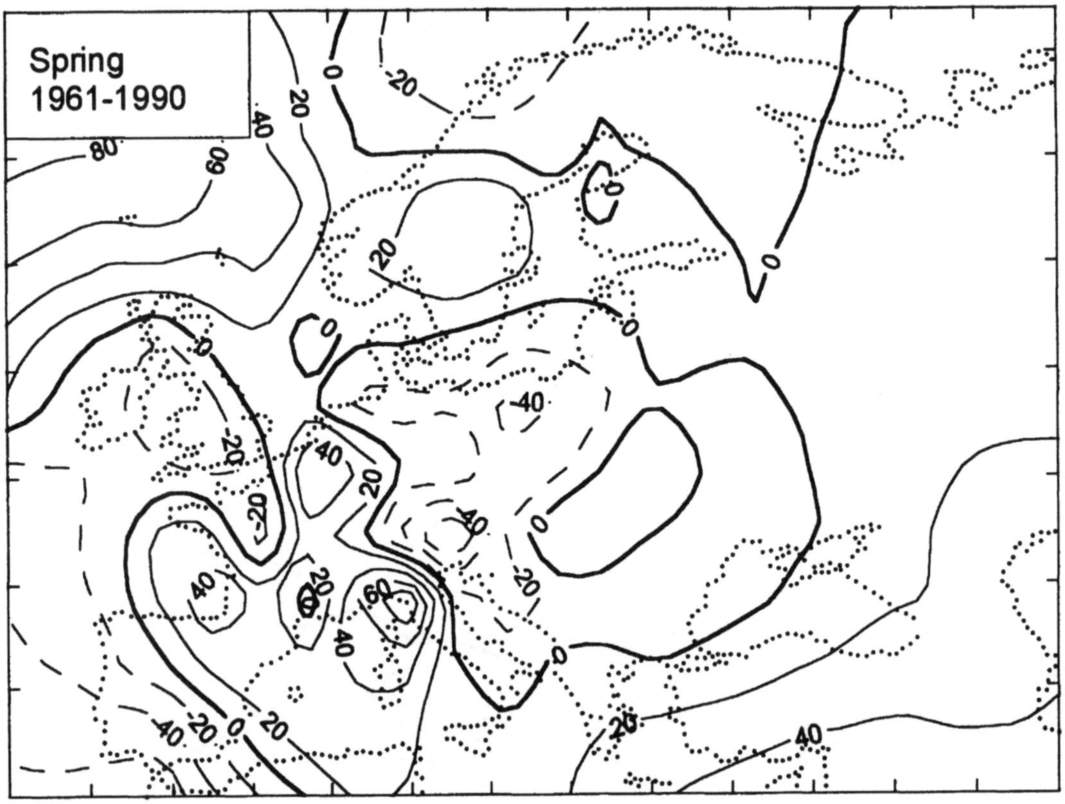

Precipitation, trend test

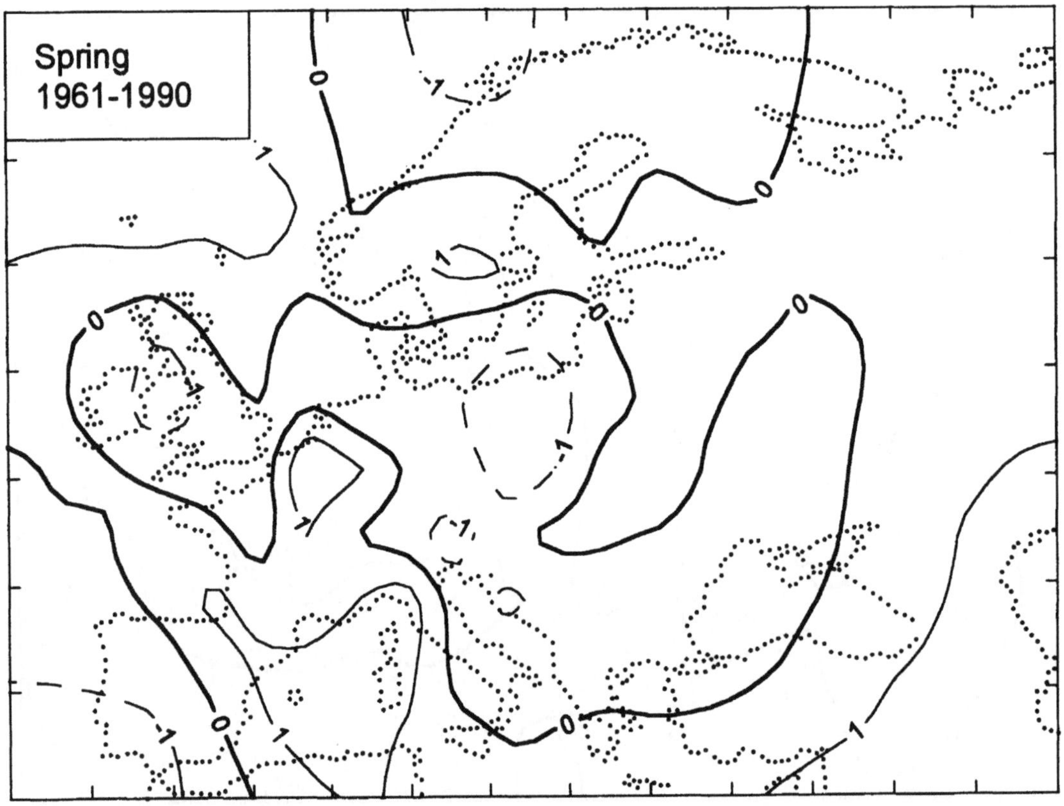

Precipitation, linear trend [mm]

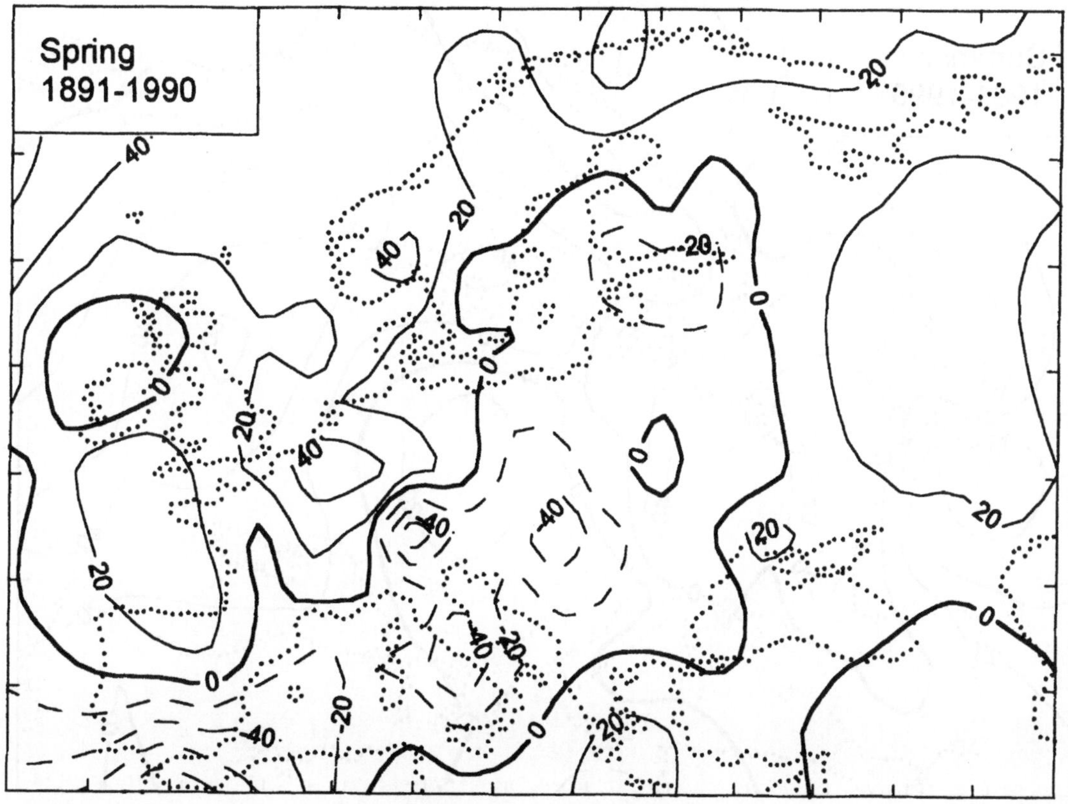

Precipitation, trend test

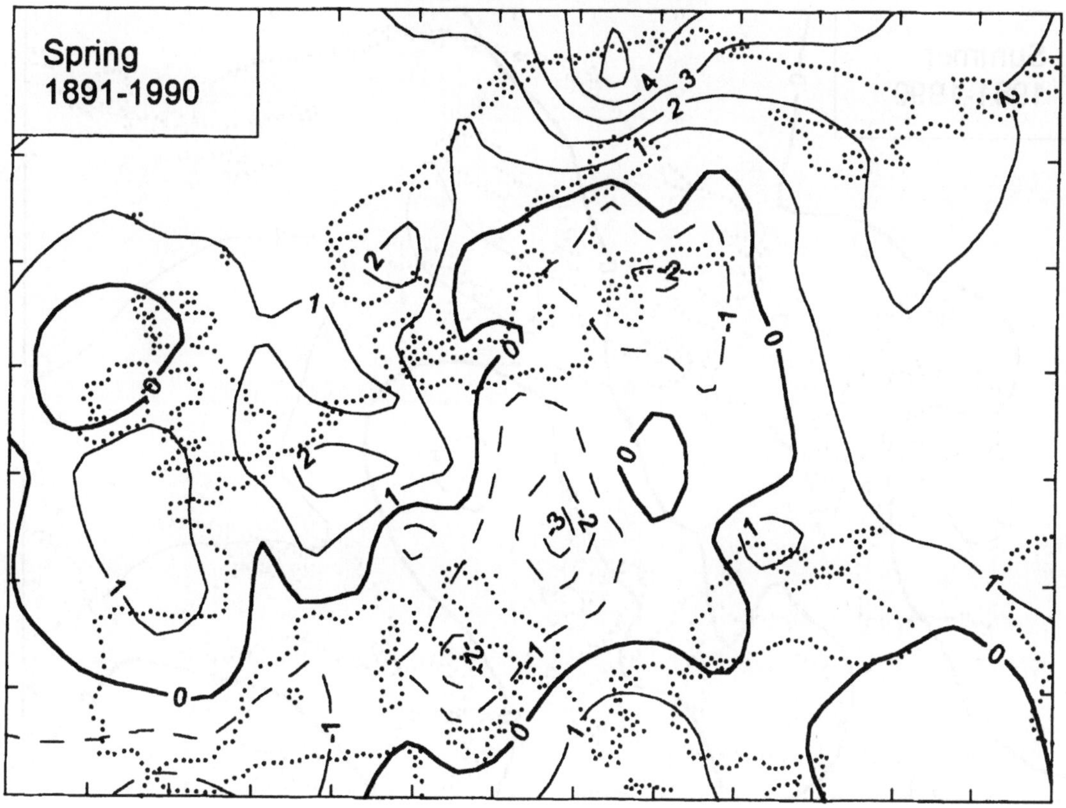

Precipitation, linear trend [mm]

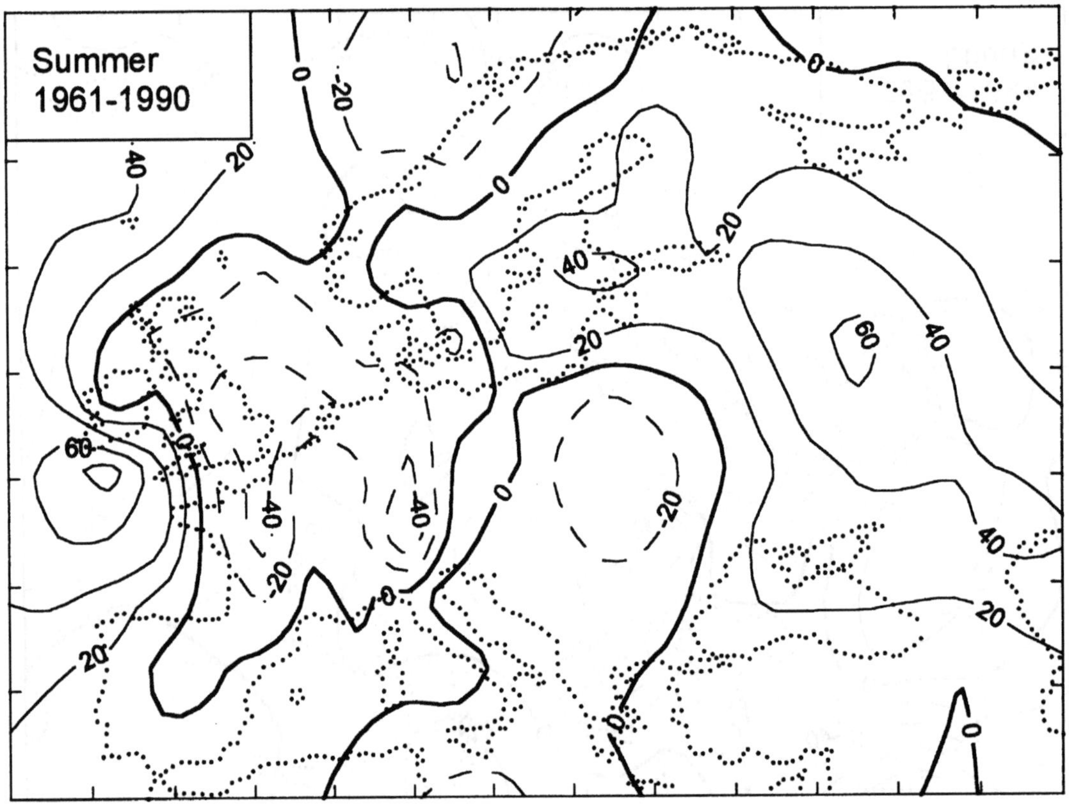

Precipitation, trend test

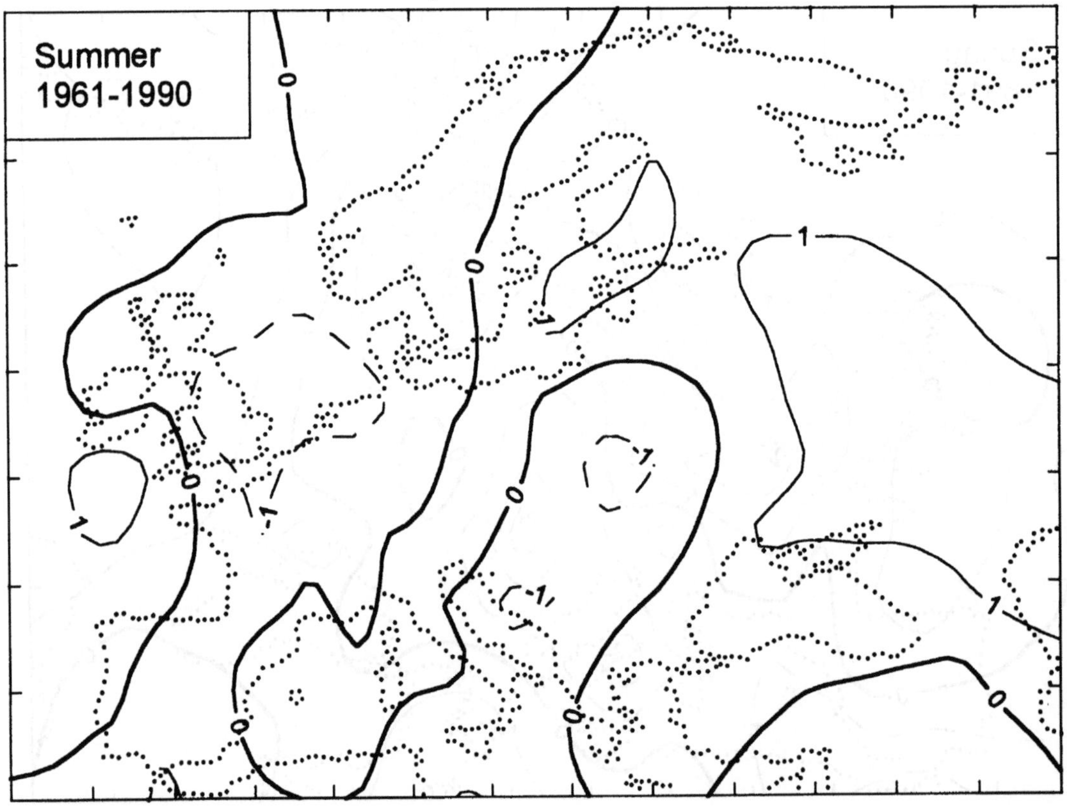

Precipitation, linear trend [mm]

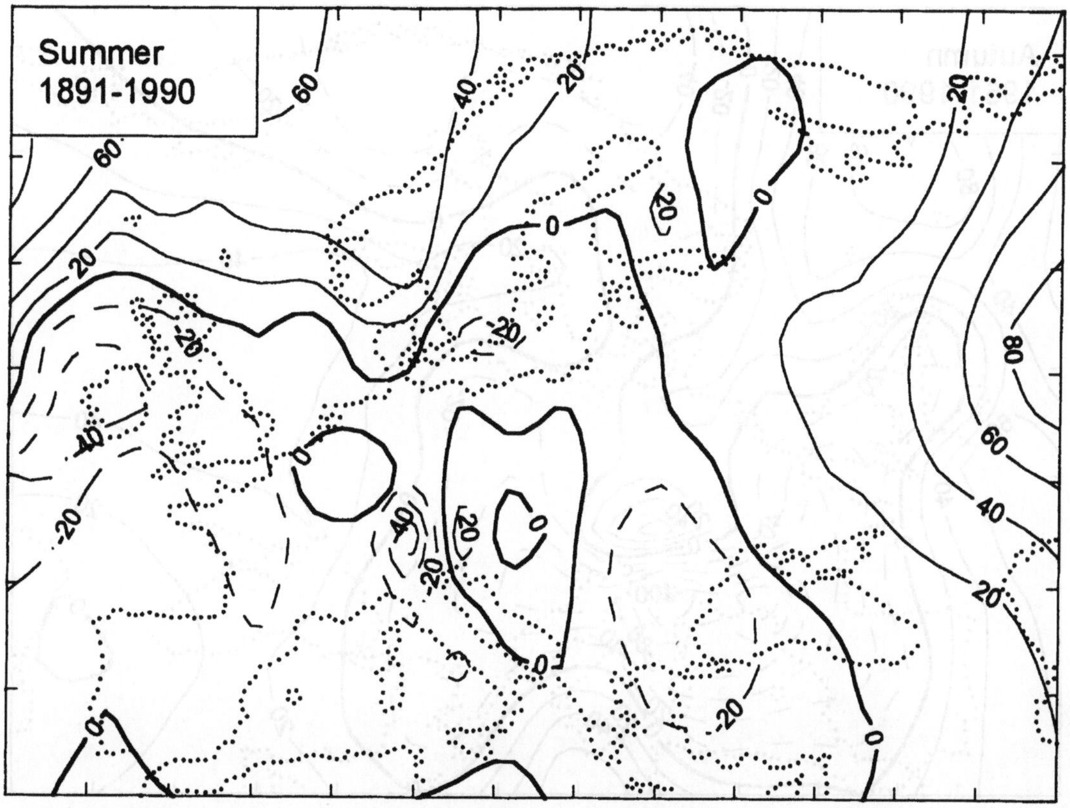

Precipitation, trend test

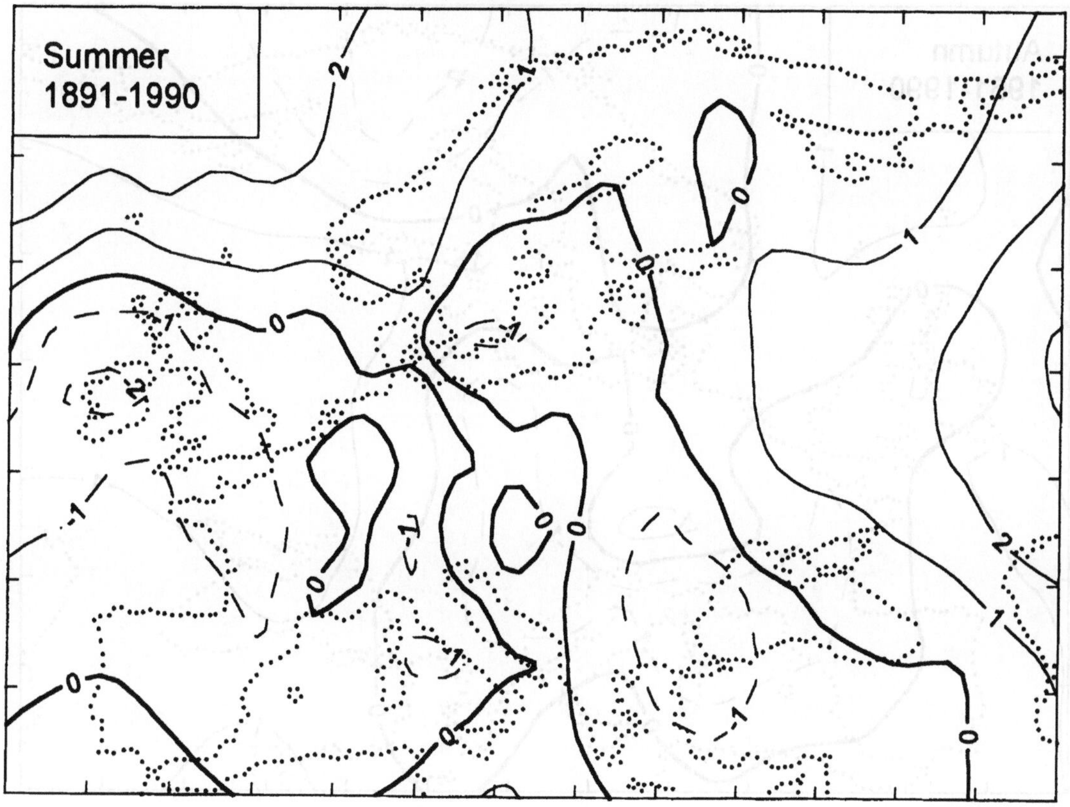

Precipitation, linear trend [mm]

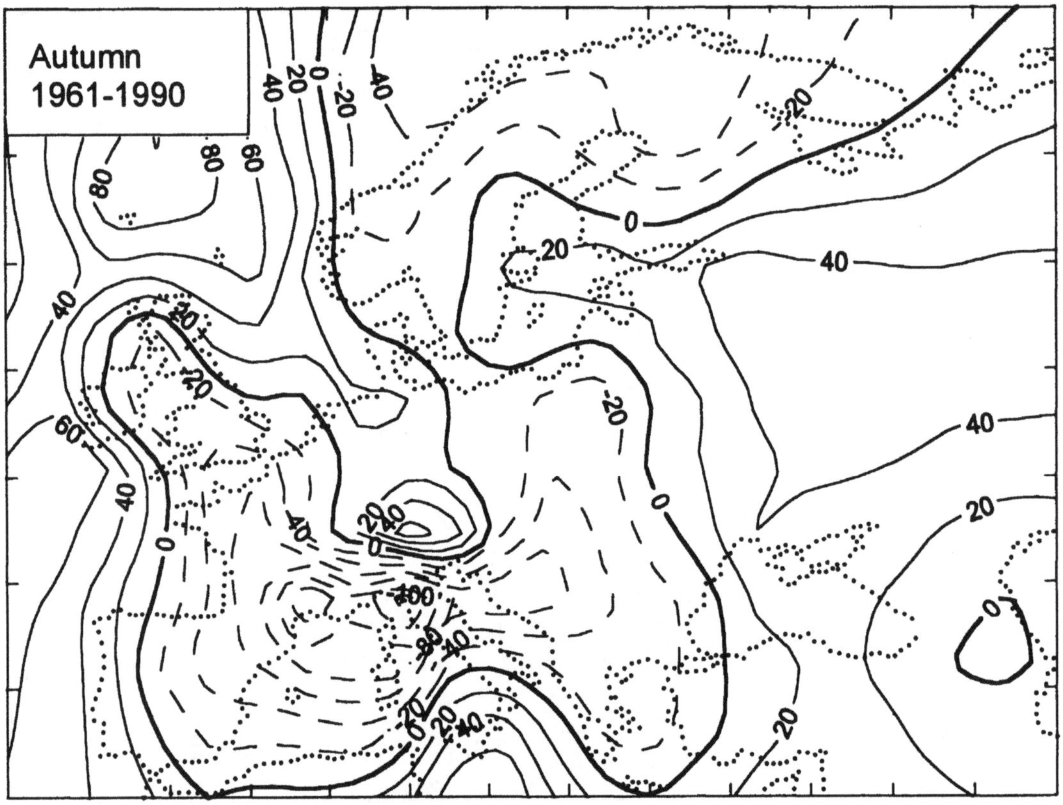

Precipitation, trend test

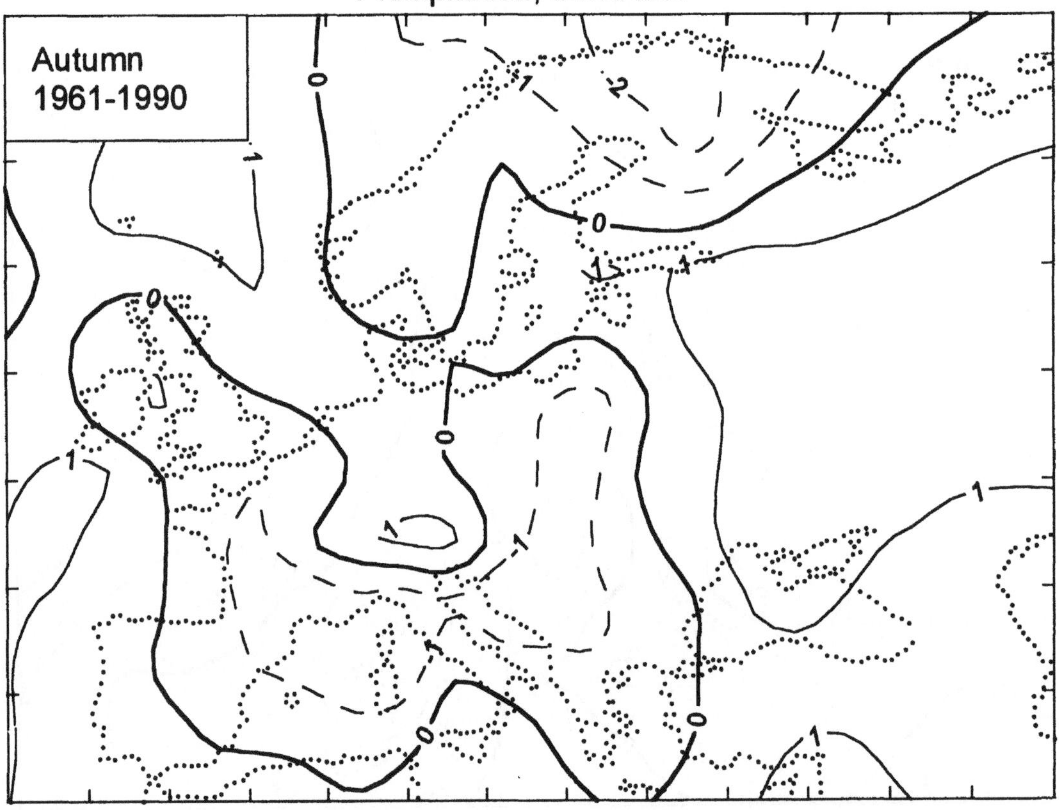

Precipitation, linear trend [mm]

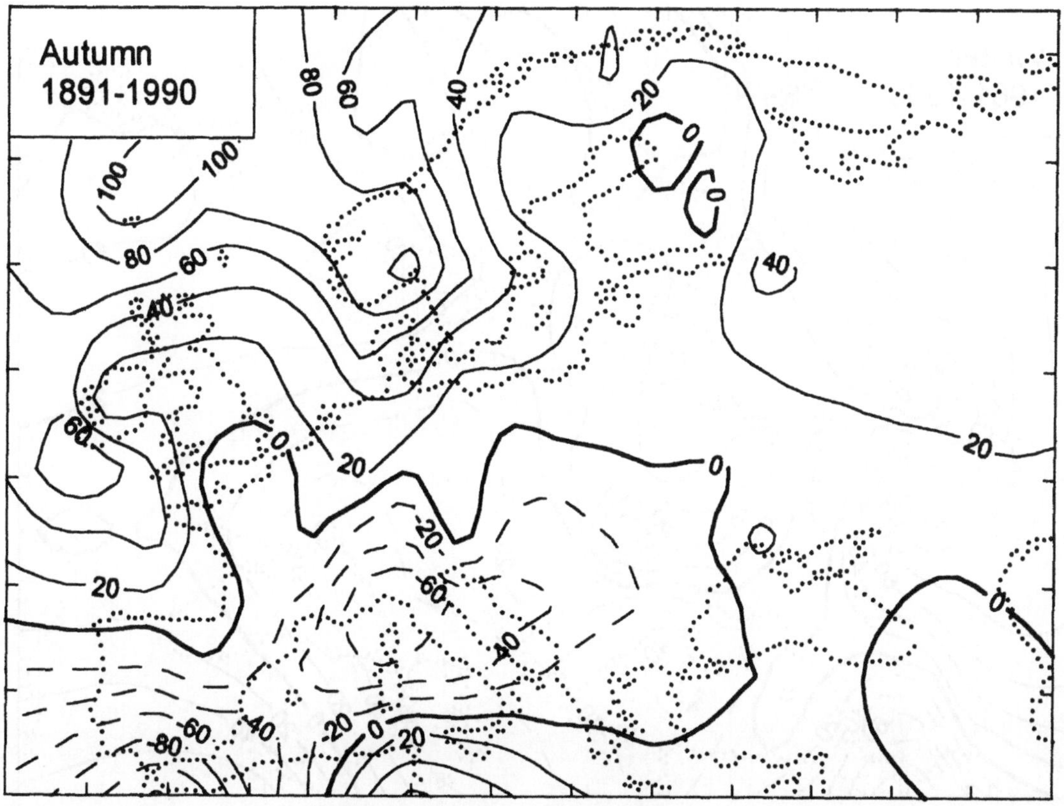

Precipitation, trend test

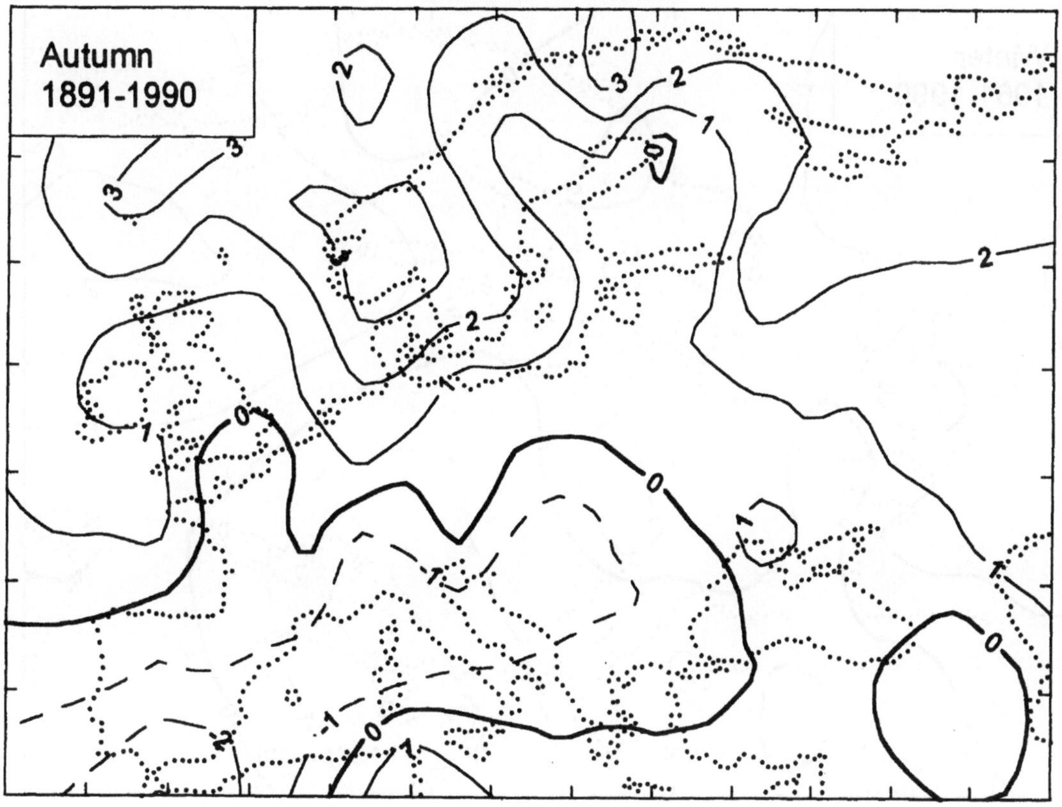

Precipitation, linear trend [mm]

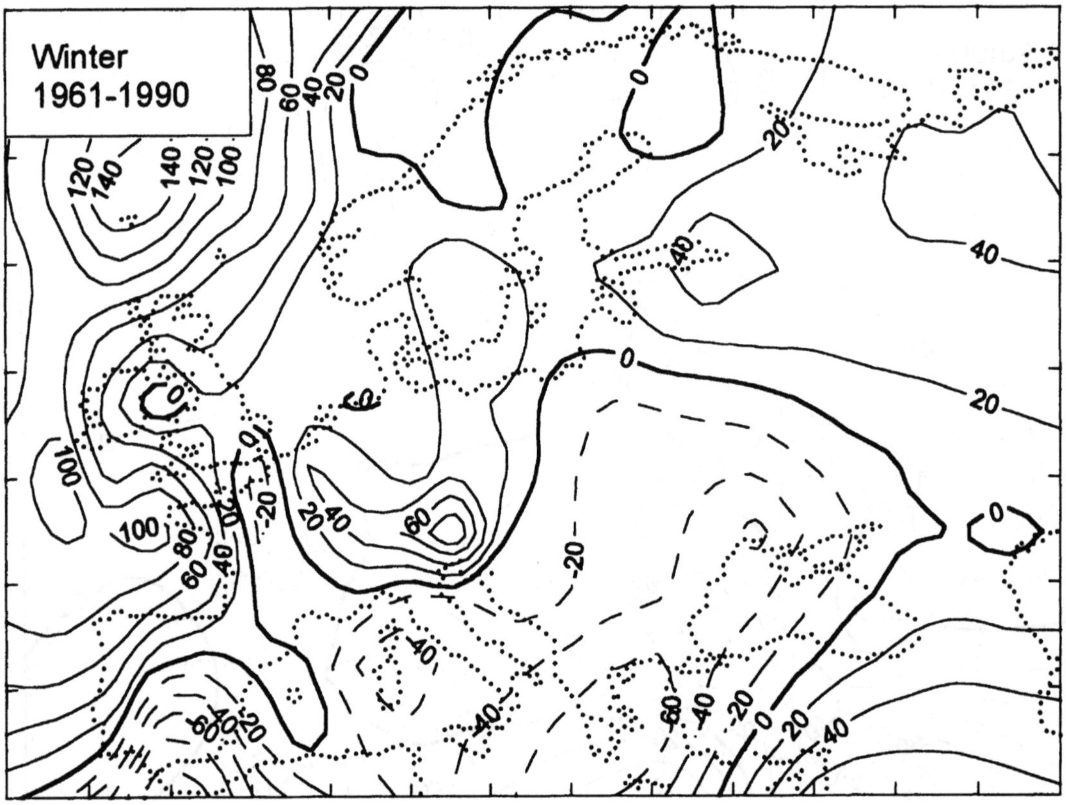

Precipitation, trend test

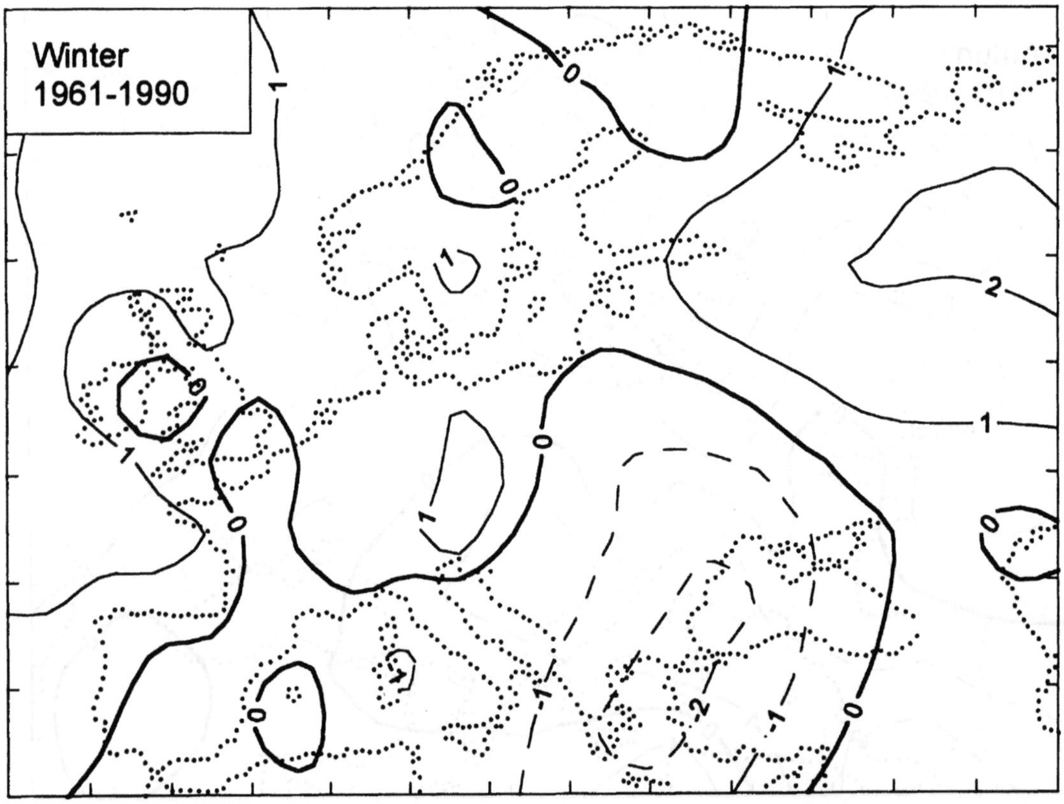

Precipitation, linear trend [mm]

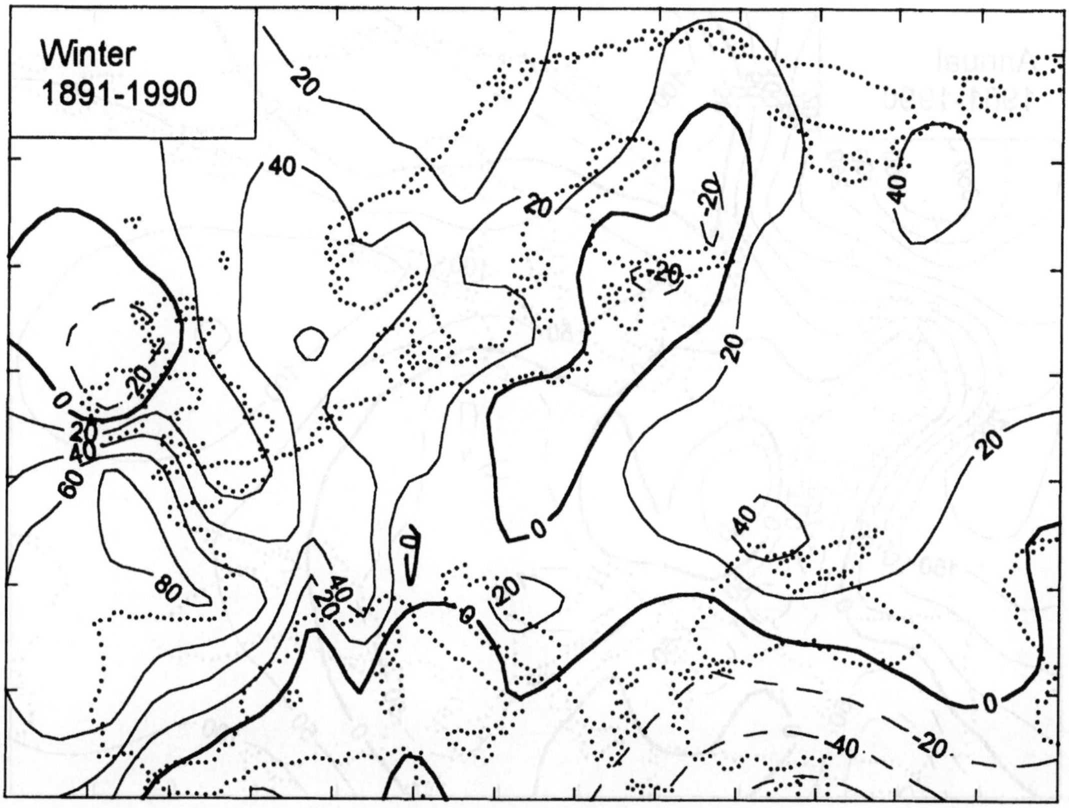

Precipitation, trend test

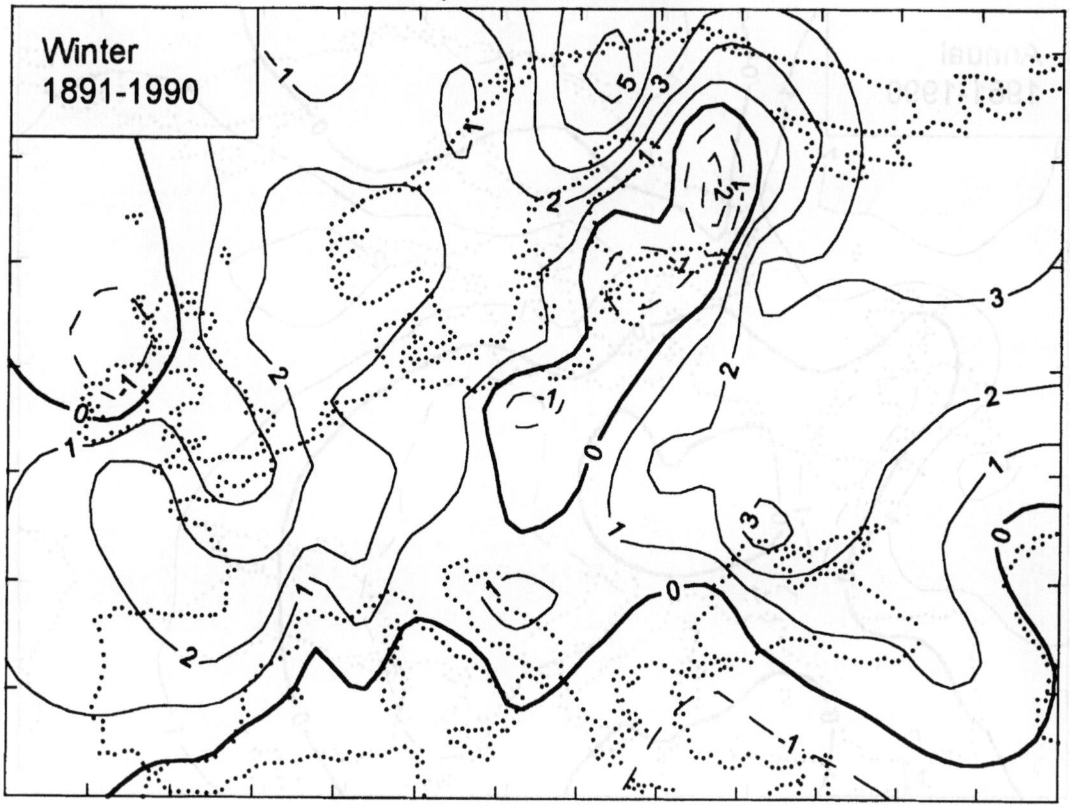

Precipitation, linear trend [mm]

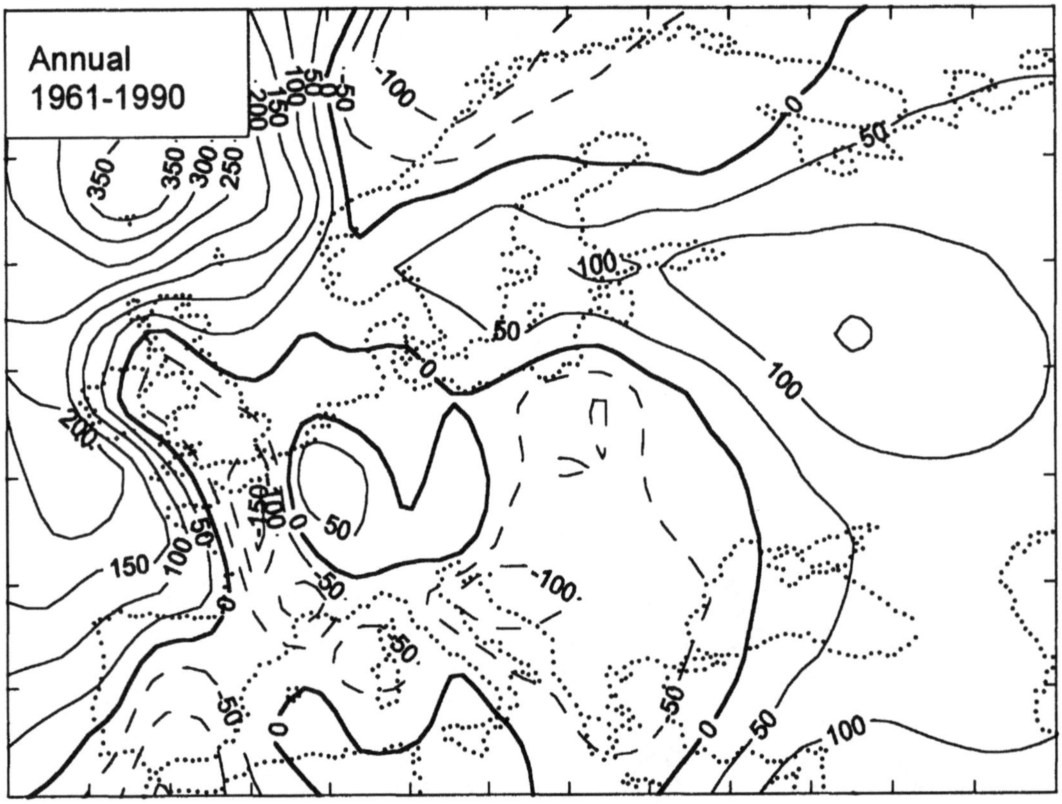

Precipitation, trend test

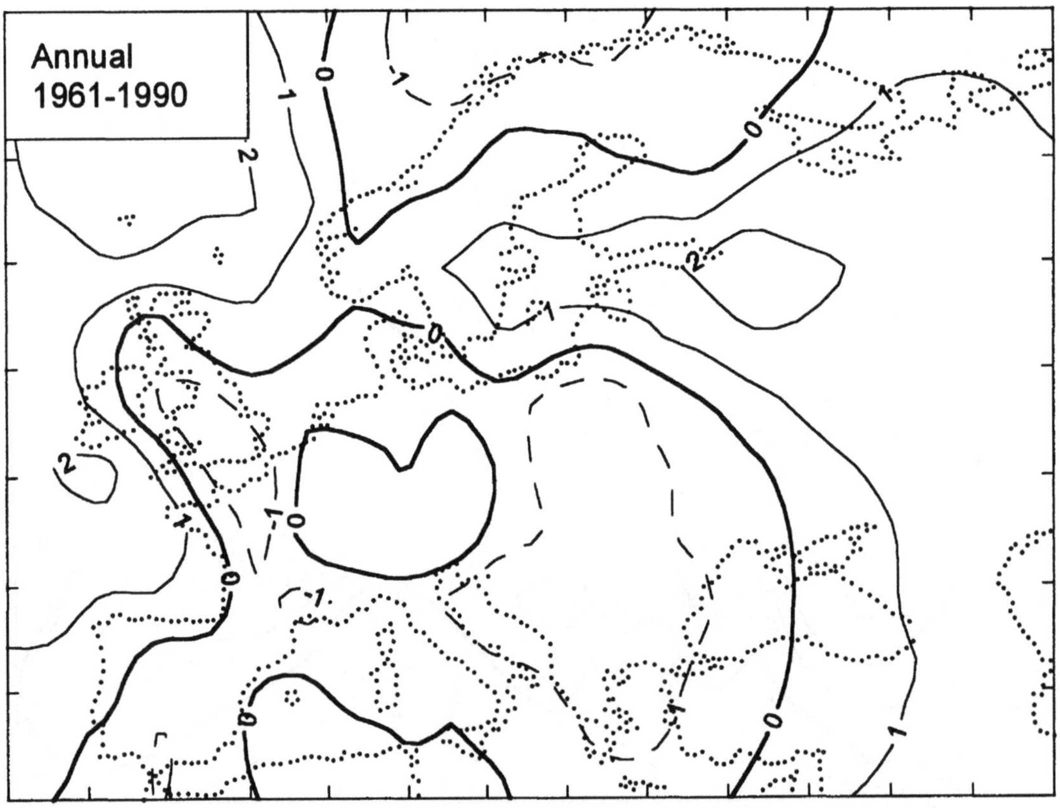

Precipitation, linear trend [mm]

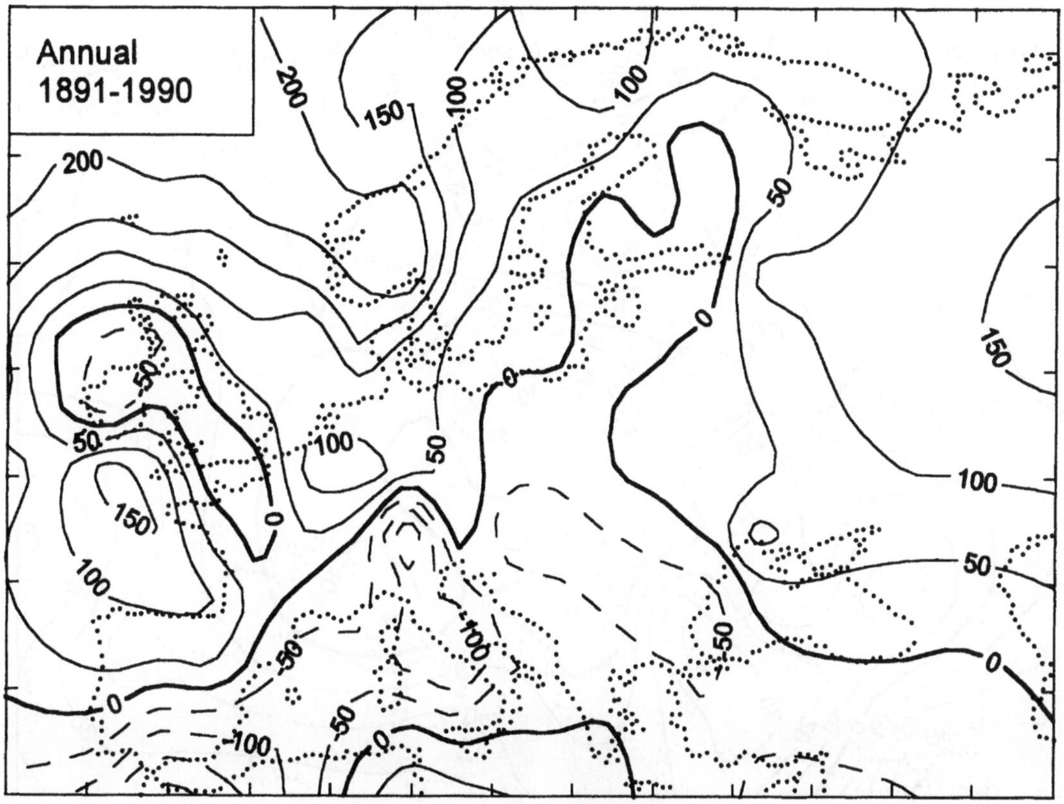

Precipitation, trend test

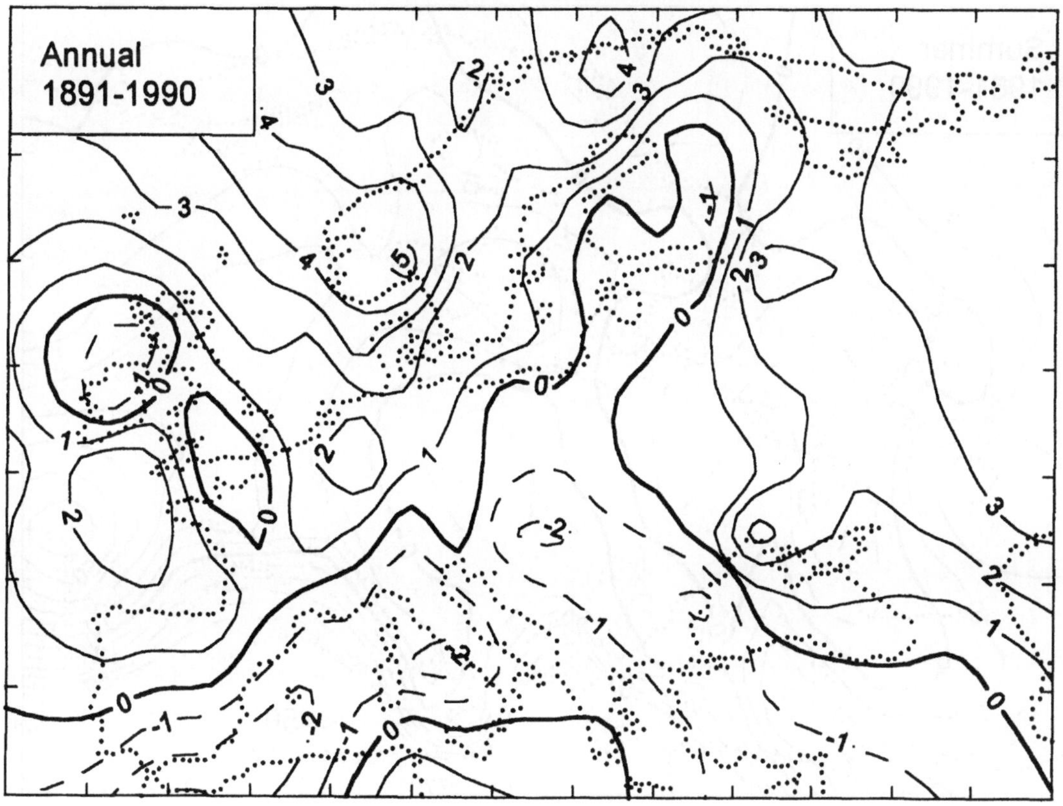

Precipitation, relative linear trend [% of the average]

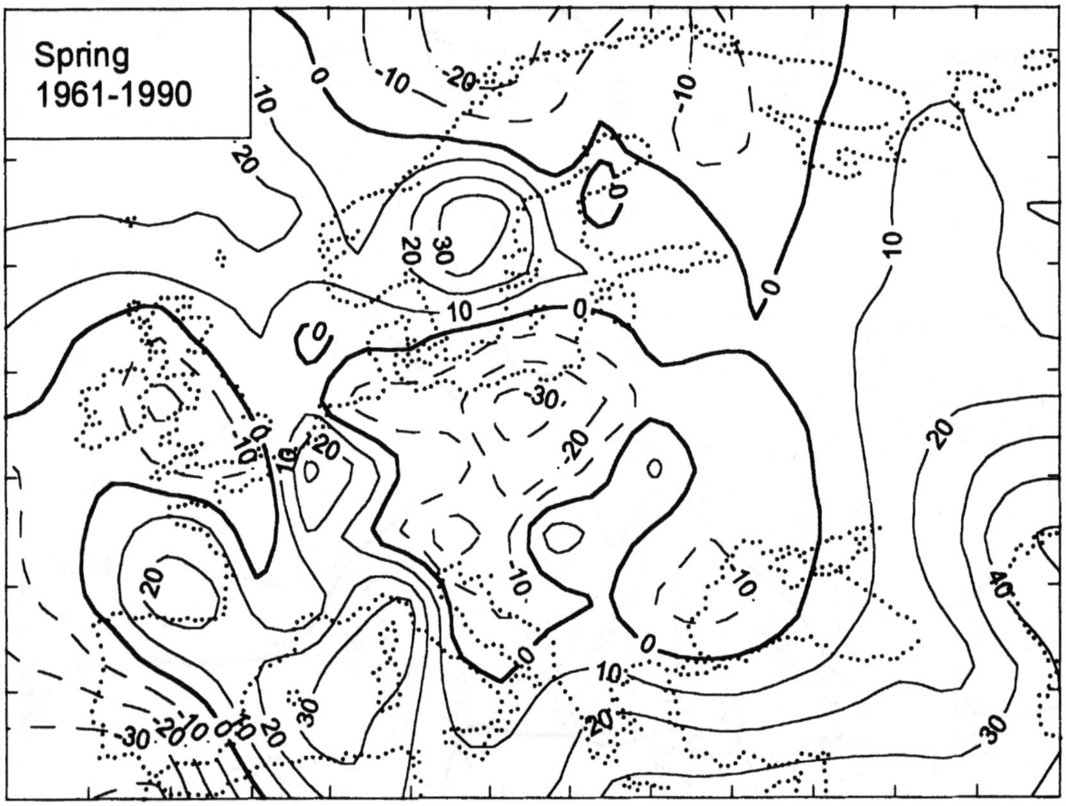

Precipitation, relative linear trend [% of the average]

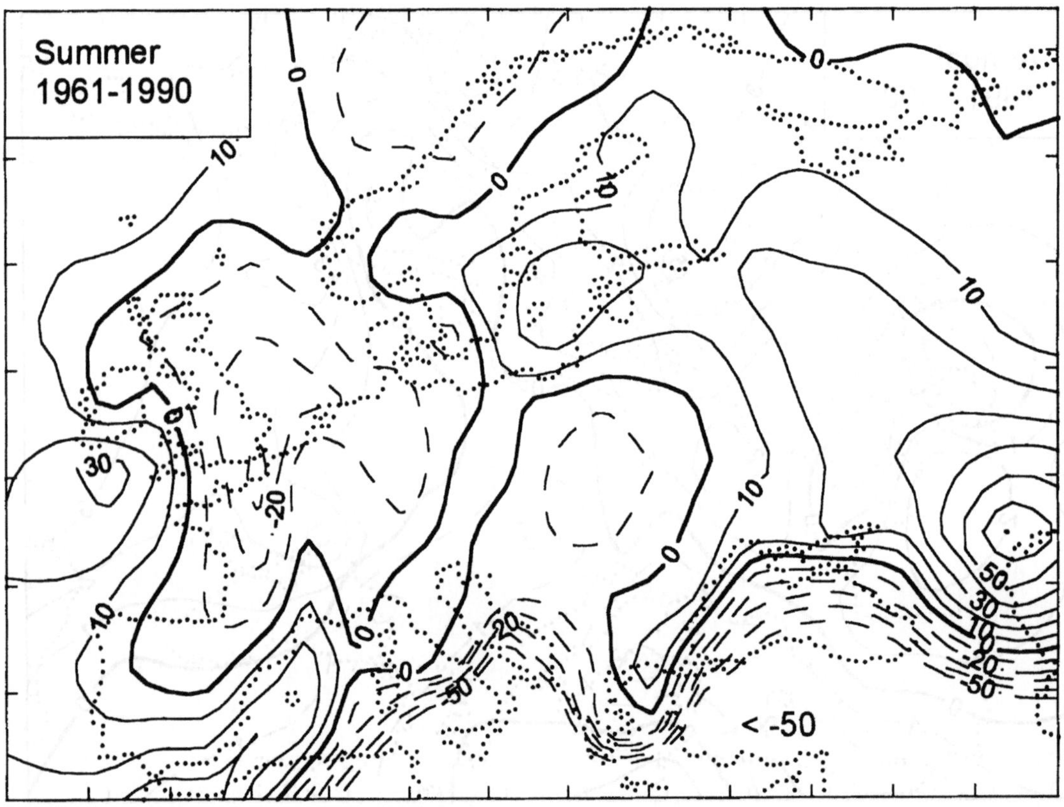

Precipitation, relative linear trend [% of the average]

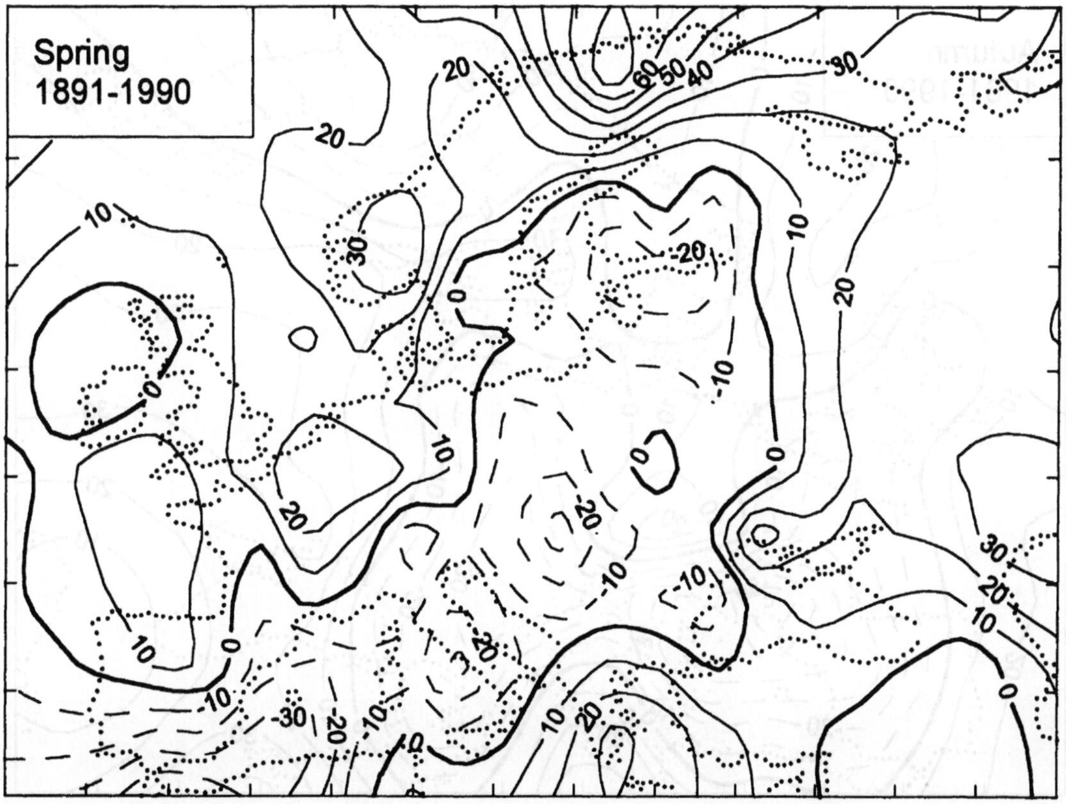

Precipitation, relative linear trend [% of the average]

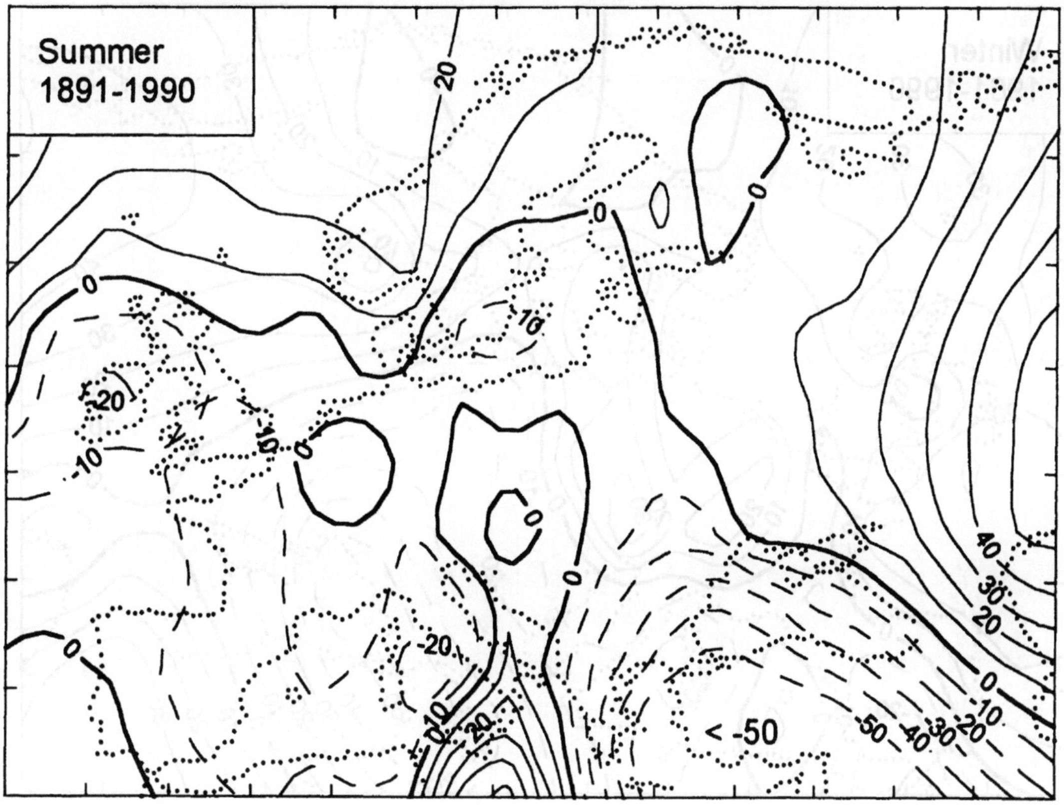

Precipitation, relative linear trend [% of the average]

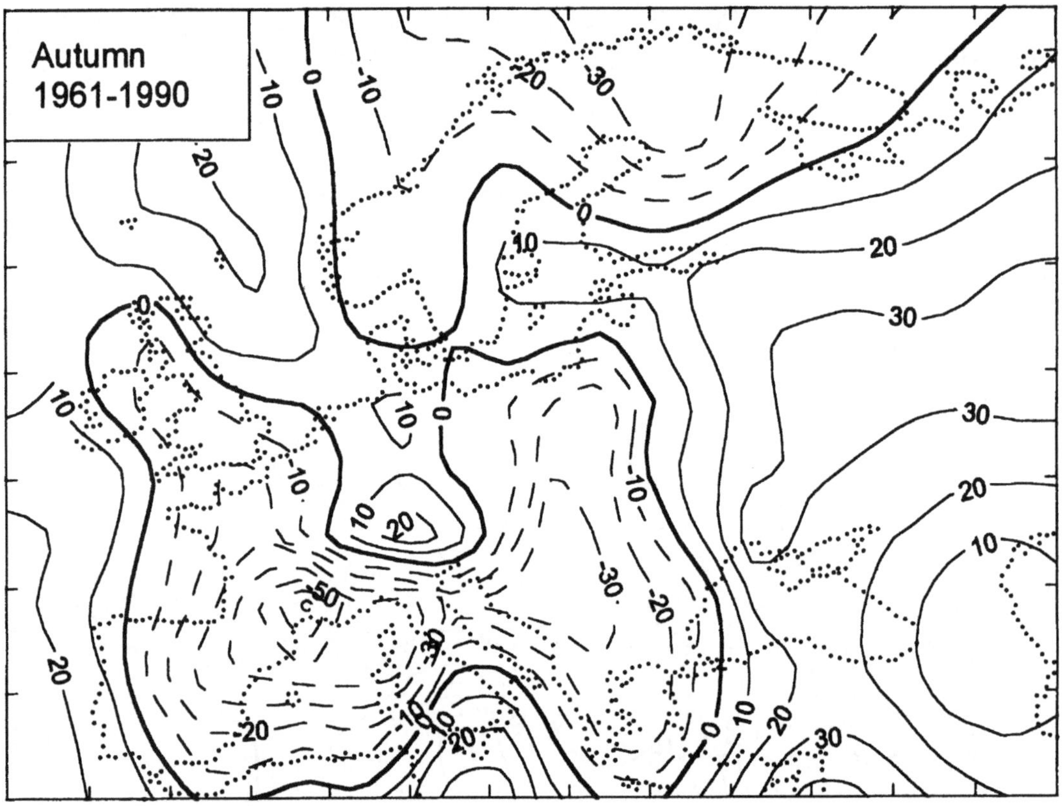

Precipitation, relative linear trend [% of the average]

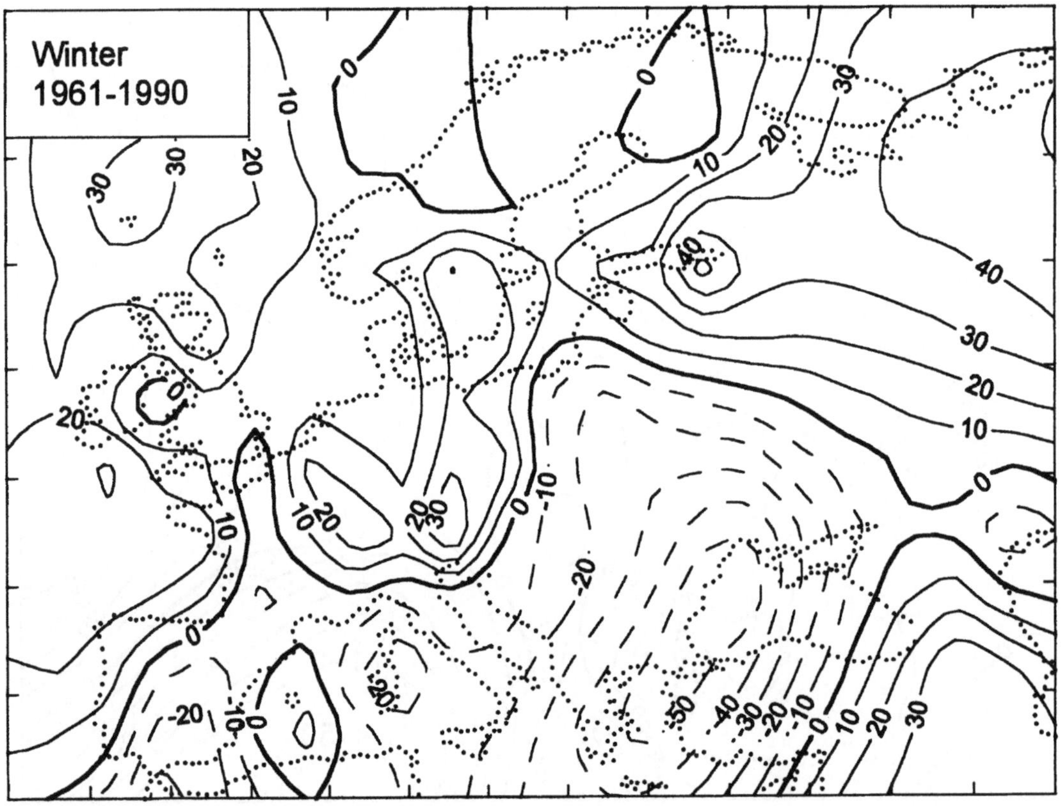

Precipitation, relative linear trend [% of the average]

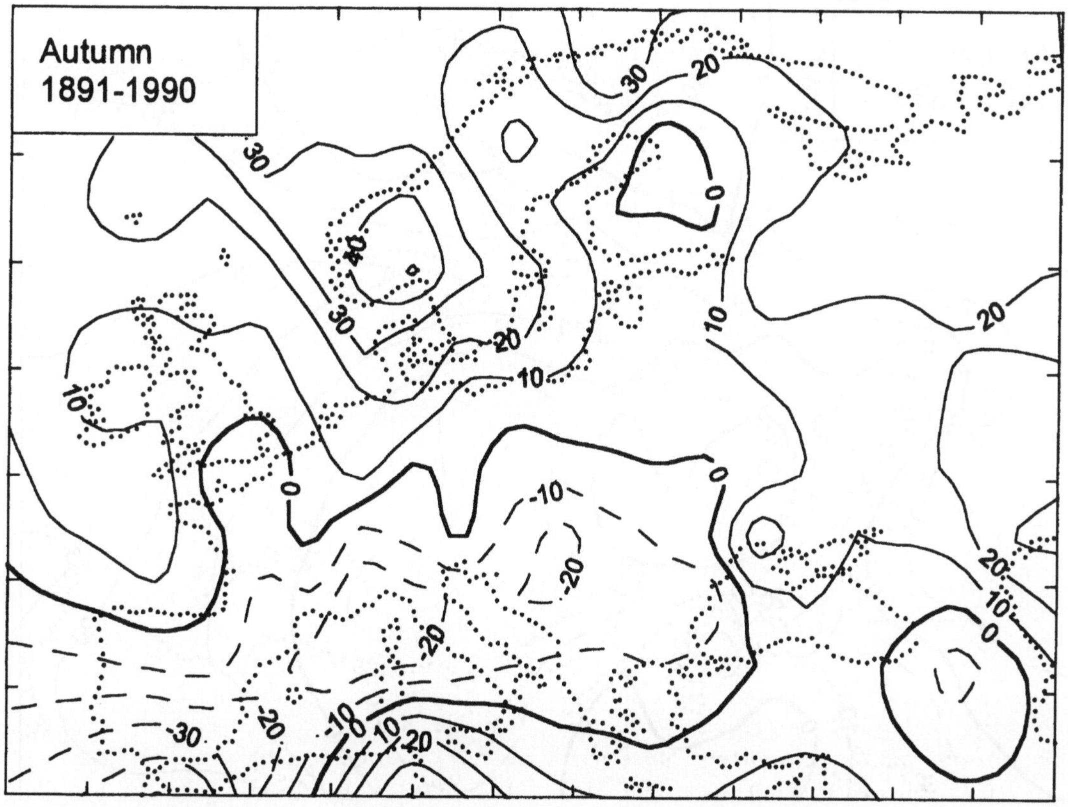

Precipitation, relative linear trend [% of the average]

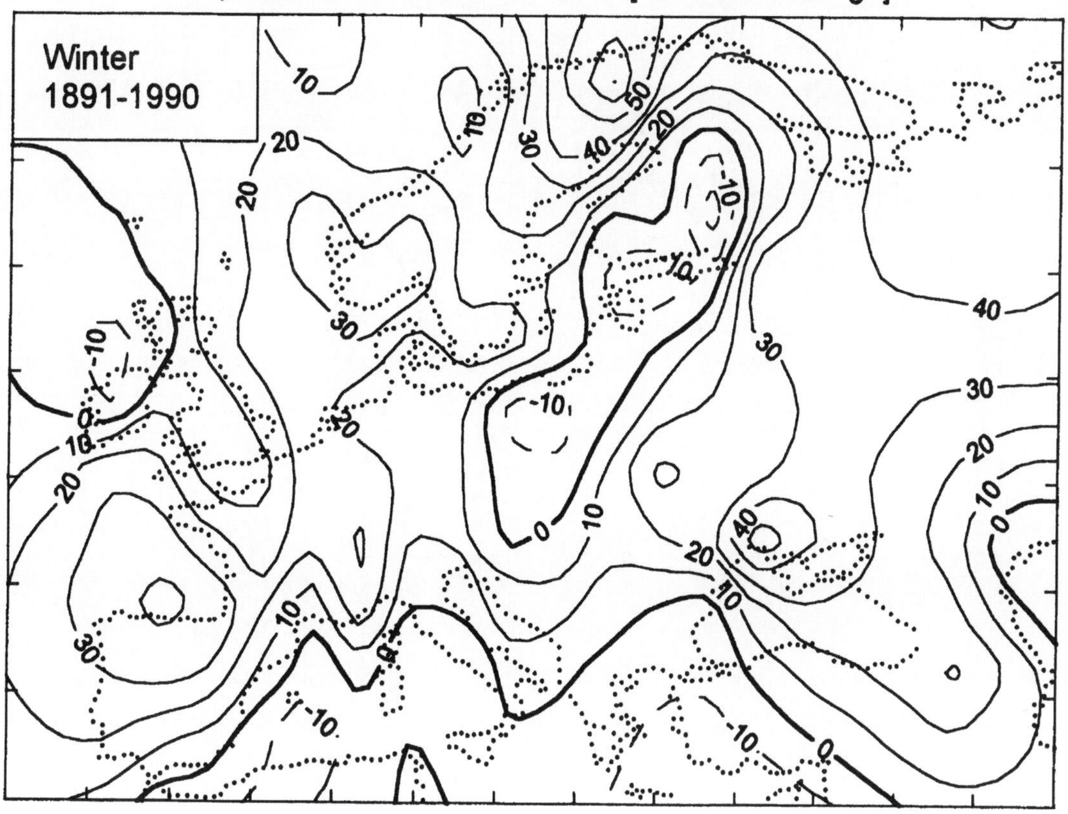

Precipitation, relative linear trend [% of the average]

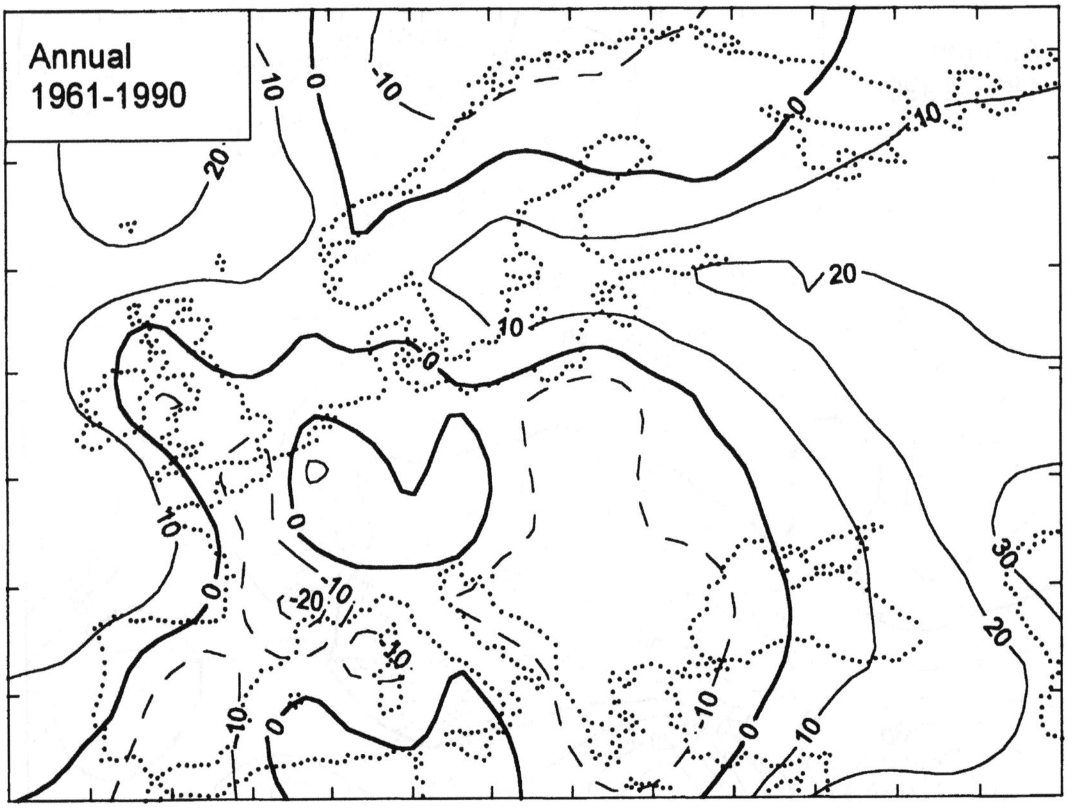

Precipitation, relative linear trend [% of the average]

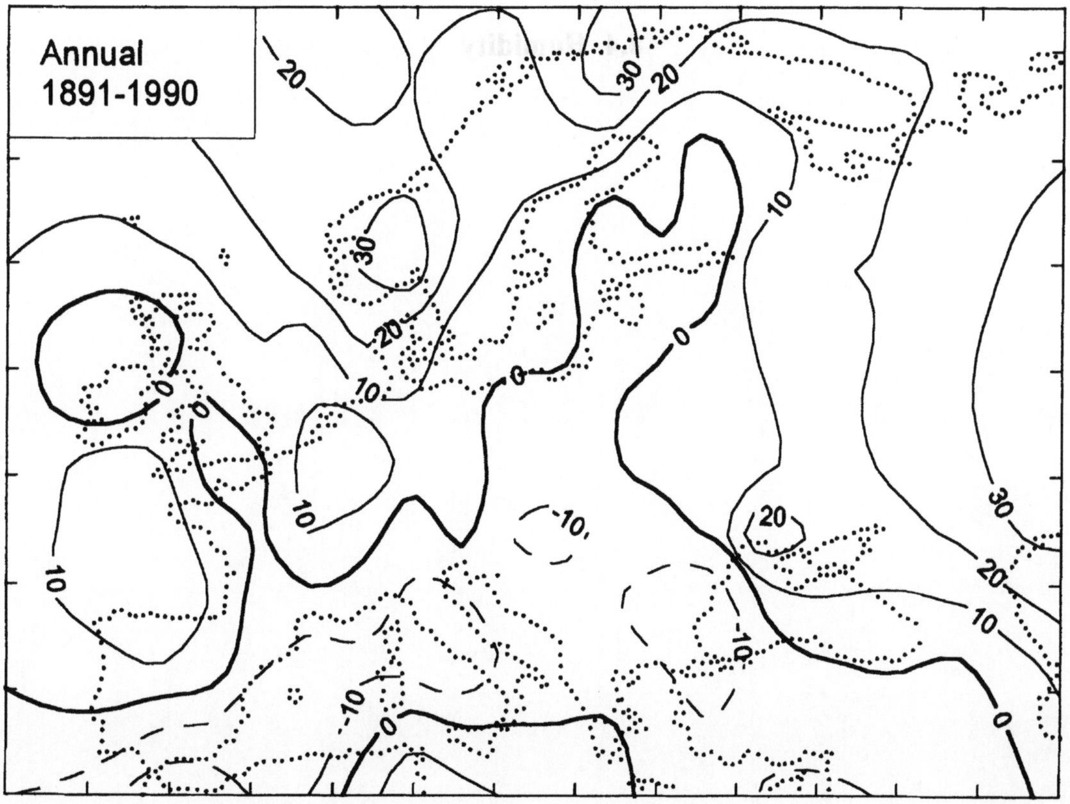

8.4. Humidity

Humidity, average [hPa]

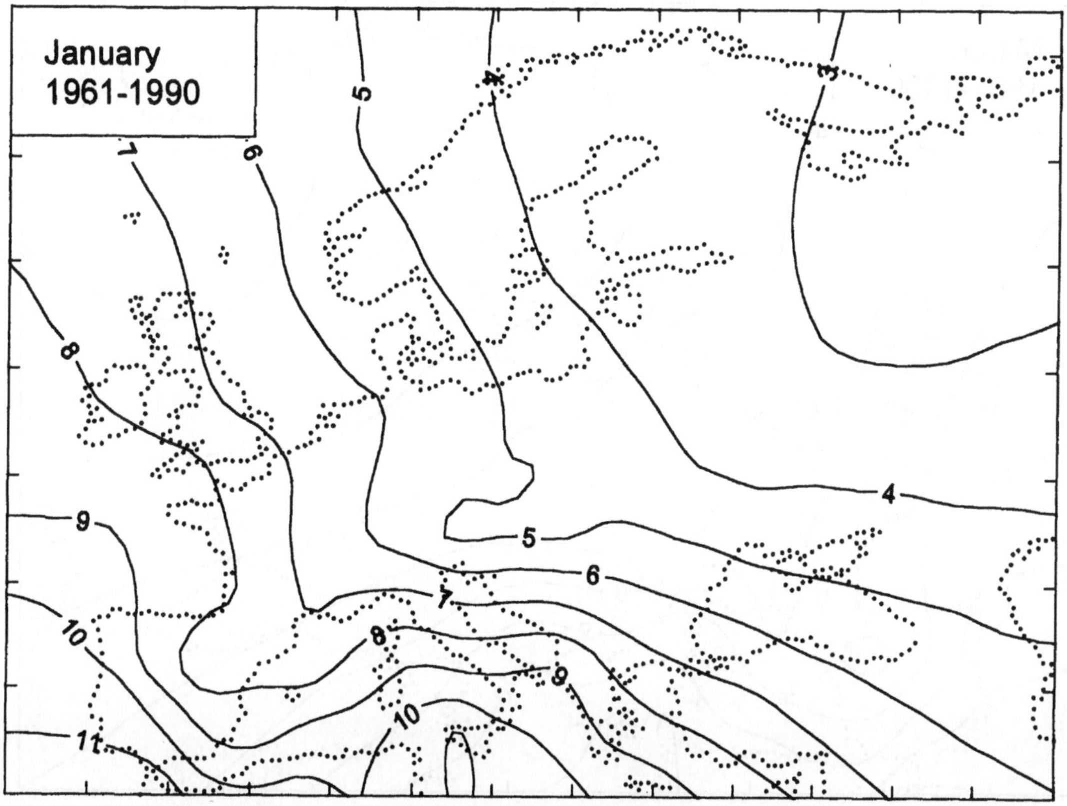

Humidity, average [hPa]

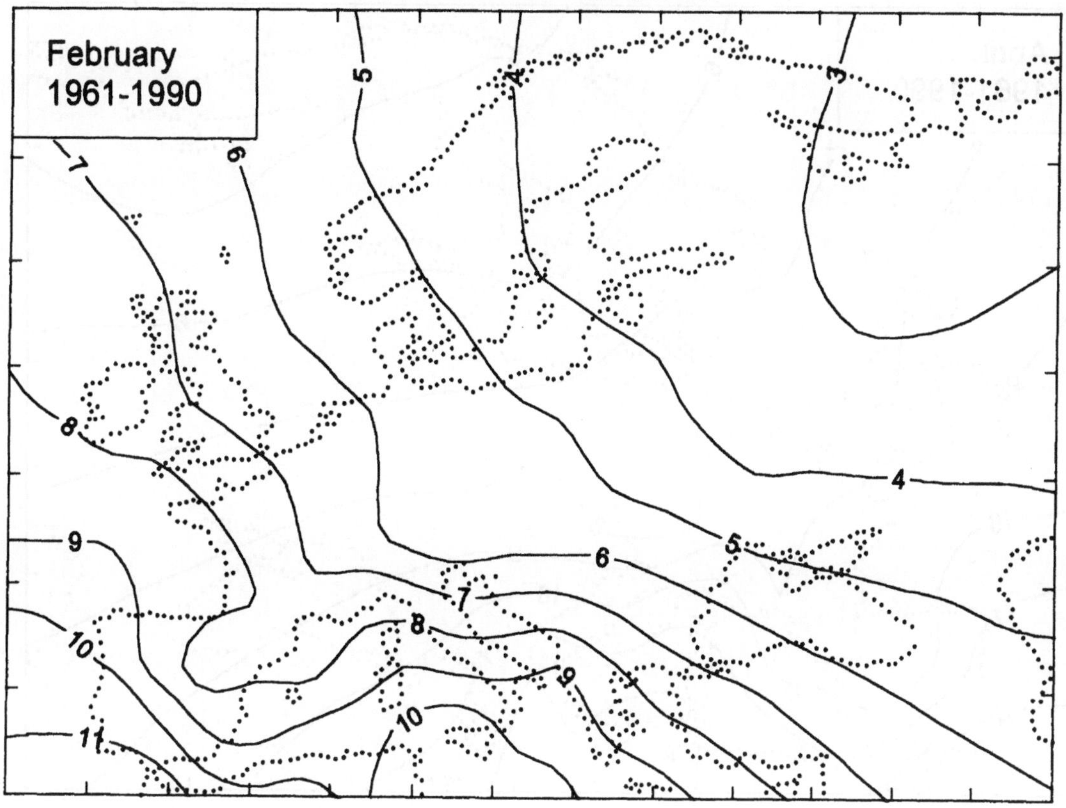

Humidity, average [hPa]

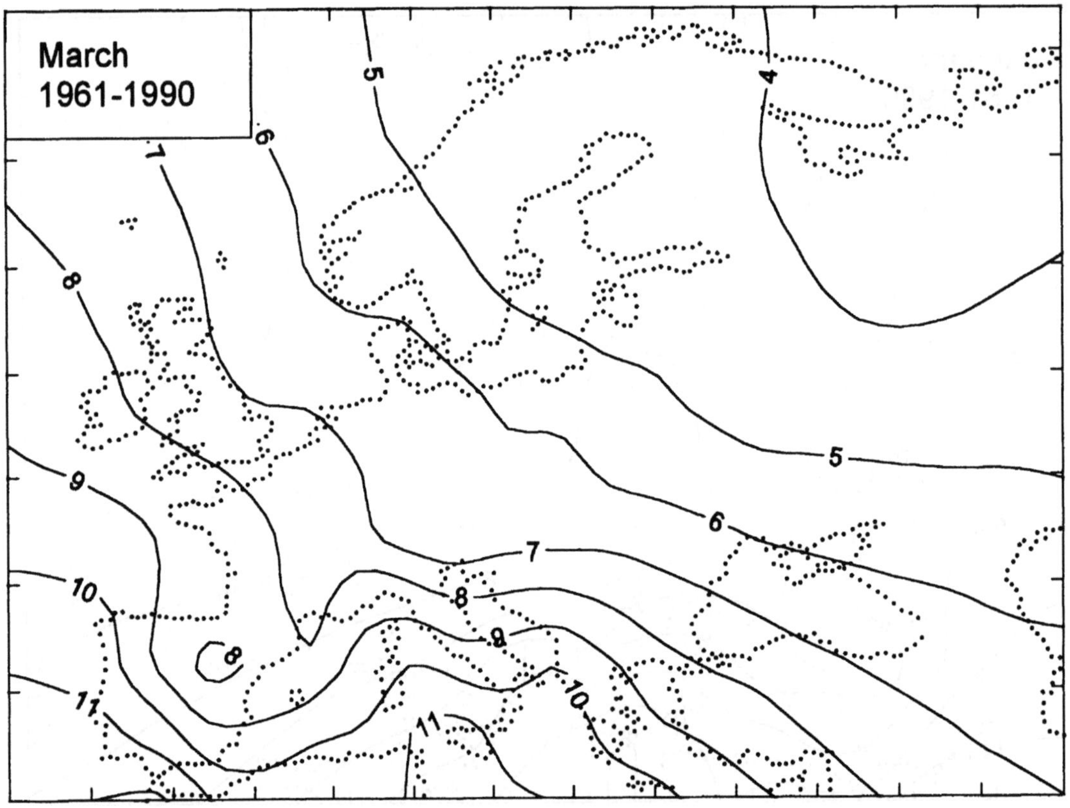

Humidity, average [hPa]

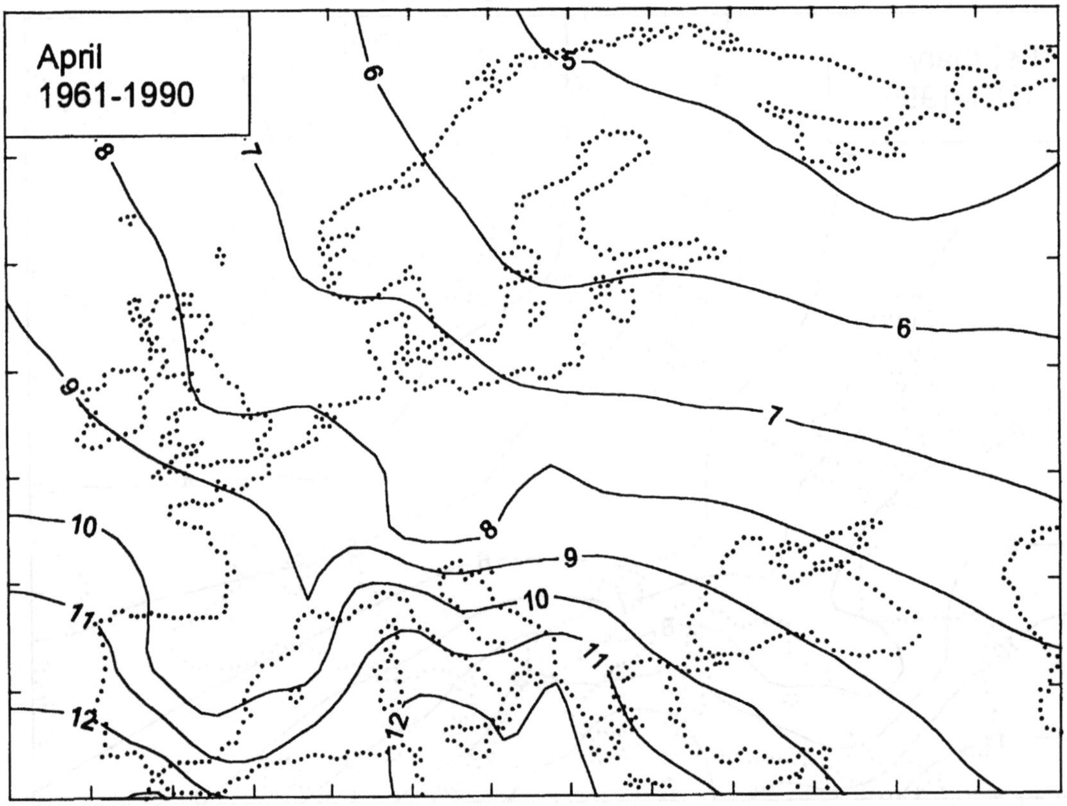

Humidity, average [hPa]

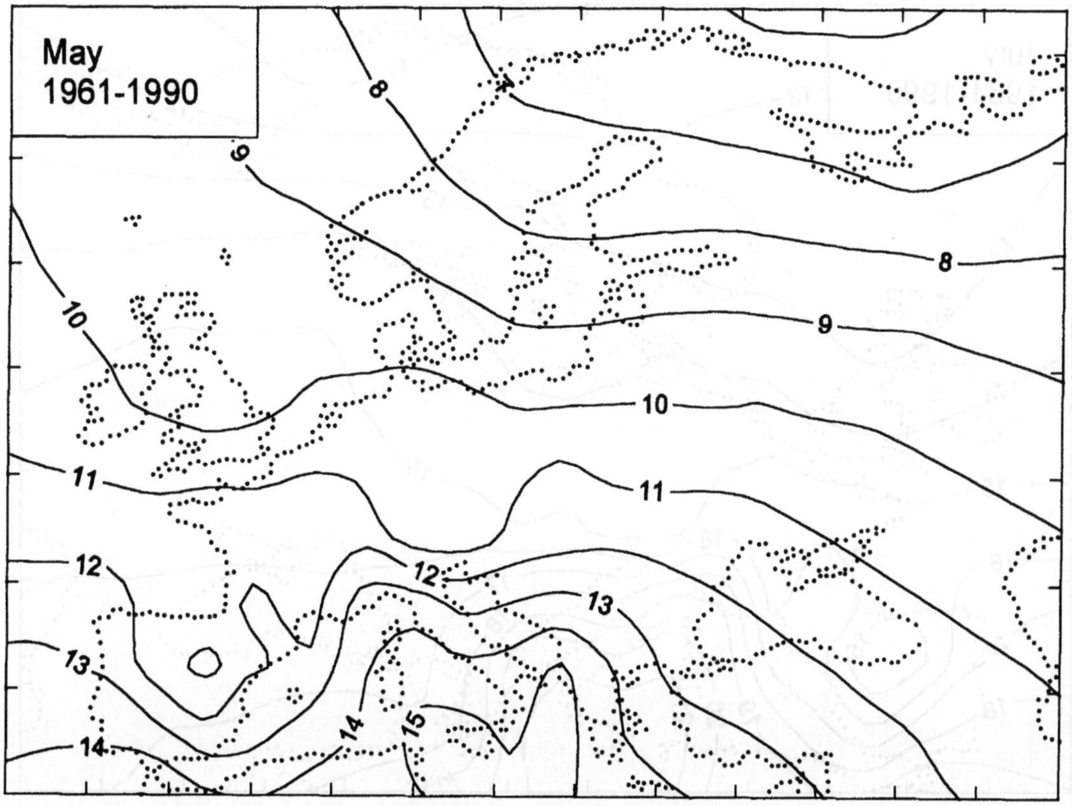

Humidity, average [hPa]

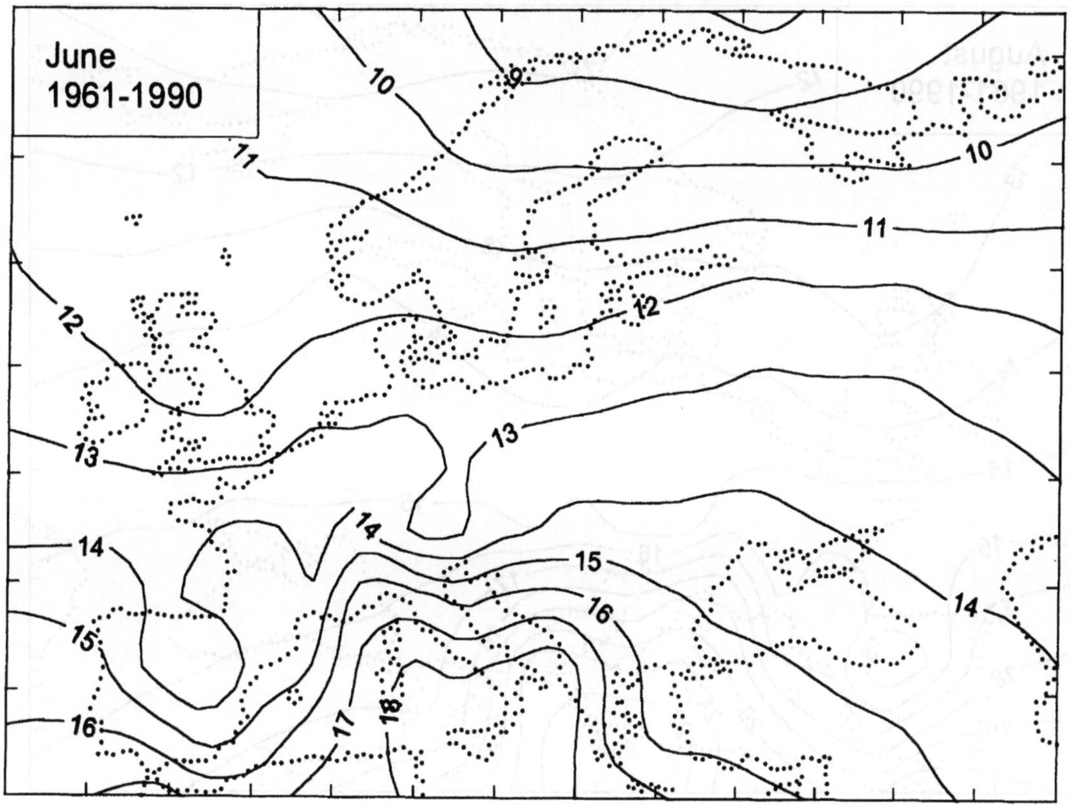

Humidity, average [hPa]

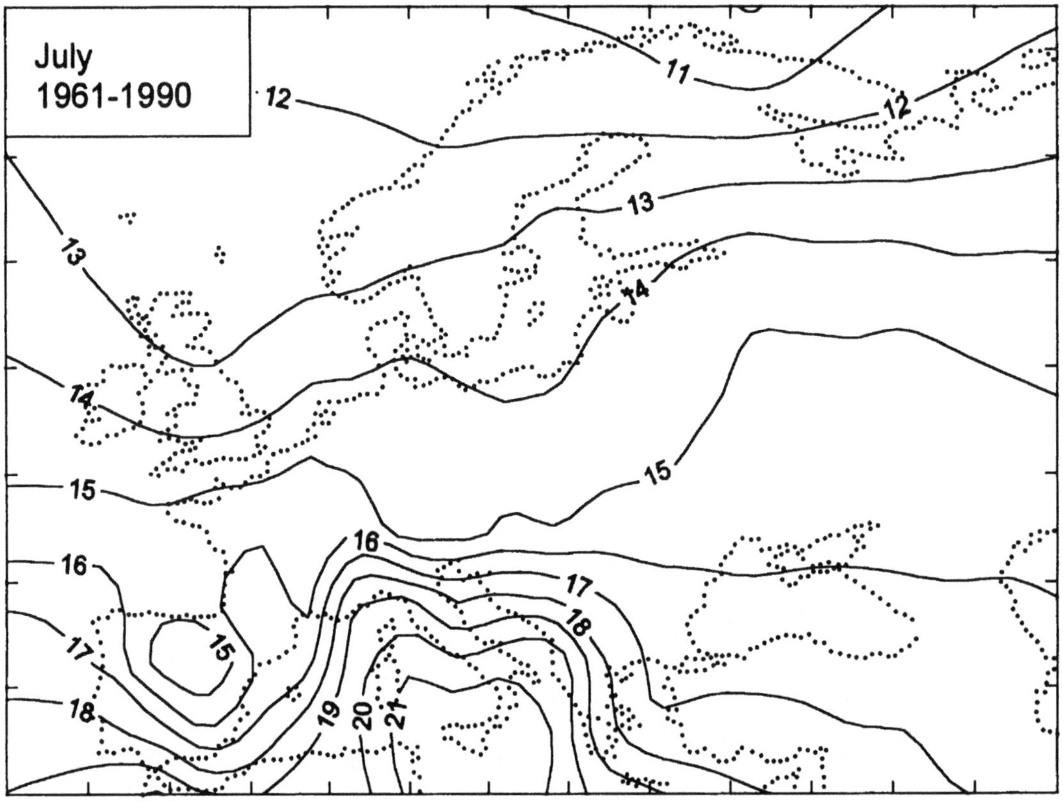

Humidity, average [hPa]

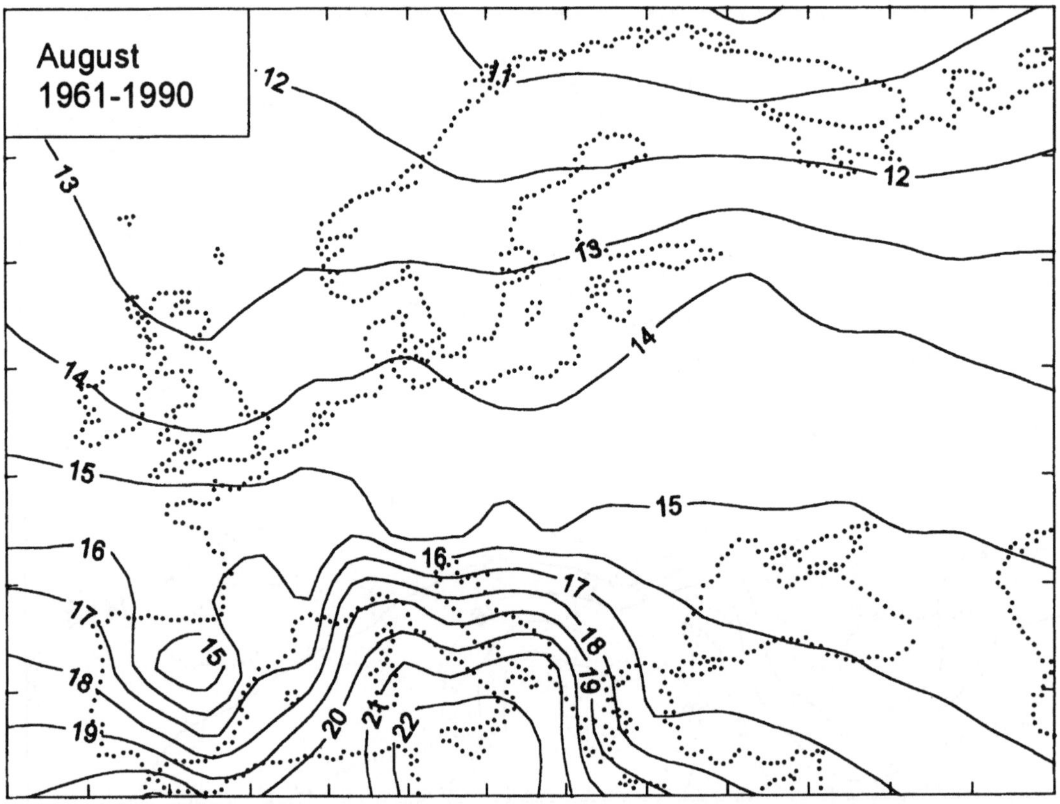

Humidity, average [hPa]

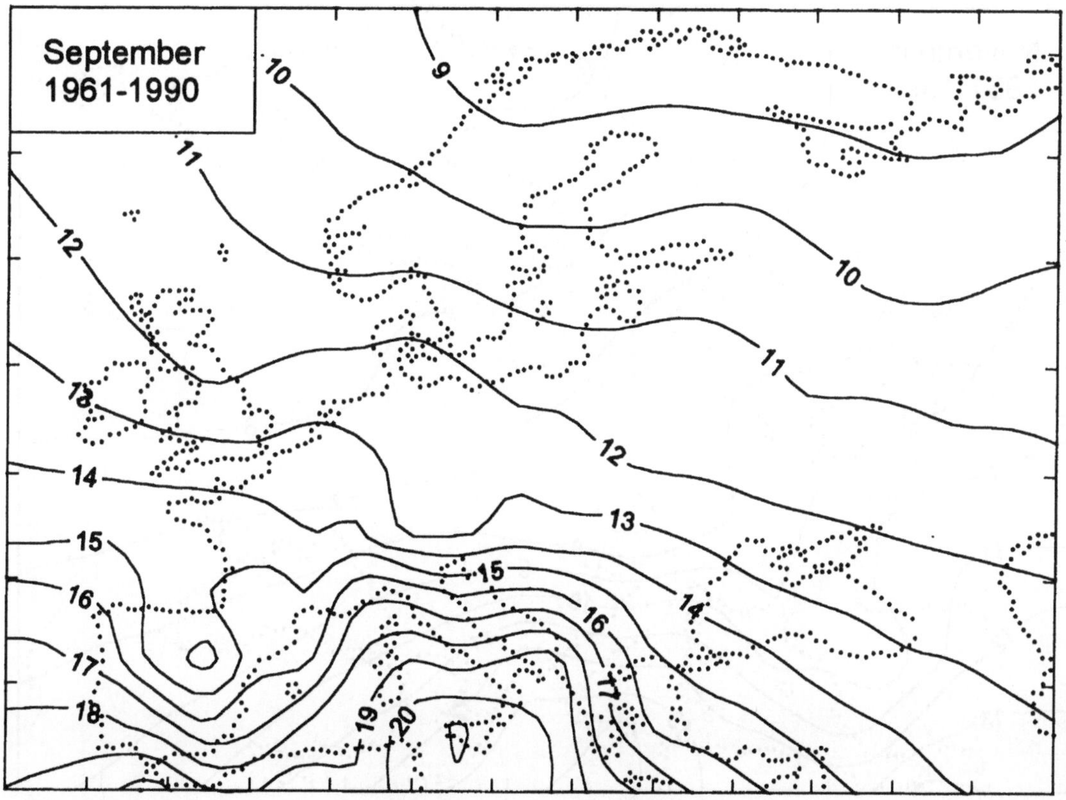

Humidity, average [hPa]

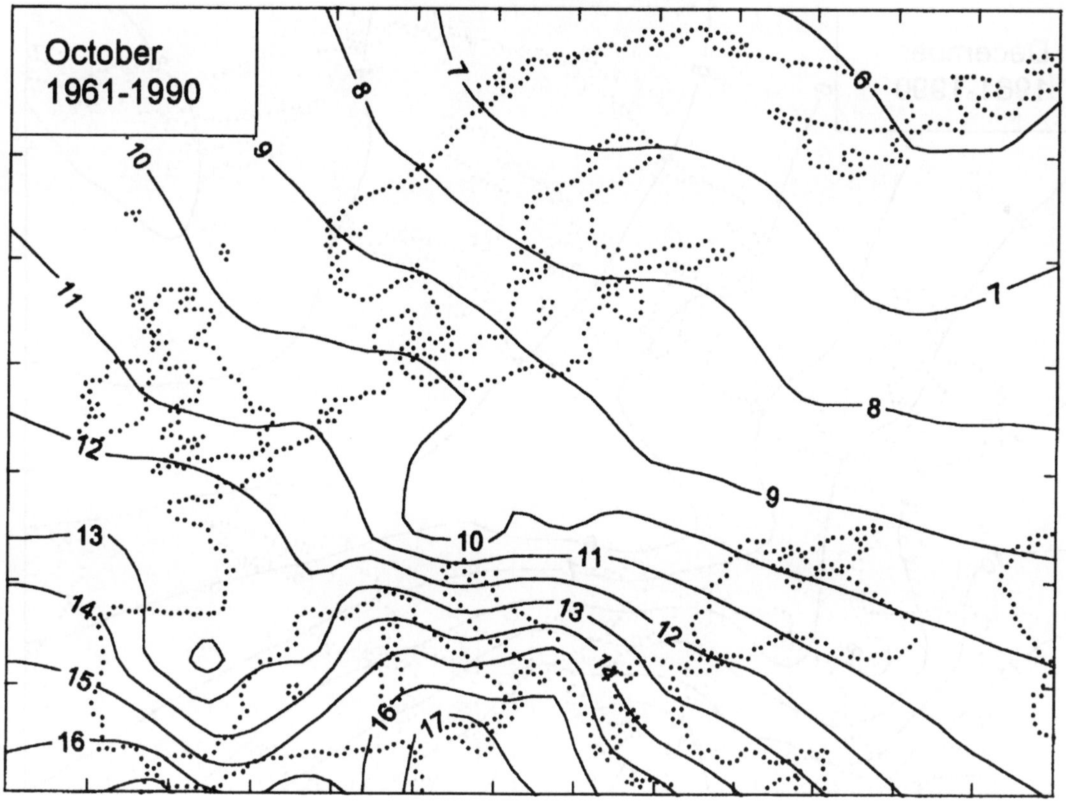

Humidity, average [hPa]

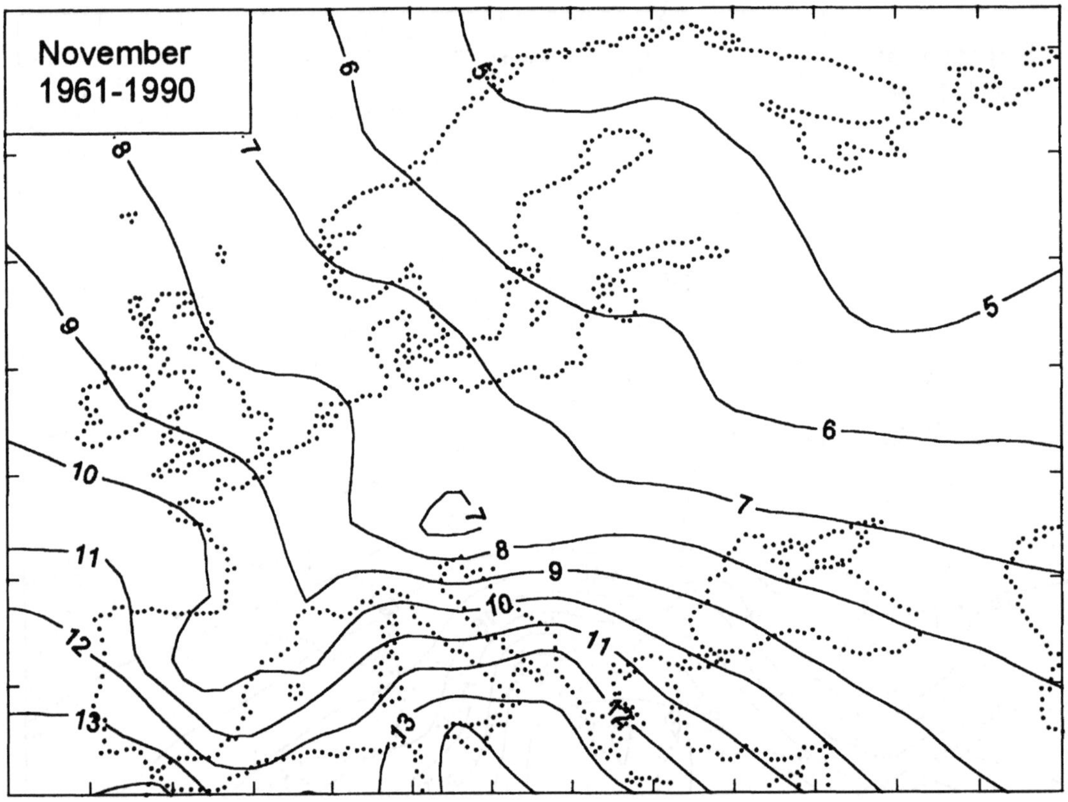

Humidity, average [hPa]

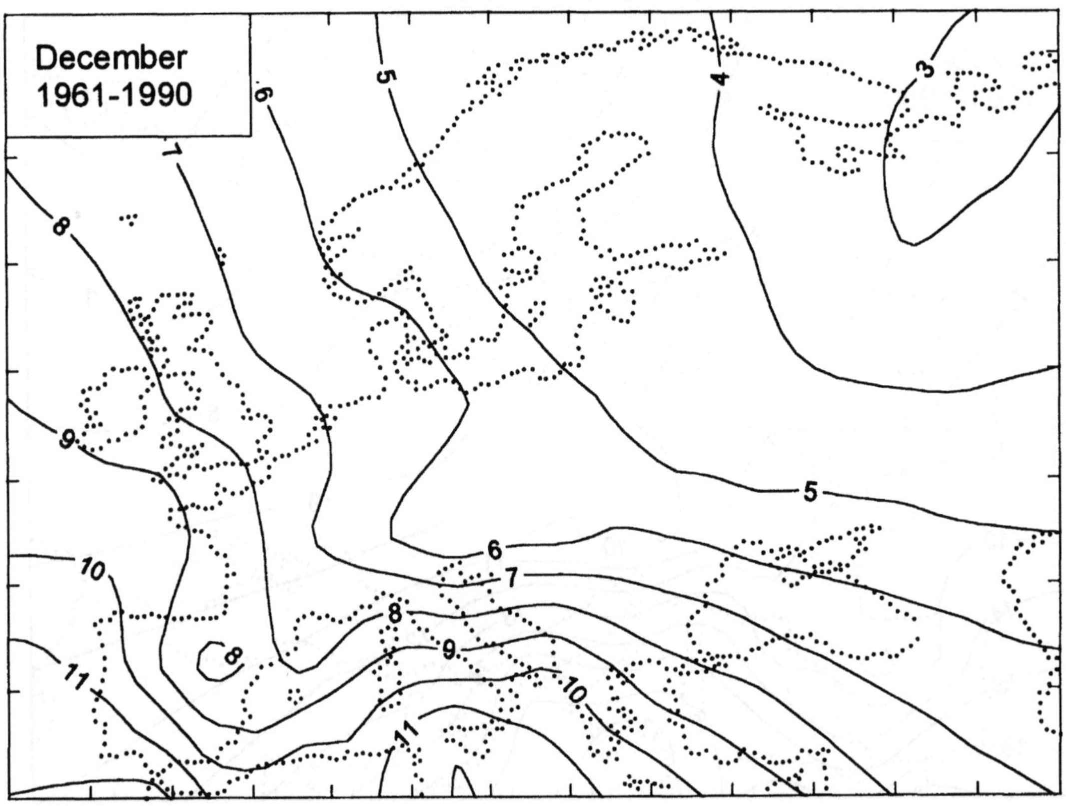

Humidity, average [hPa]

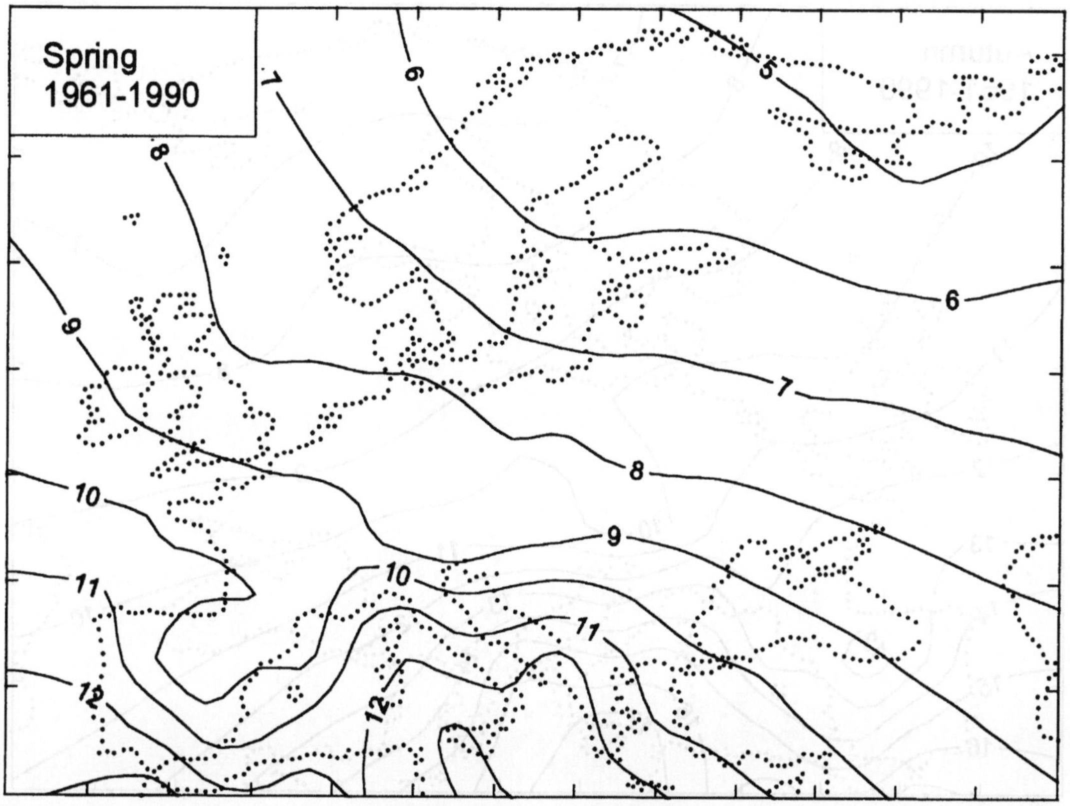

Humidity, average [hPa]

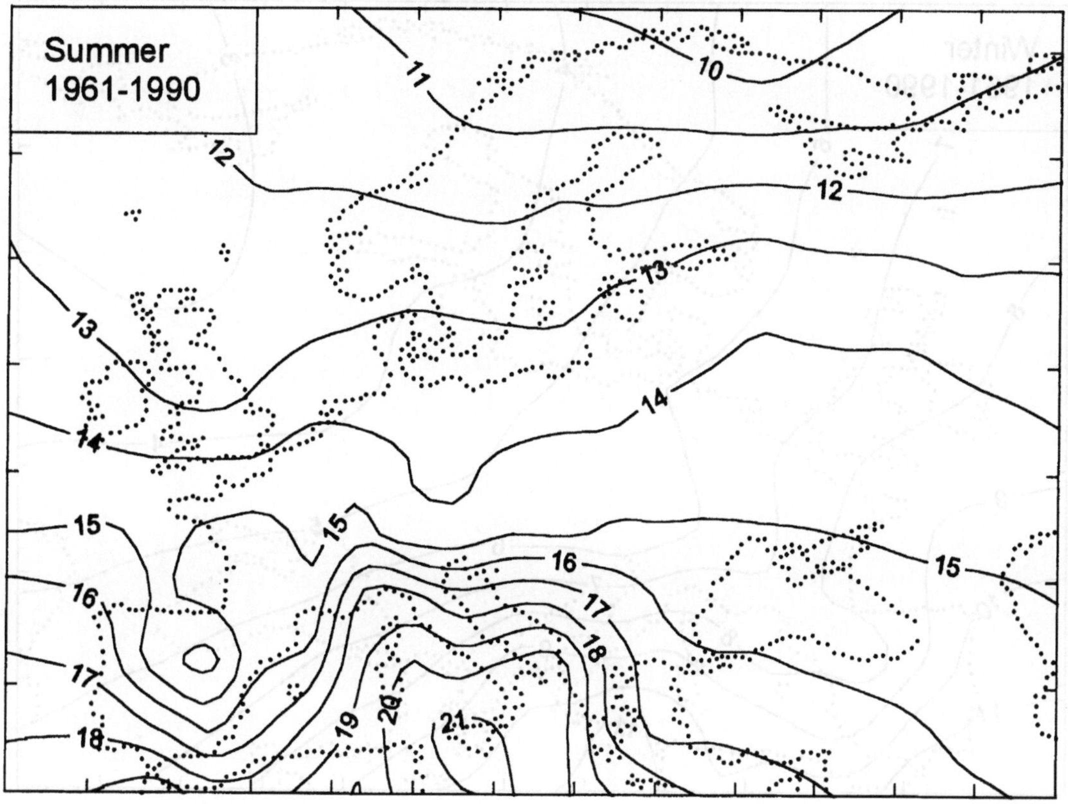

Humidity, average [hPa]

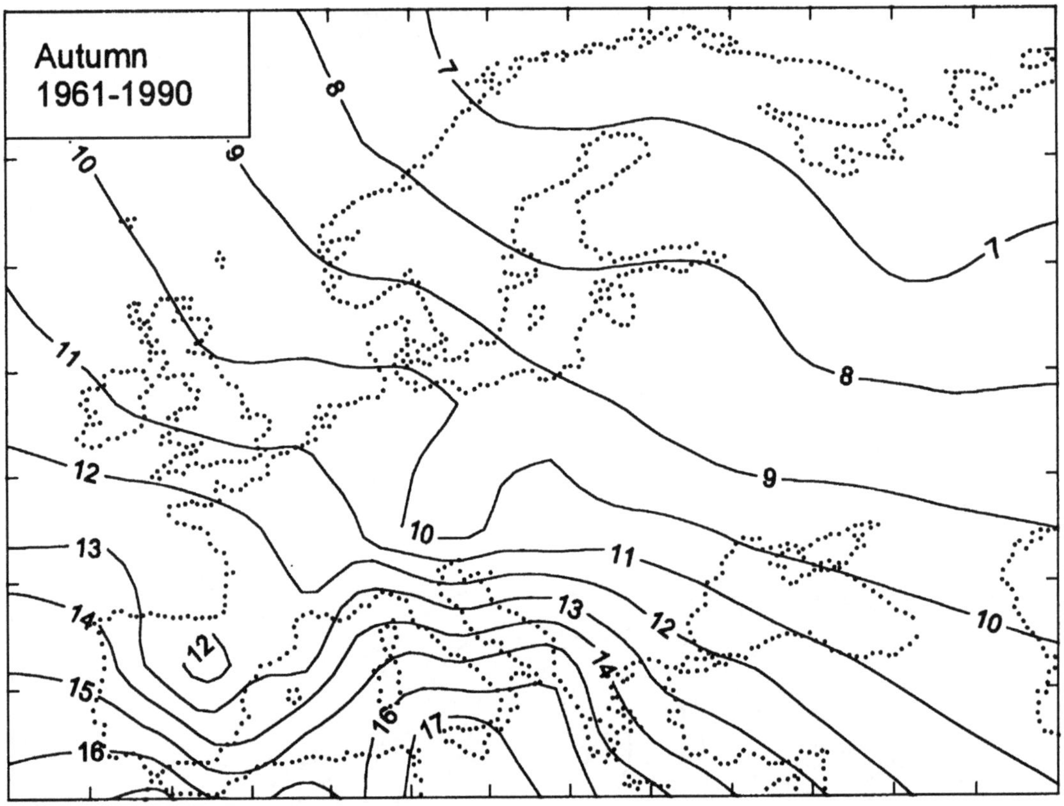

Humidity, average [hPa]

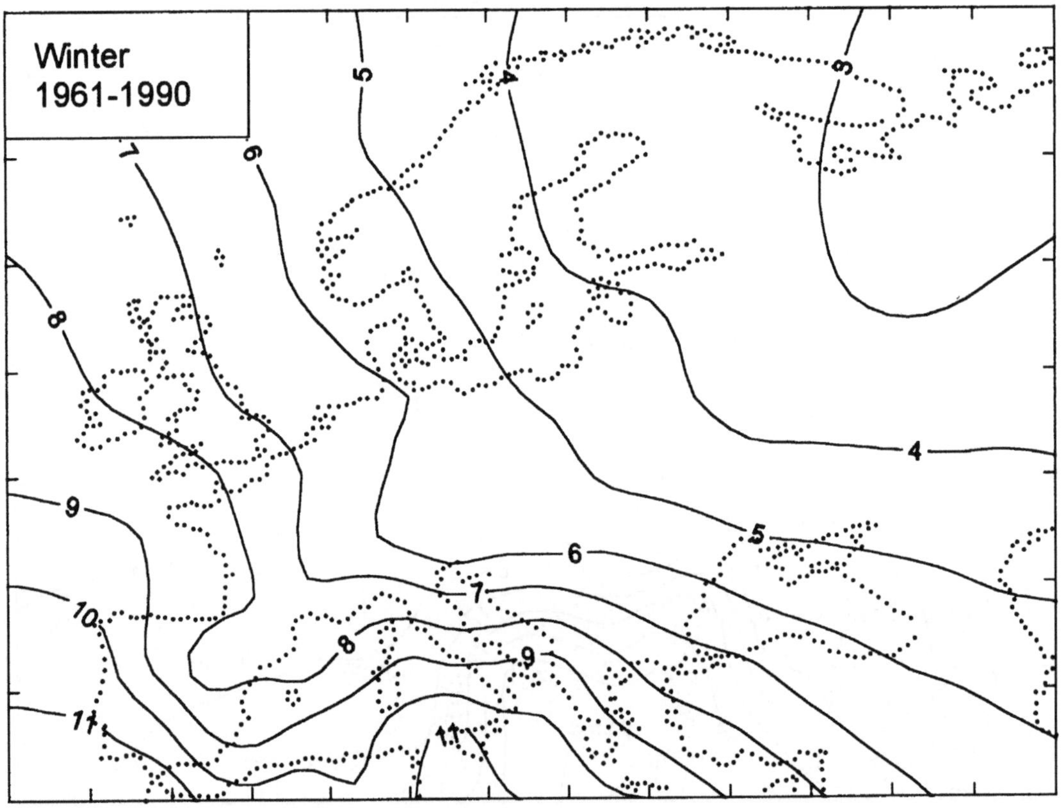

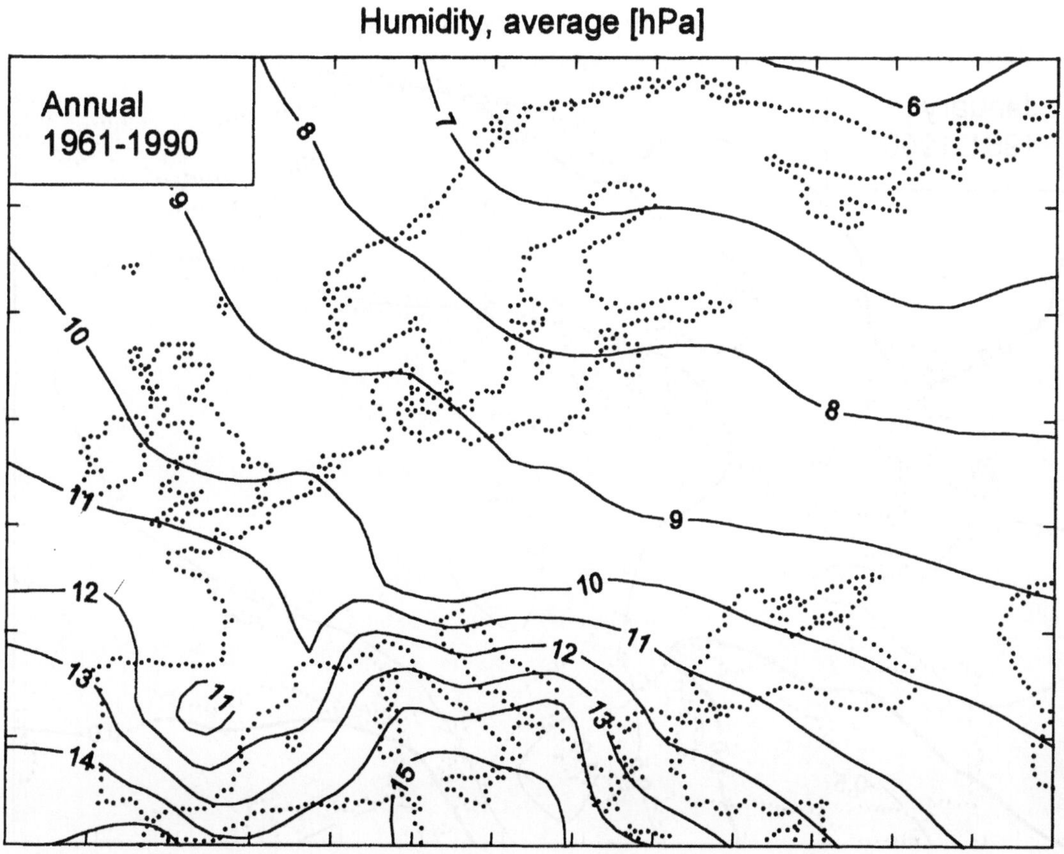
Humidity, average [hPa]
Annual
1961-1990
6
7
8
9
10
11
12
13
14
15
8
9
10
11
12
13
11

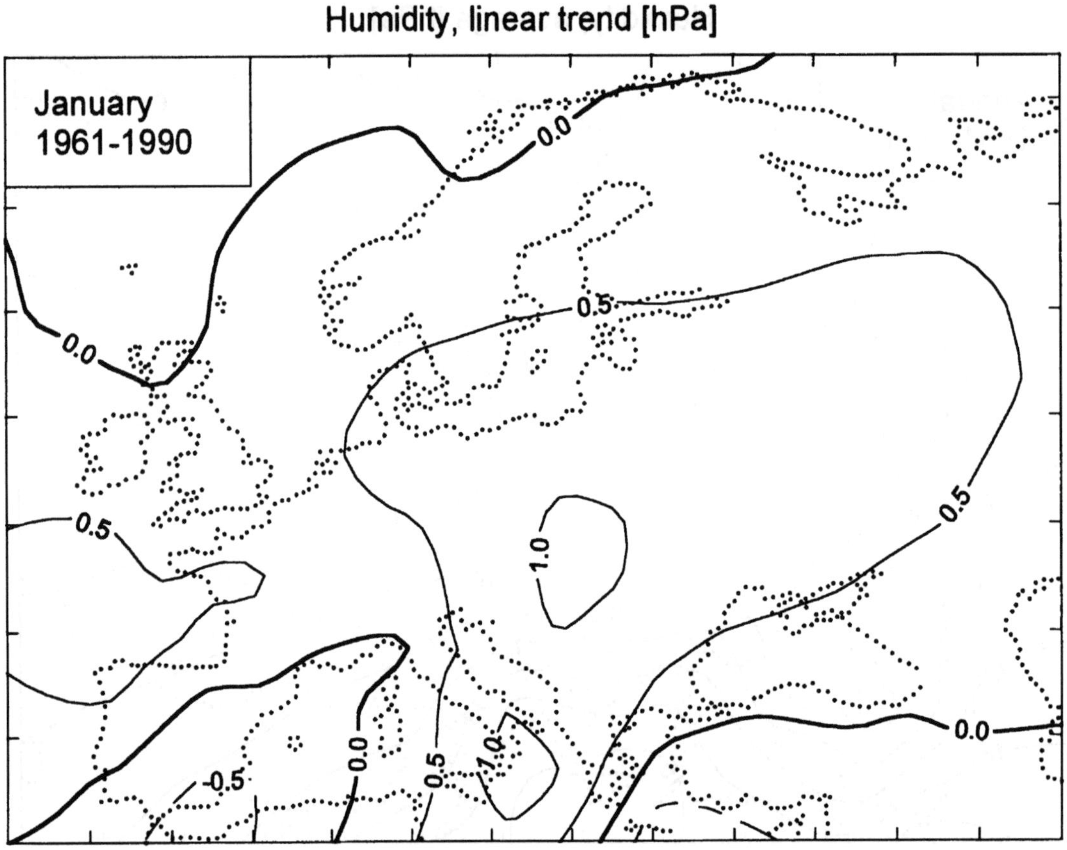
Humidity, linear trend [hPa]
January
1961-1990
0.0
0.5
1.0
-0.5

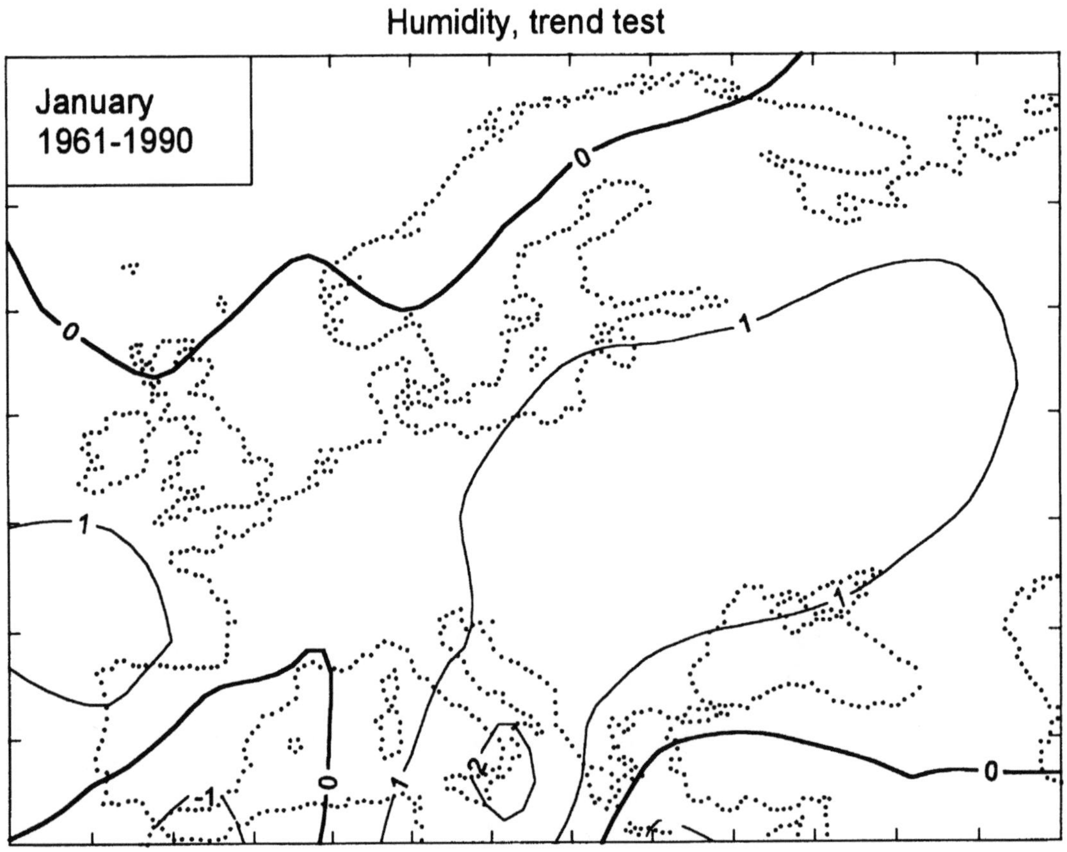
Humidity, trend test
January
1961-1990
0
1
2
-1

Humidity, linear trend [hPa]

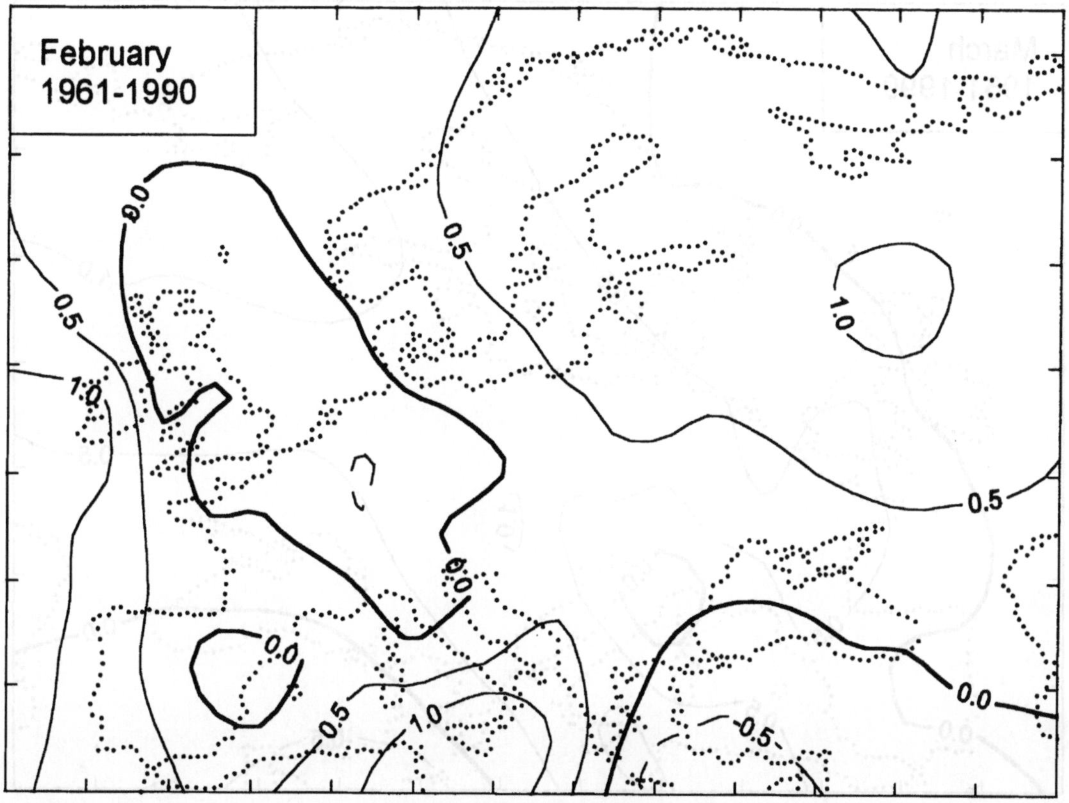

Humidity, trend test

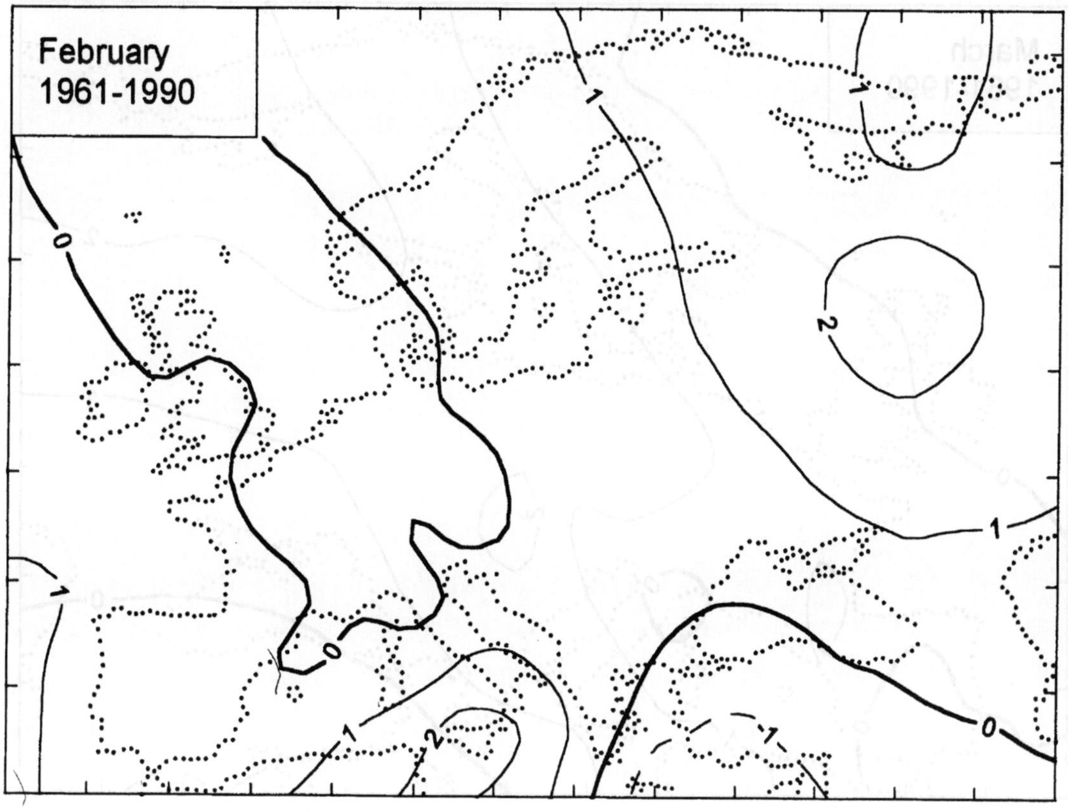

Humidity, linear trend [hPa]

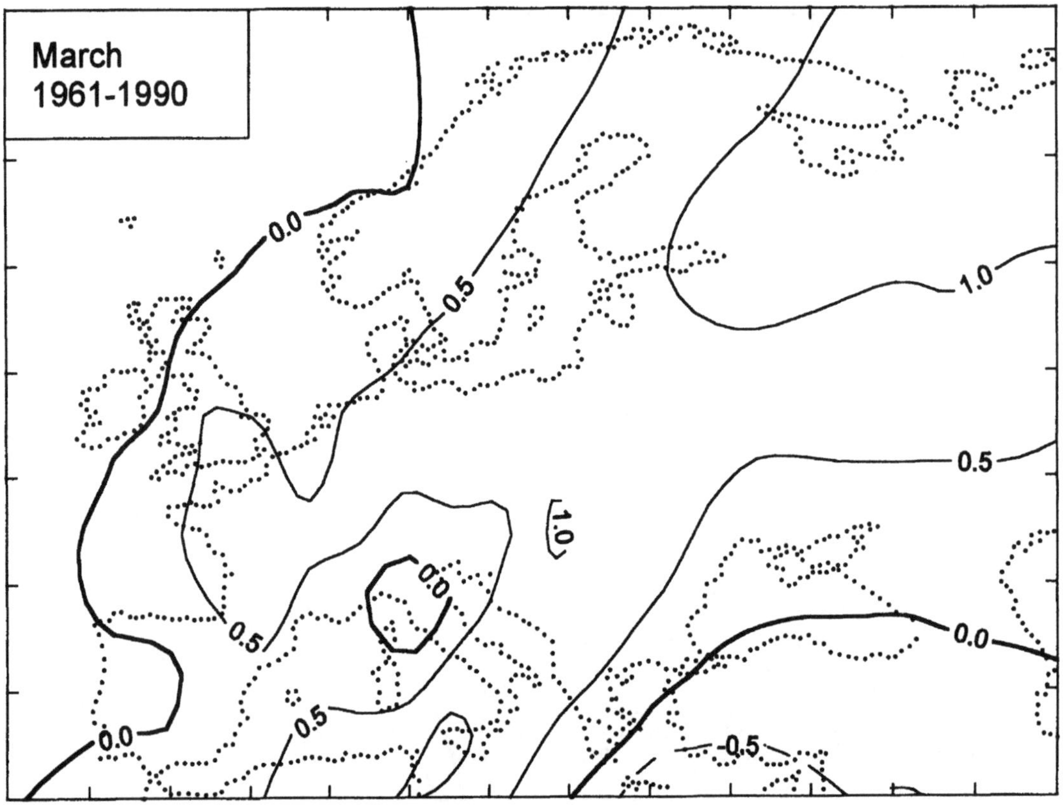

Humidity, trend test

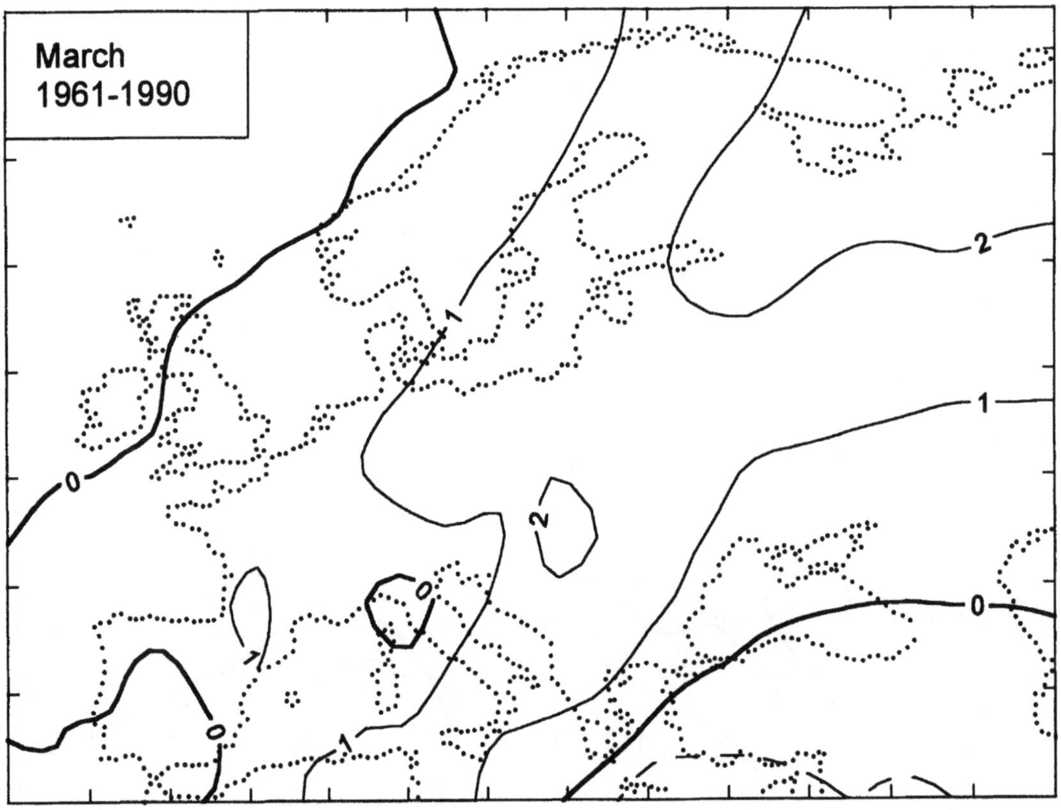

Humidity, linear trend [hPa]

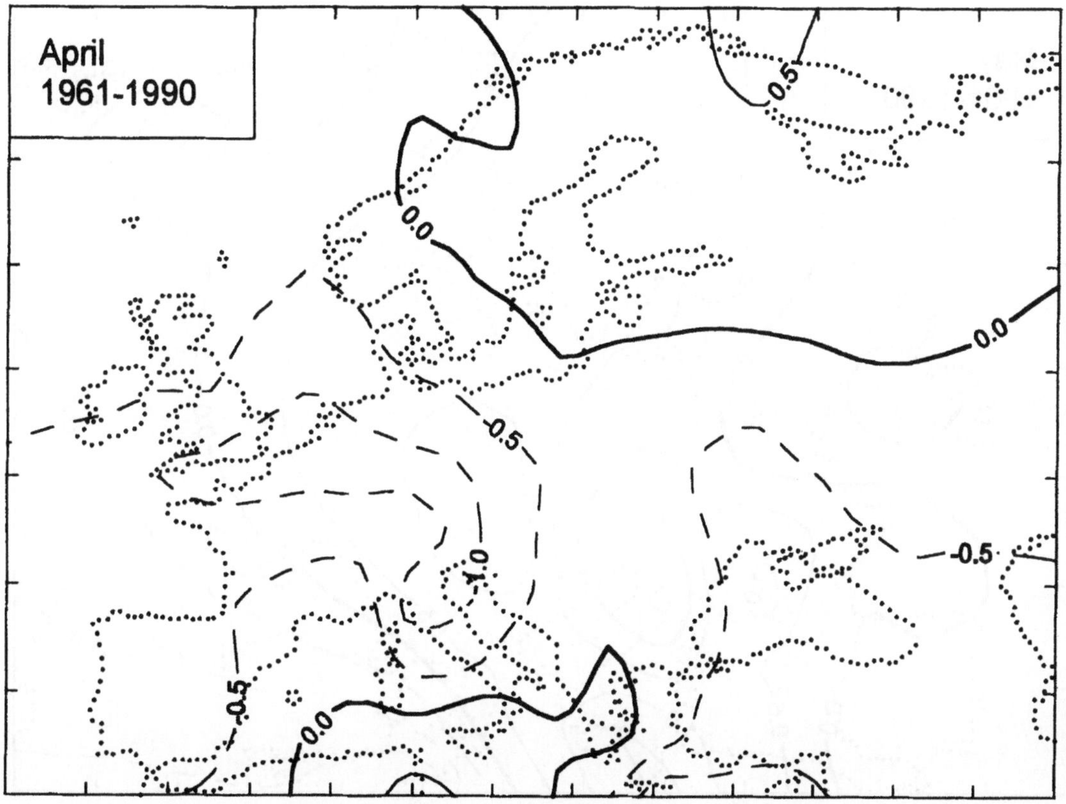

Humidity, trend test

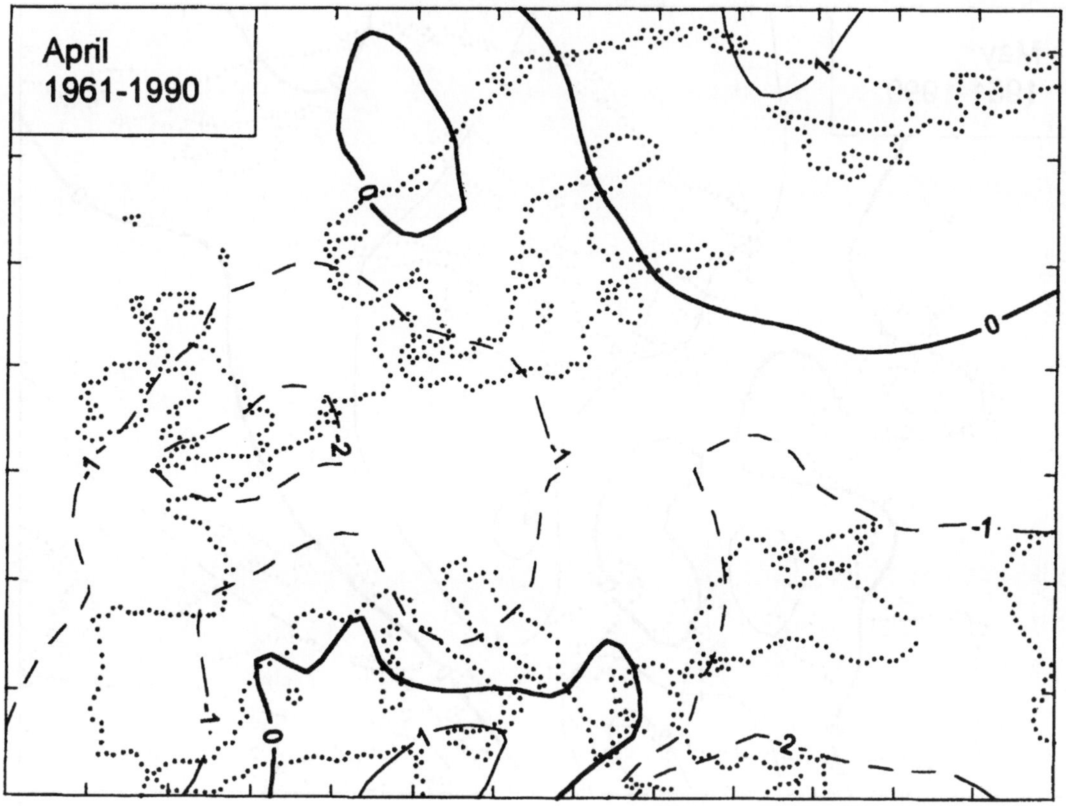

Humidity, linear trend [hPa]

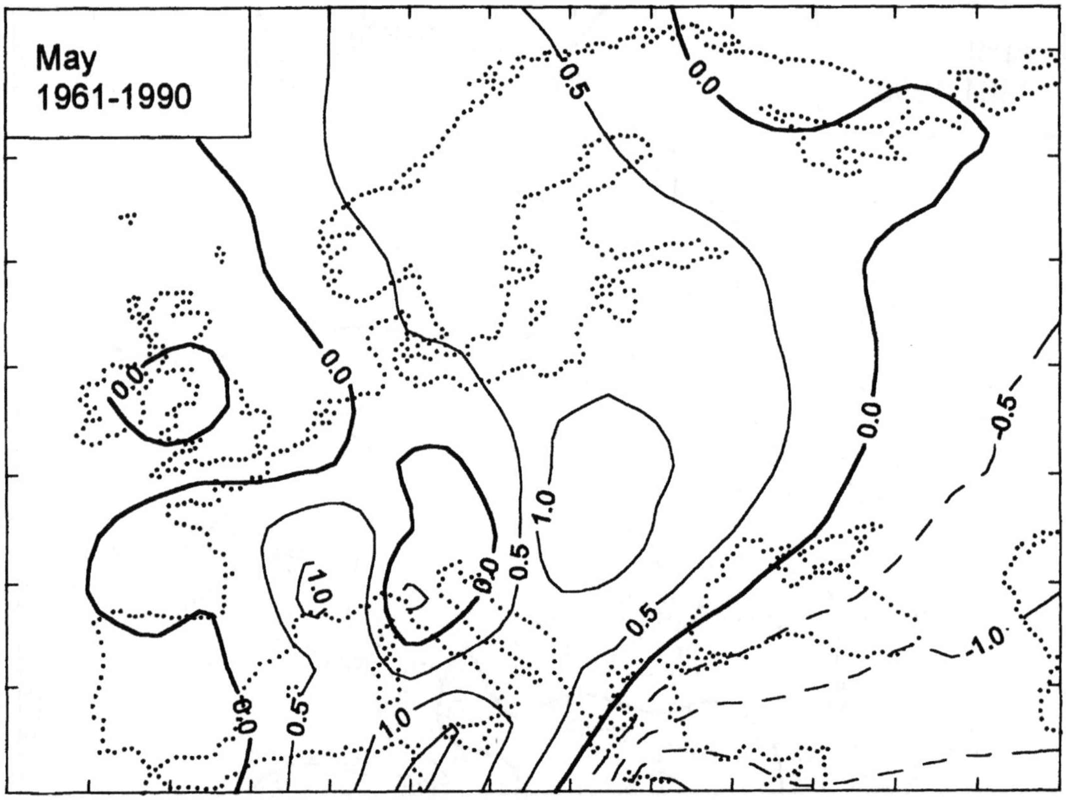

Humidity, trend test

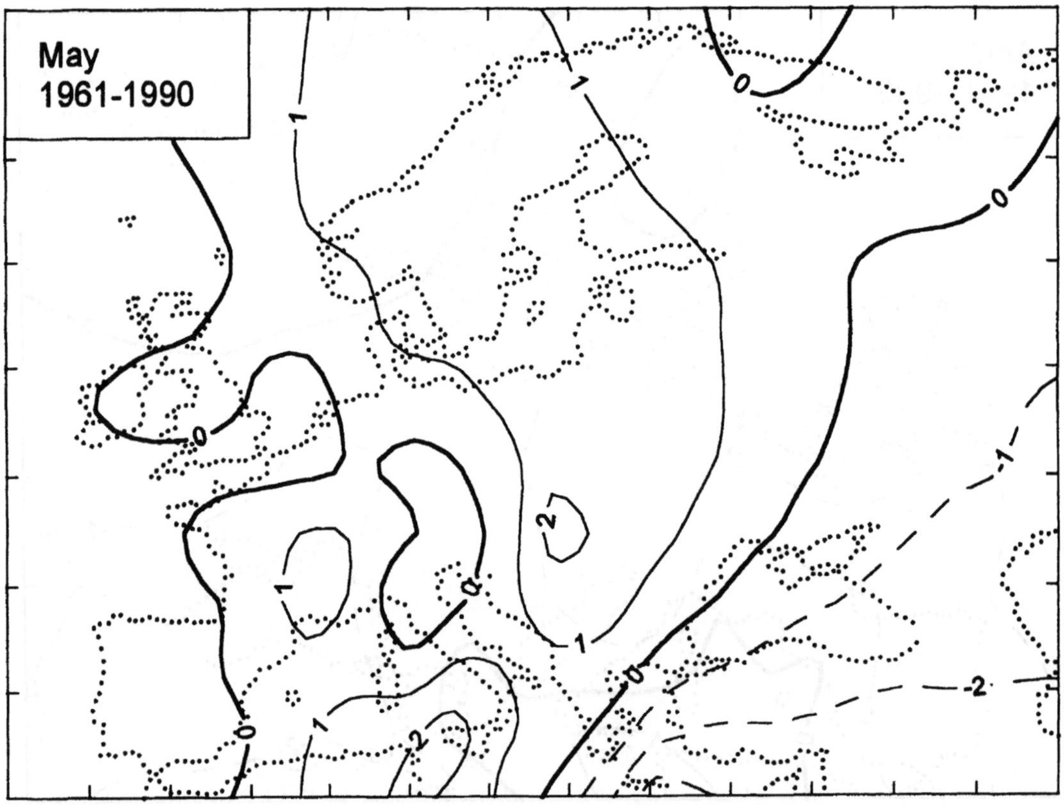

Humidity, linear trend [hPa]

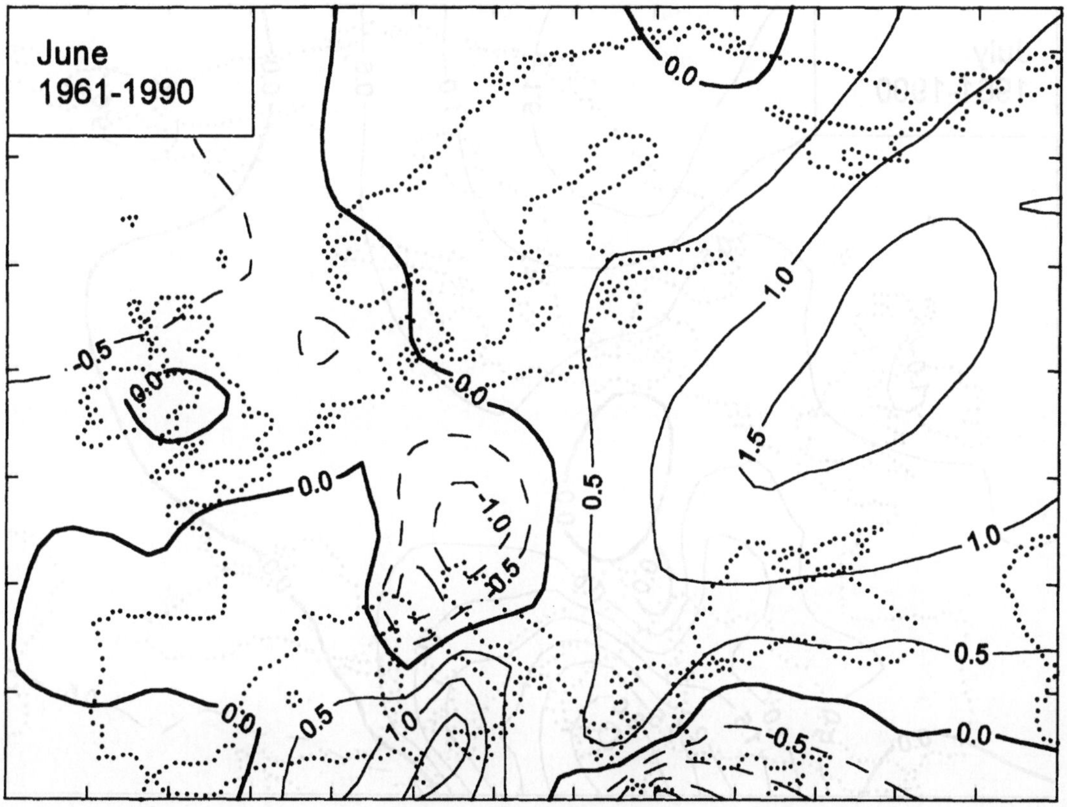

Humidity, trend test

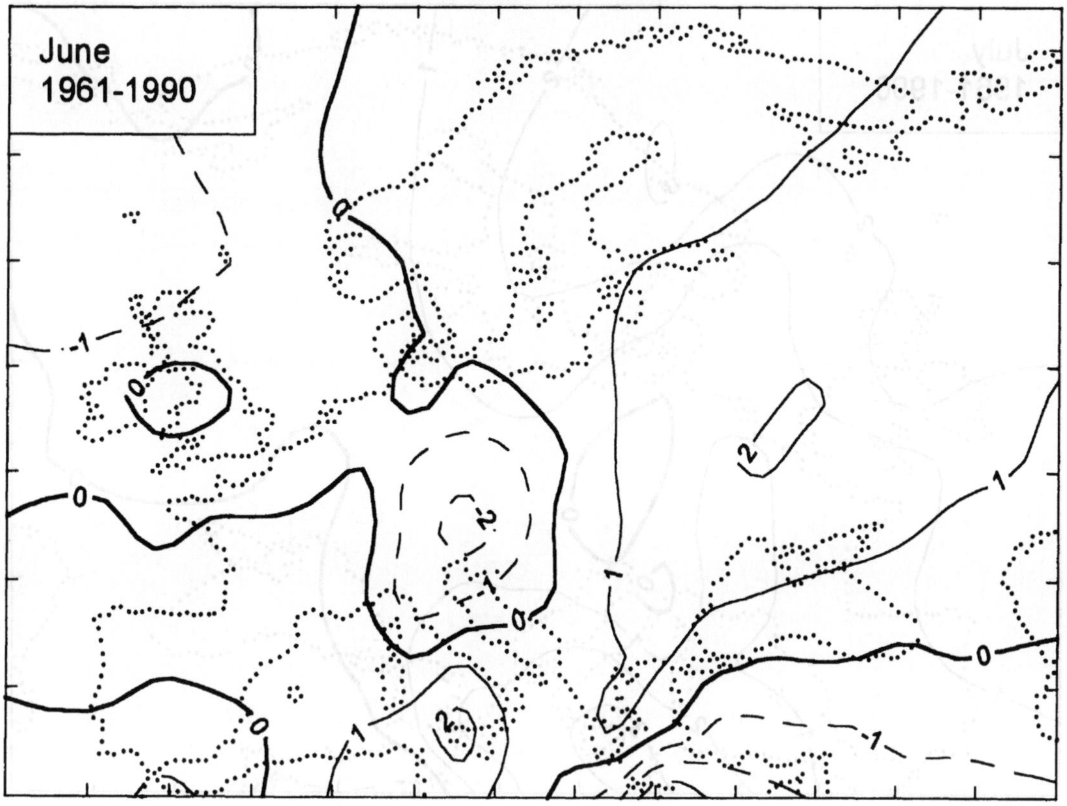

Humidity, linear trend [hPa]

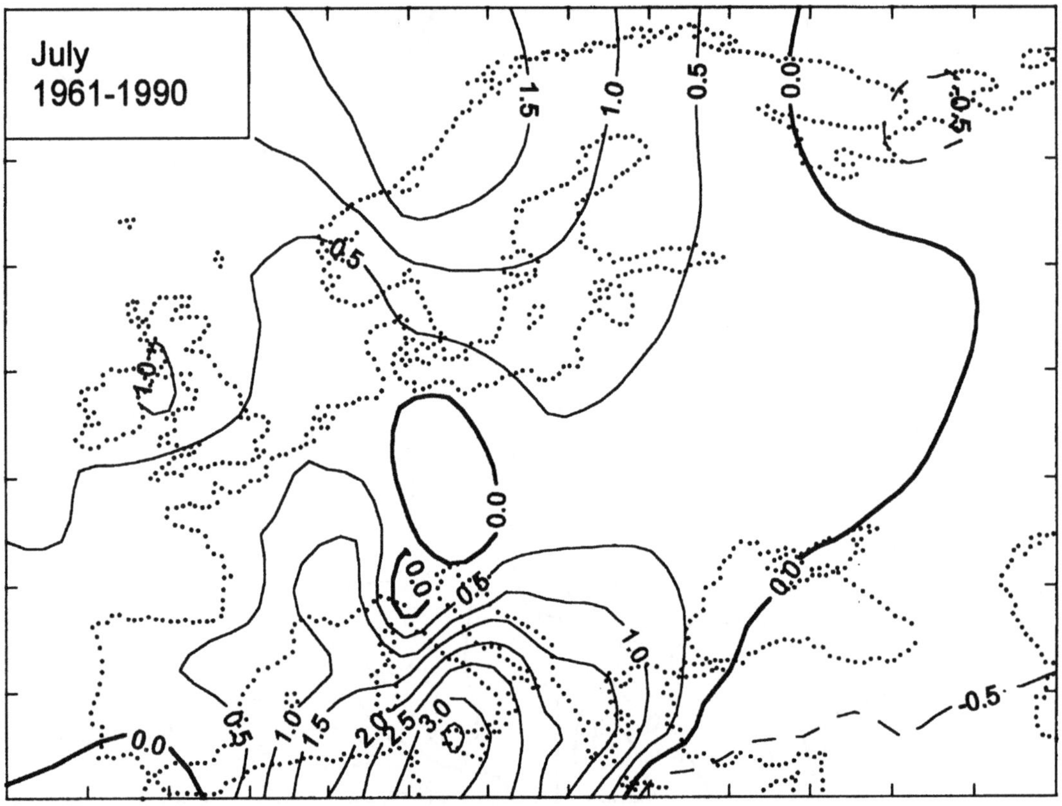

Humidity, trend test

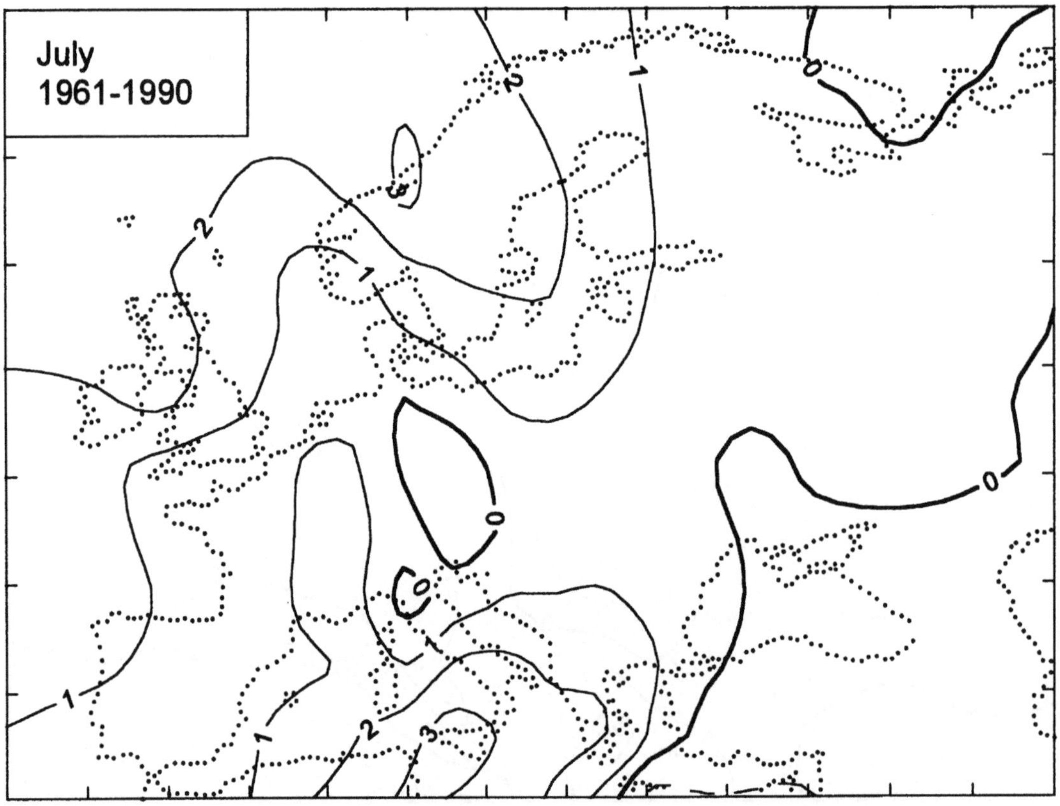

Humidity, linear trend [hPa]

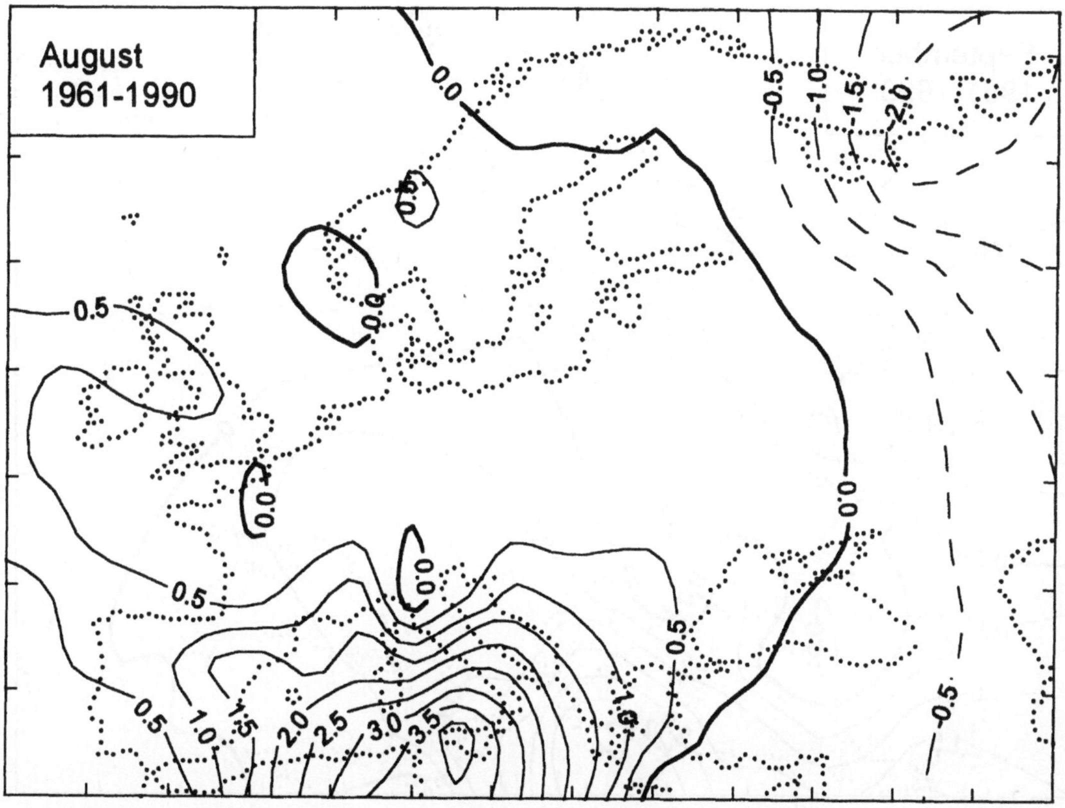

Humidity, trend test

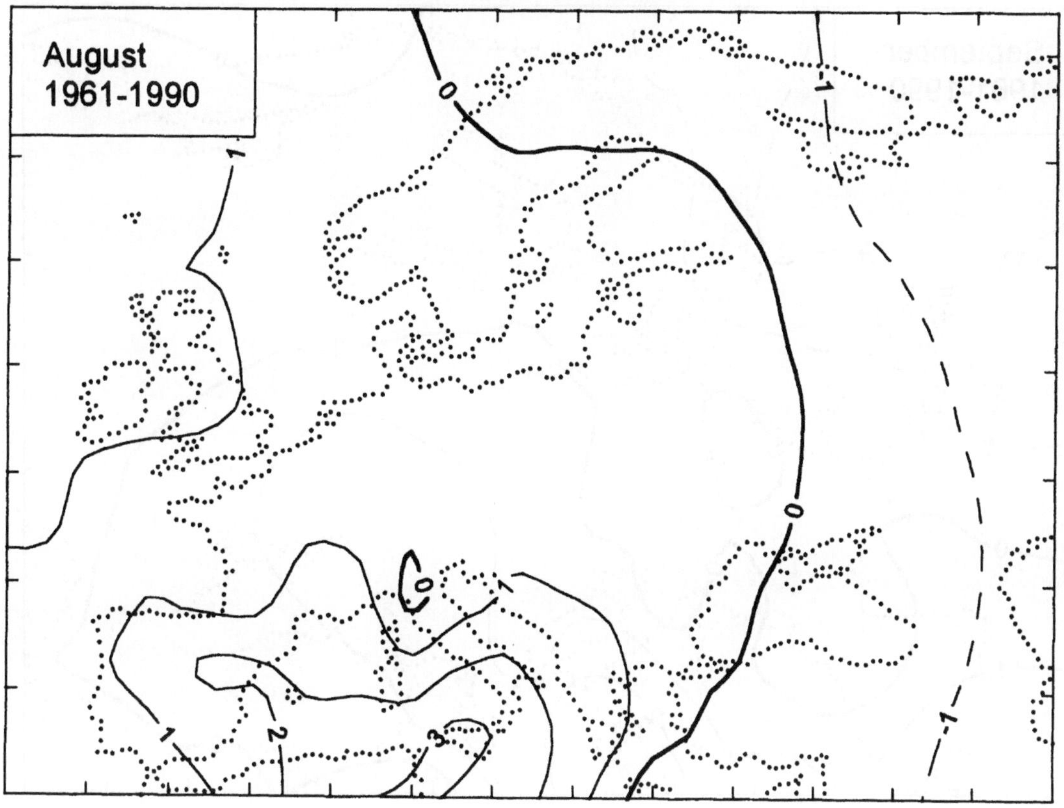

Humidity, linear trend [hPa]

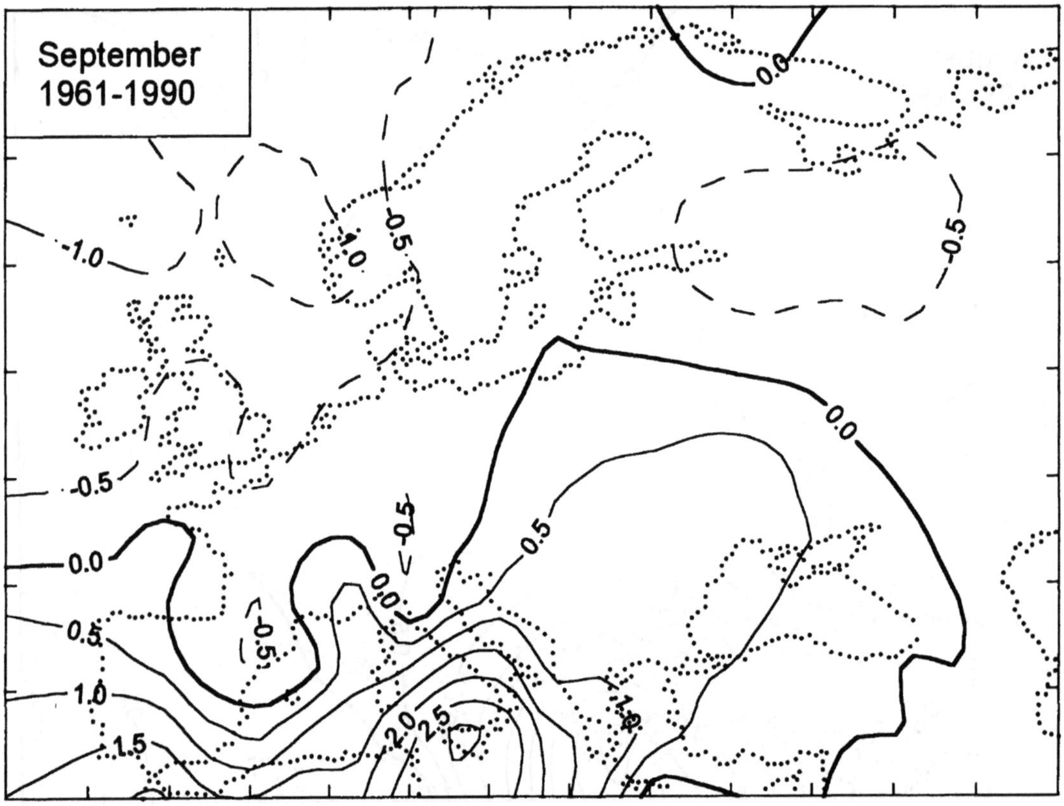

Humidity, trend test

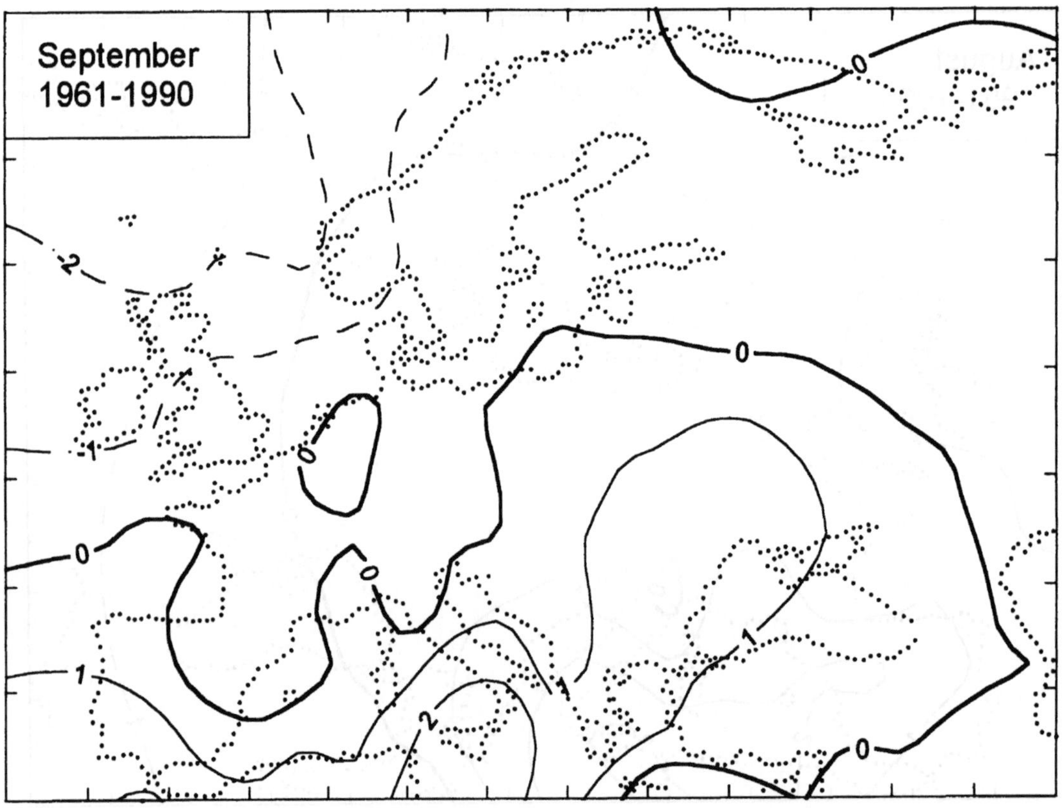

Humidity, linear trend [hPa]

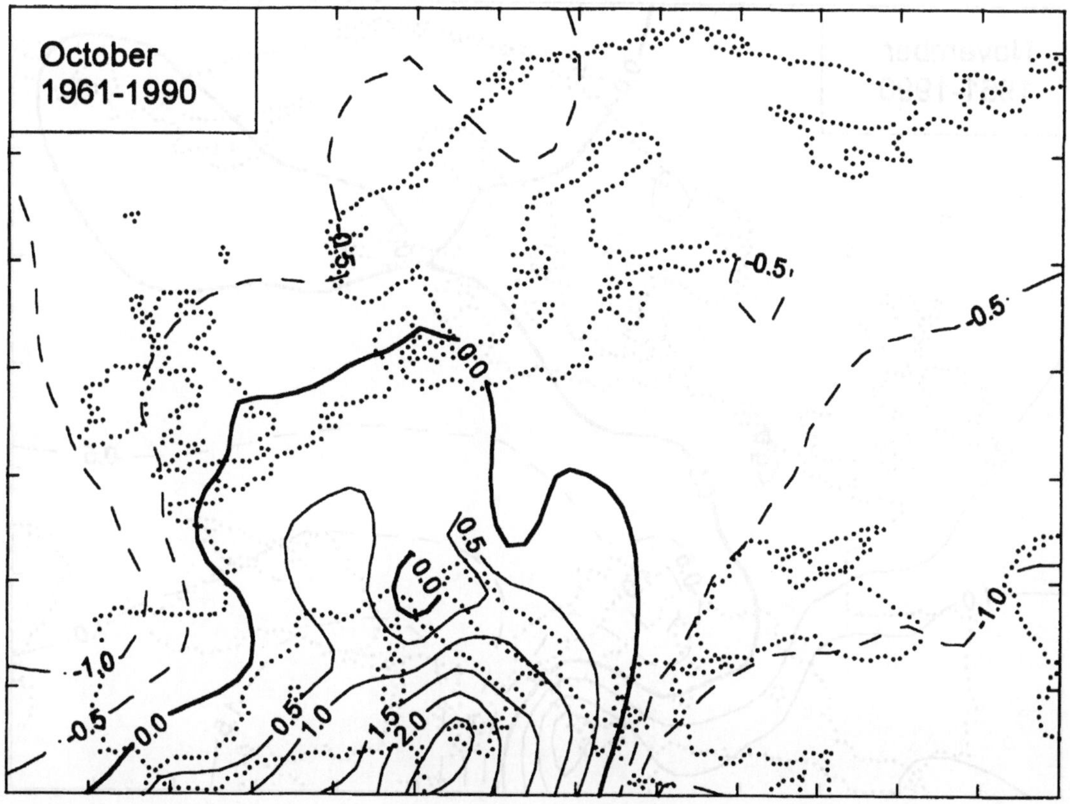

Humidity, trend test

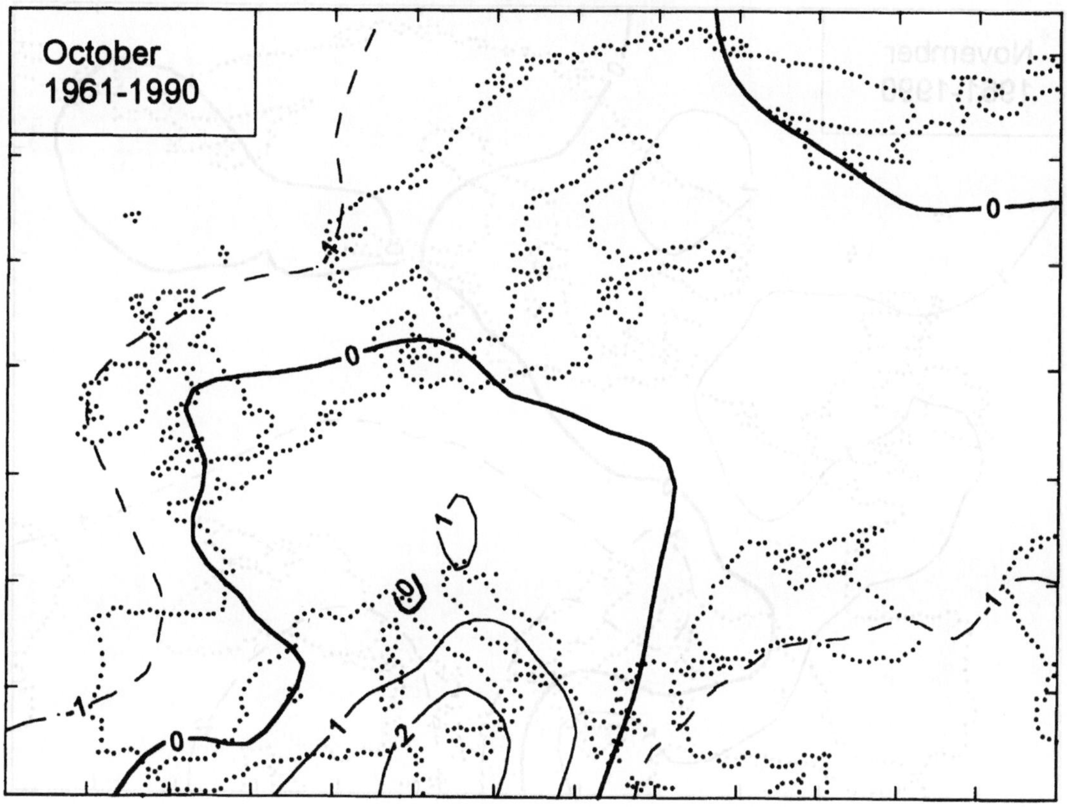

Humidity, linear trend [hPa]

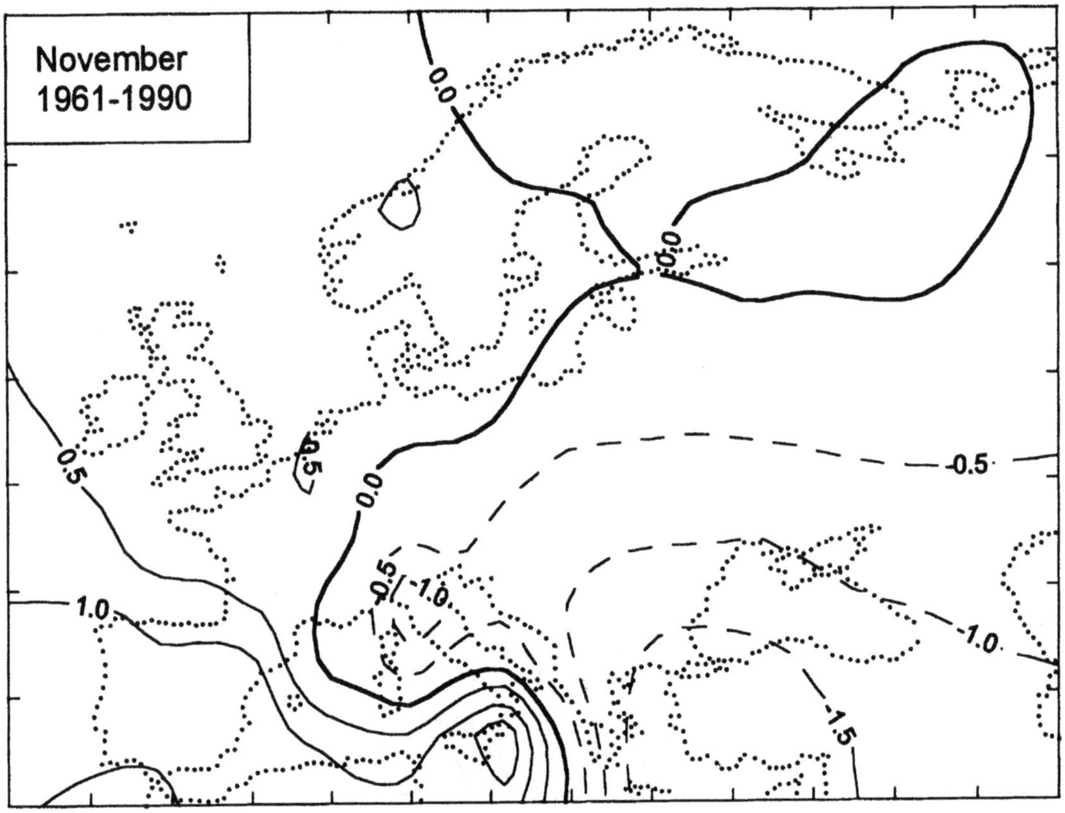

Humidity, trend test

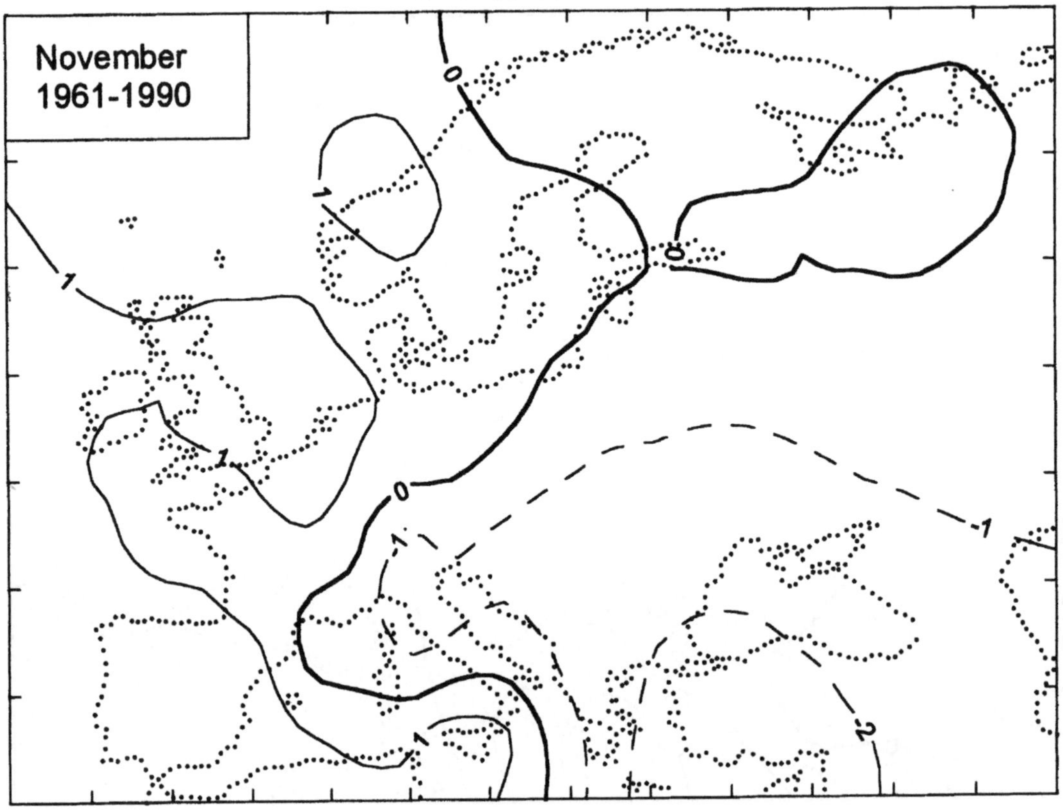

Humidity, linear trend [hPa]

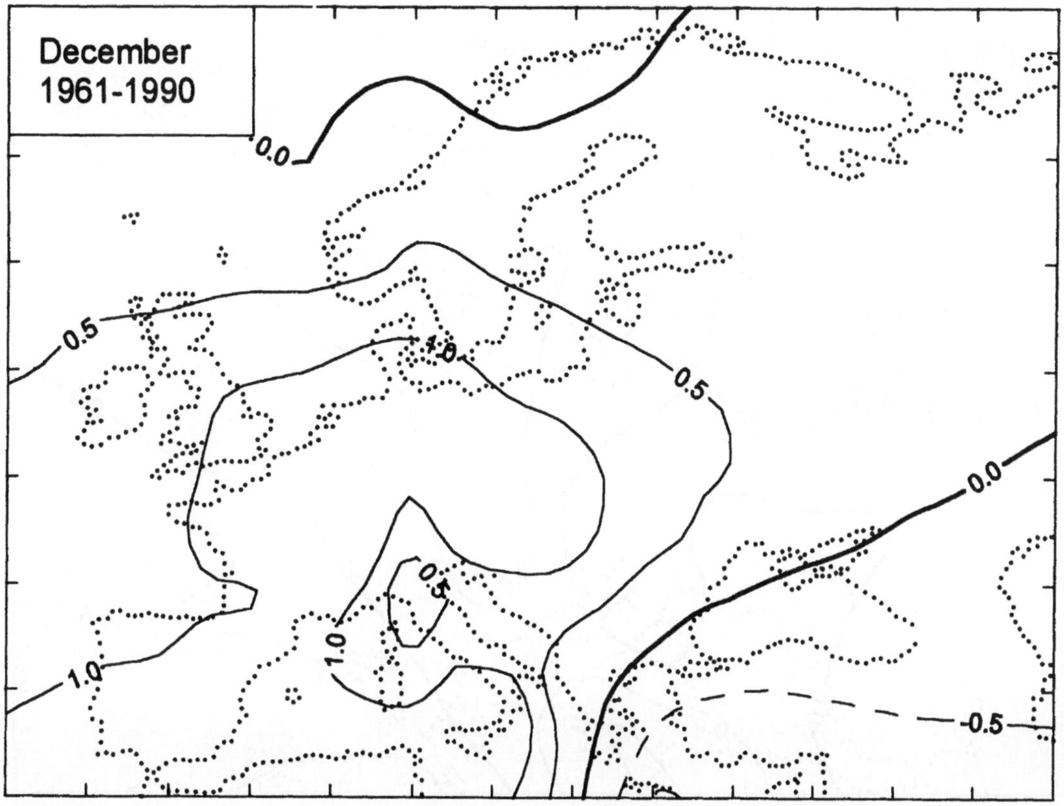

Humidity, trend test

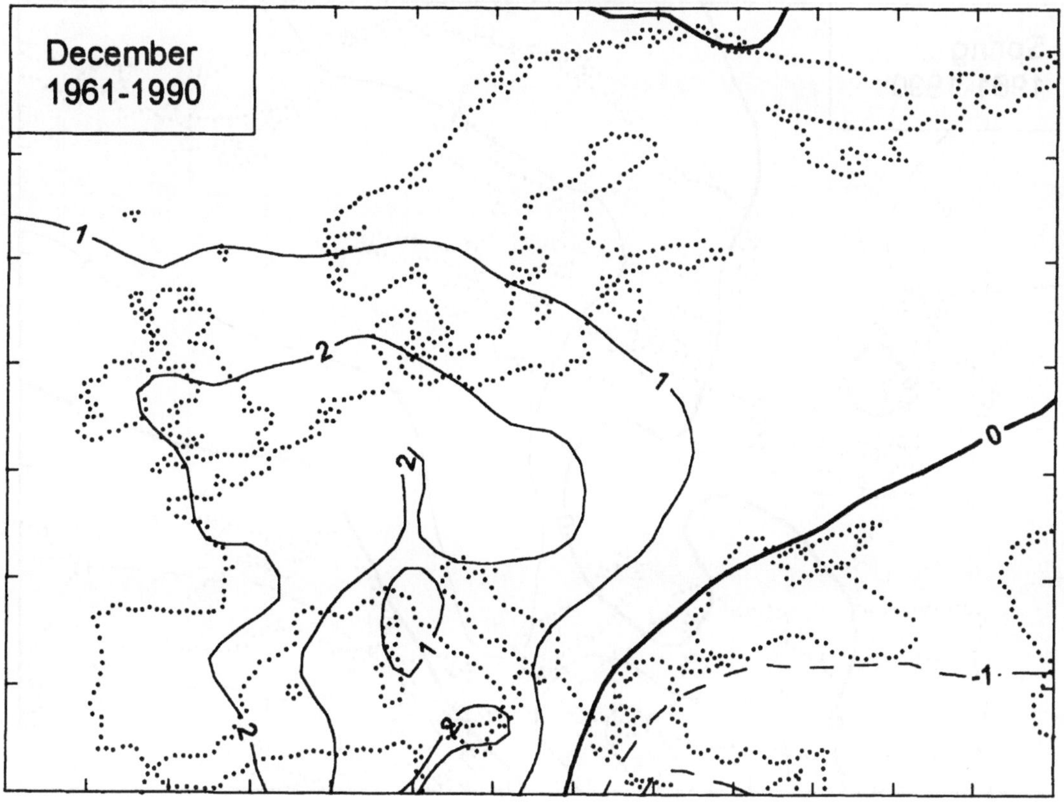

Humidity, linear trend [hPa]

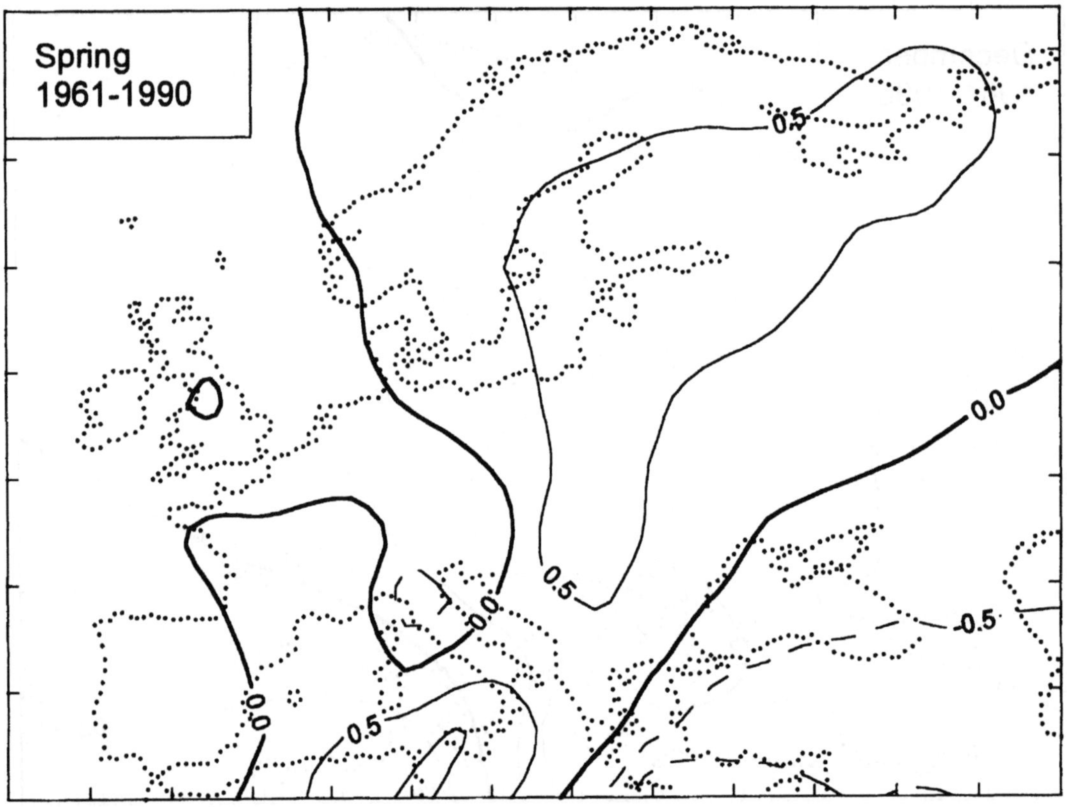

Humidity, trend test

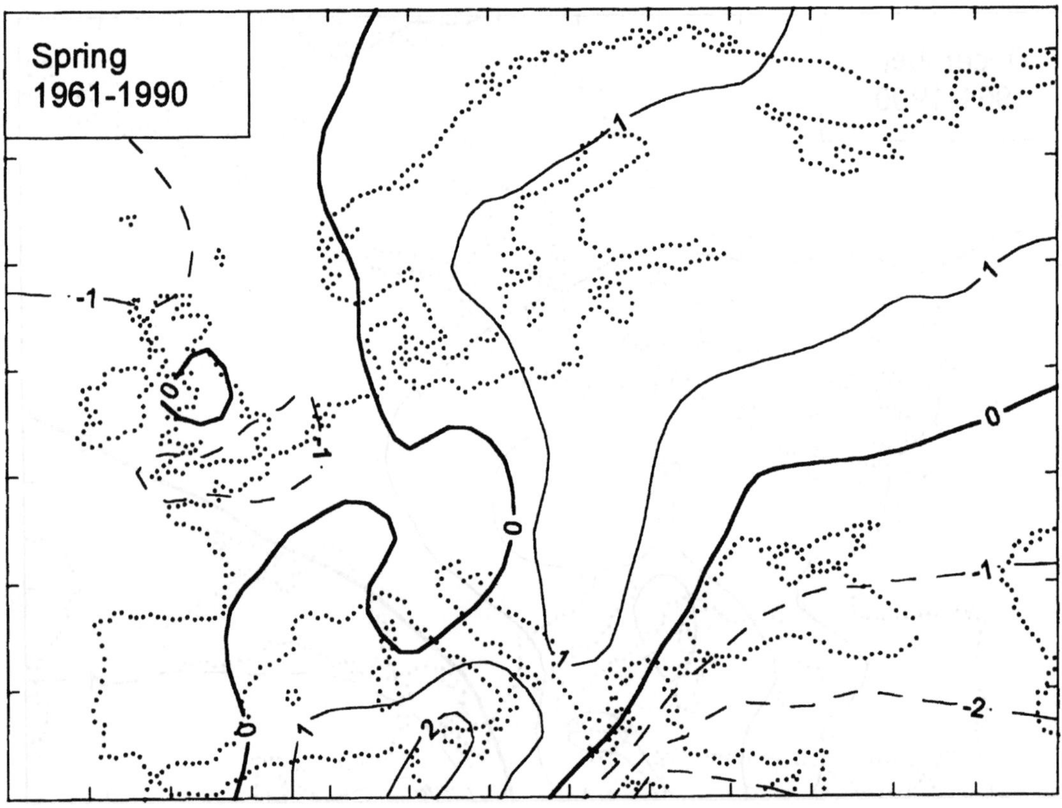

Humidity, linear trend [hPa]

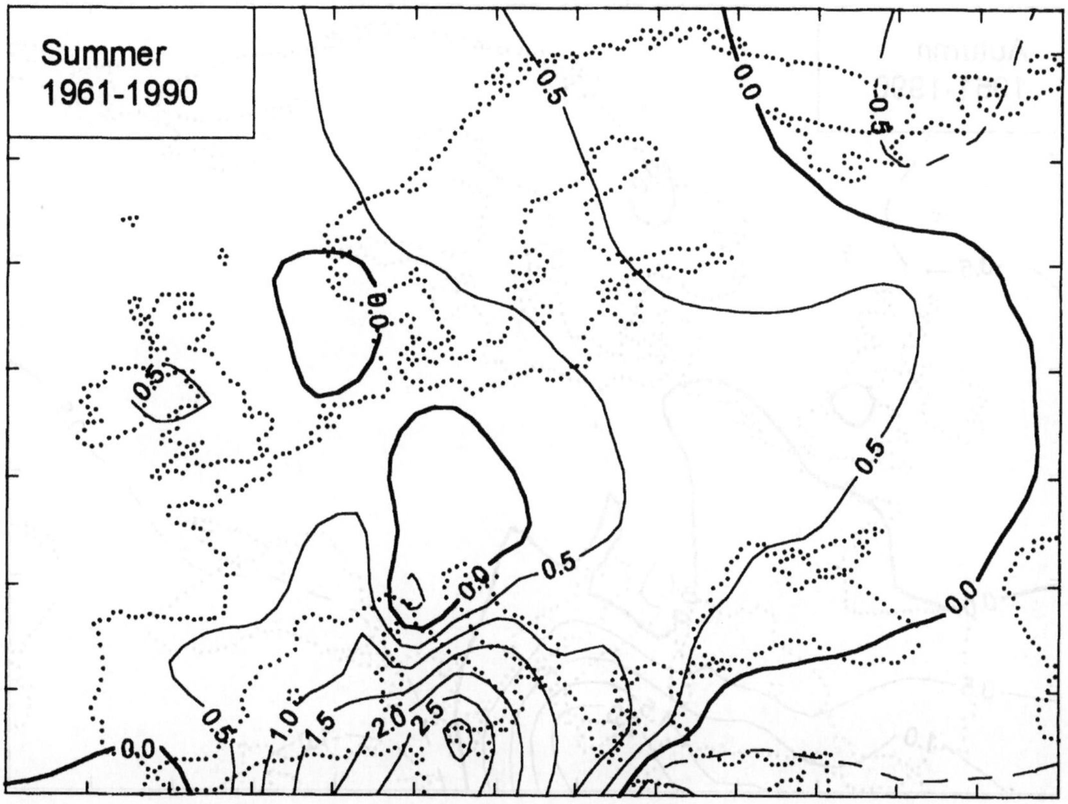

Humidity, trend test

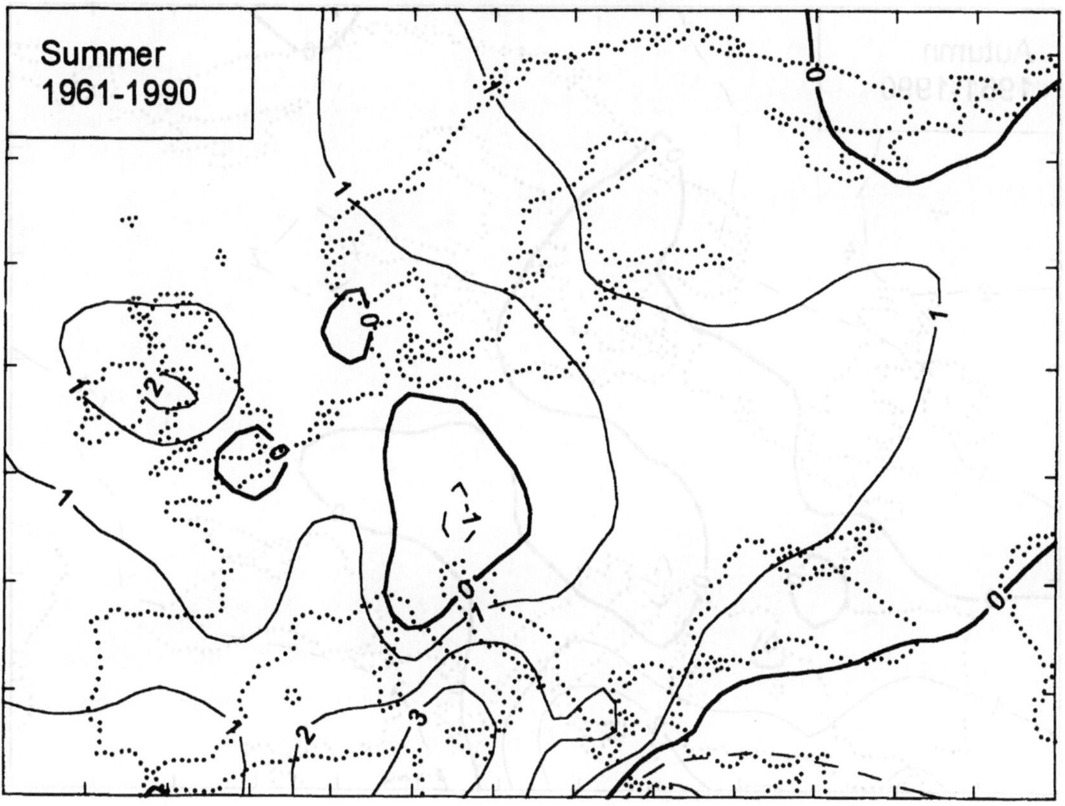

Humidity, linear trend [hPa]

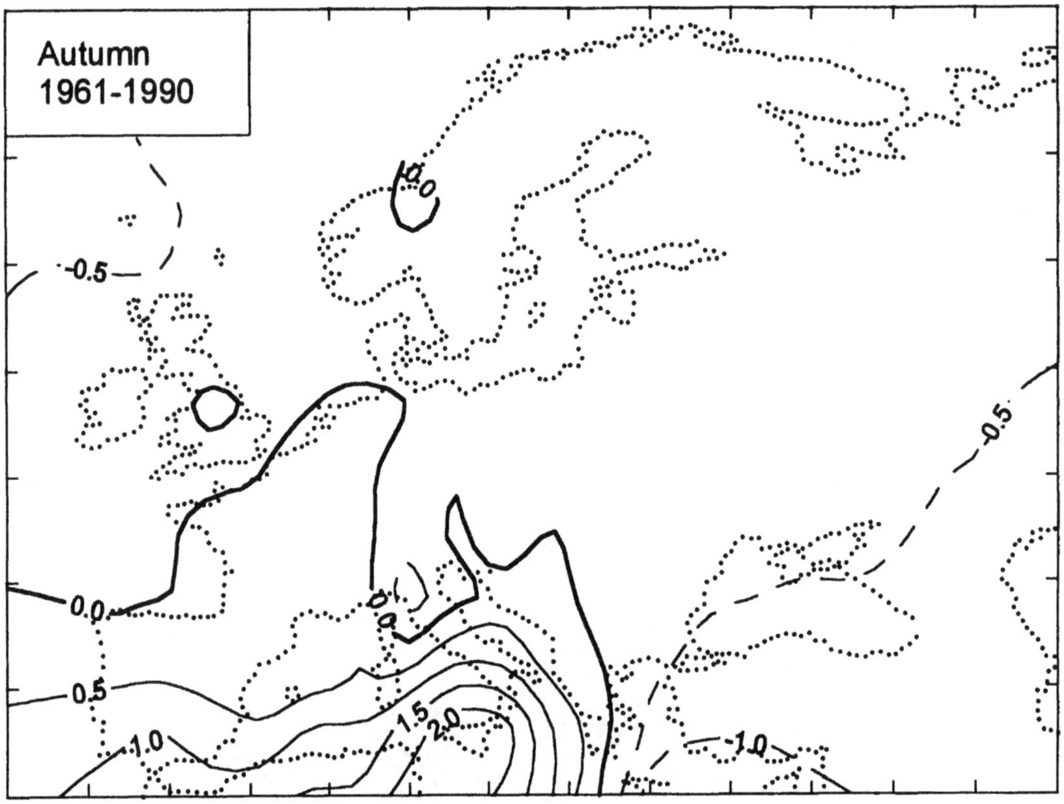

Humidity, trend test

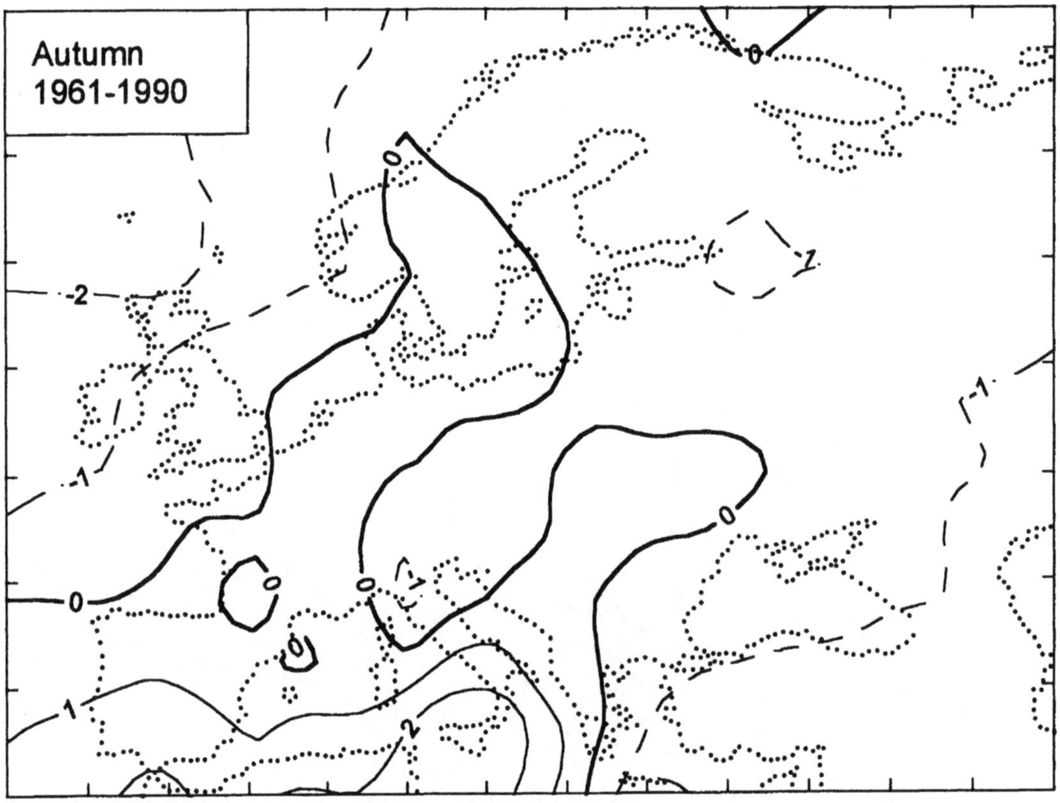

Humidity, linear trend [hPa]

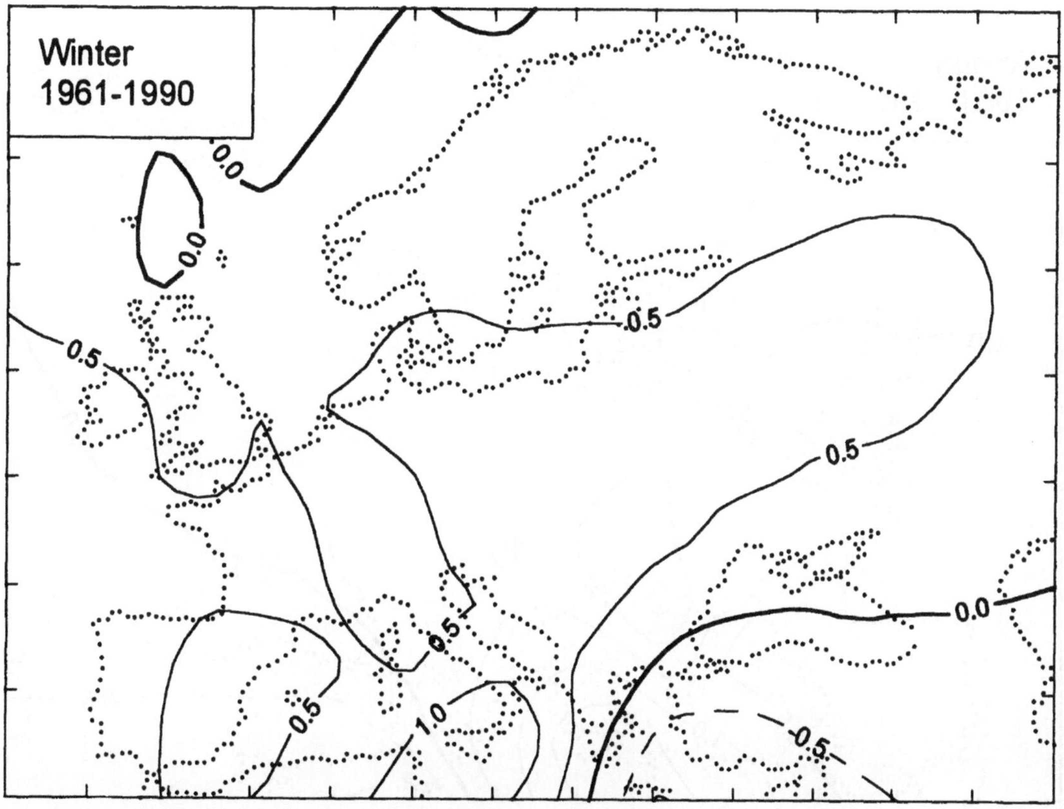

Humidity, trend test

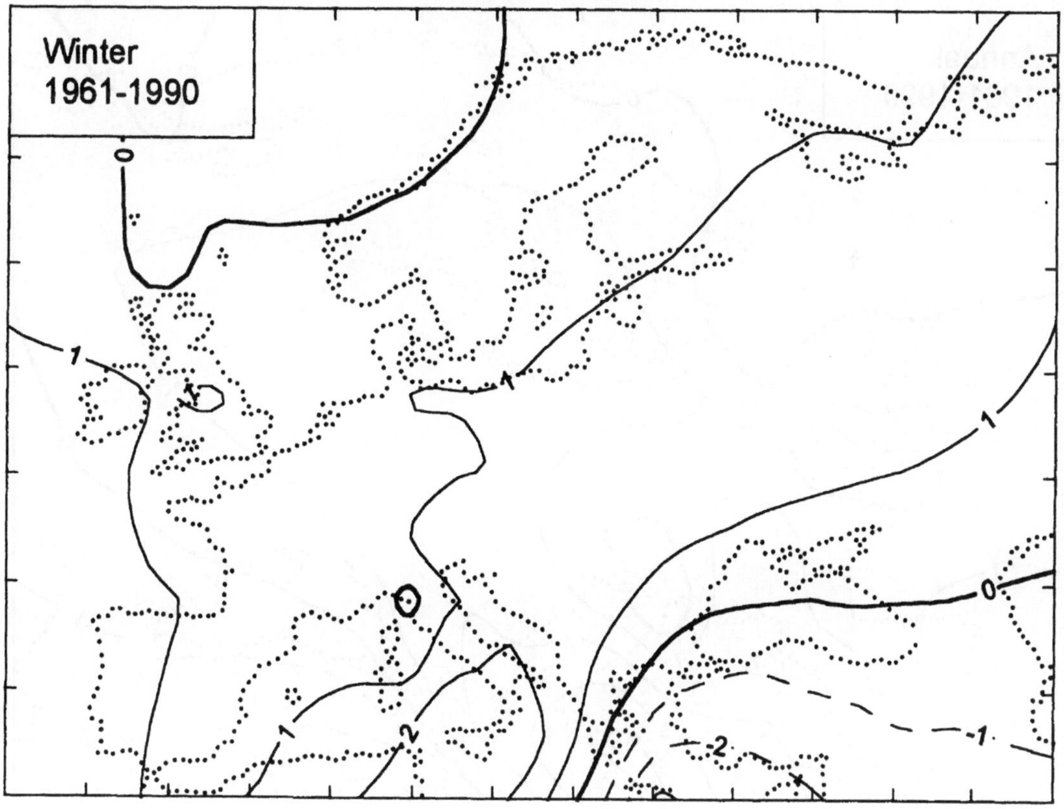

Humidity, linear trend [hPa]

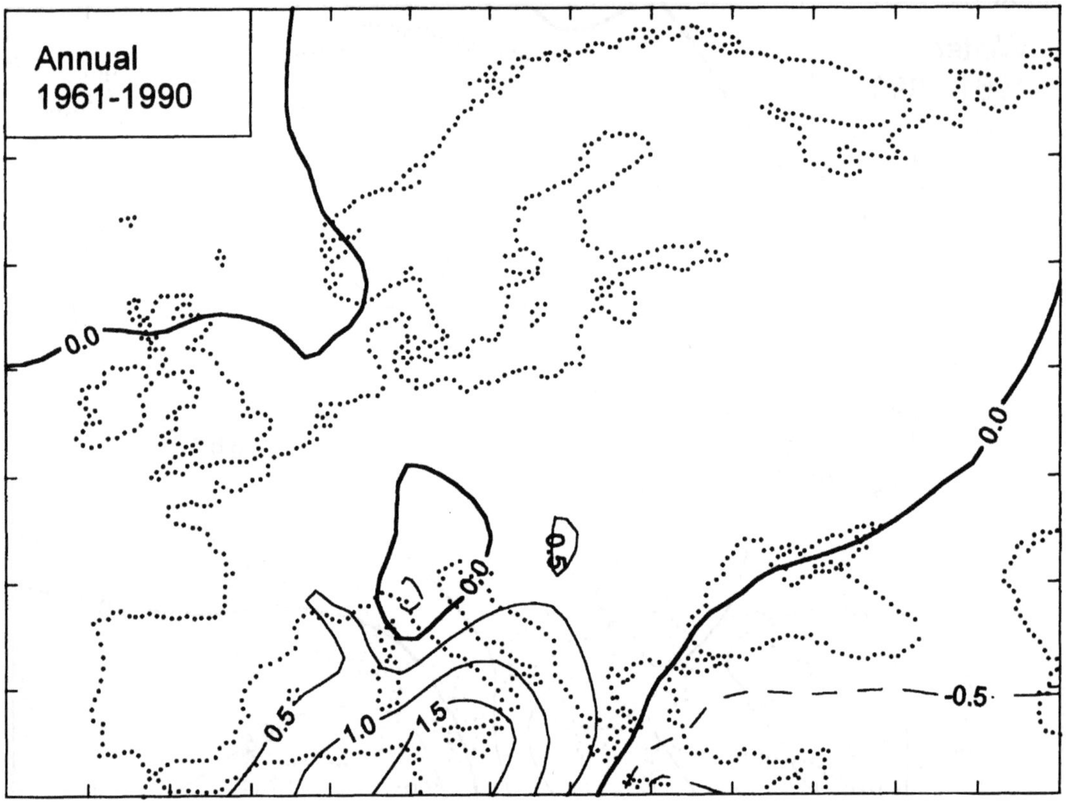

Humidity, trend test

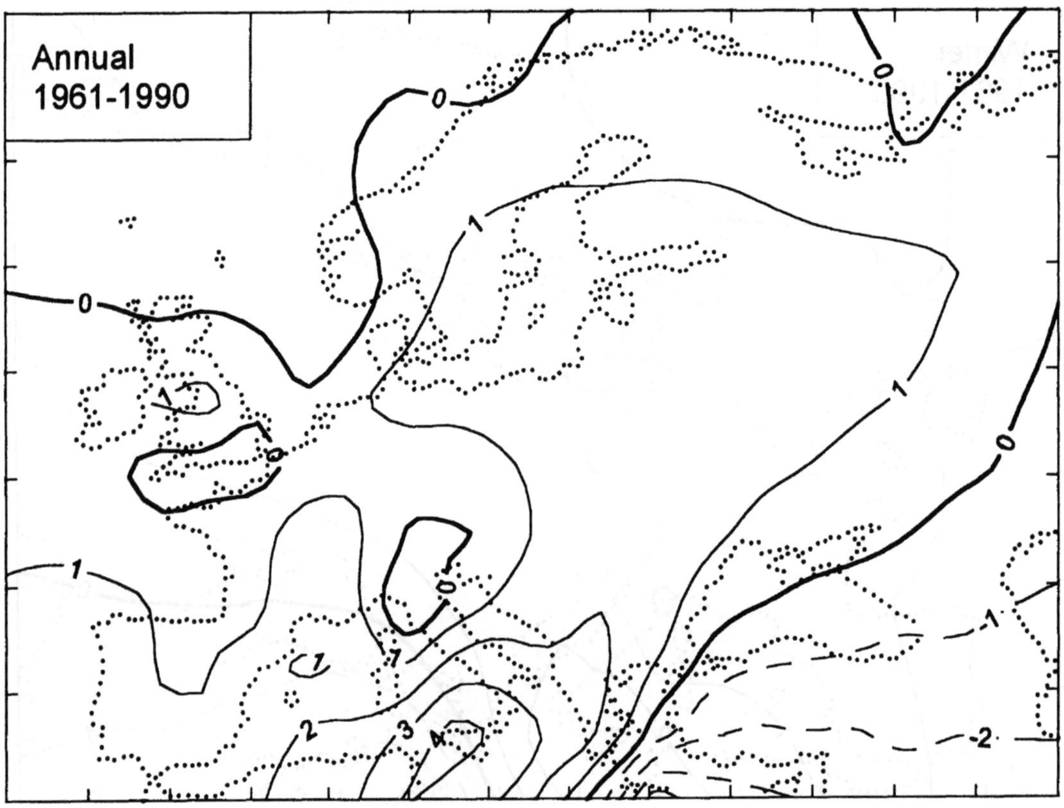

APPENDIX A

Station and related homogeneity list

Remarks:

1) T kor = corrected air temperature time series,
P kor = corrected MSL pressure time series.

2) T = surface air temperature time series,
P = MSL pressure time series,
R = precipitation time series,
H = humidity time series;
1st number = 1891–1990, 2nd number = 1961–1990,
homogeneity classes see Table 3.2. (page 20).

Name of station 1)	WMO-No.	Geographical coordinates in °		Altitude in m	Climate element and homogeneity class 2)			
Aberdeen (T kor)	030910	57.2N	2.2W	59	T 1,1	R	P 4,1	
Ajaccio	077610	41.9N	8.8E	9		R		
Akureyri	040630	65.4N	18.1W	27	T 5,1		P 3,3	
Arhangelsk	225500	64.6N	40.5E	13	T 2,1	R	P 4,5	H
Astrahan	348800	46.4N	48.0E	18	T 2,1	R	P 5,5	
Athinai	167140	38.0N	23.7E	107	T 3,3	R	P 2,1	H
Augsburg	108520	48.4N	10.9E	486		R		
Bad Gleichenberg	112450	46.5N	15.5E	294		R		
Basel	066450	47.6N	7.6E	318	T 3,1	R	P 5,5	
Belfast	039170	54.4N	6.1W	81	T 3,1	R		
Bergen	013160	60.4N	5.3E	44	T 1,1		P 4,1	H
Berlin/Dahlem	103810	52.6N	13.4E	58	T 5,1	R	P 4,1	
Berlin/Tempelhof	103840	52.5N	13.4E	49				H
Bleiberg/Kreuth	112127	46.6N	13.7E	904		R		
Bodoe	011520	67.3N	14.4E	13	T 1,1	R	P 2,1	H
Bordeaux	075100	44.8N	0.7W	61		R		H
Bremen	102240	53.0N	8.5E	3	T 5,5	R		
Brest	071100	48.4N	4.4W	103				H
Brindisi	163200	40.7N	17.9E	10				H
Bucuresti	154220	44.4N	26.1E	82	T 4,3		P 5,4	
Budapest	128400	47.5N	19.0E	129	T 2,3	R	P 1,3	H
Central-England	03.......	52.8N	2.5W	10	T 3,1			

Name of station [1])	WMO-No.	Geographical coordinates in °		Alti-tude in m	Climate element and homogeneity class [2])			
Cherbourg	070210	49.6N	1.5E	138		R		
Clausthal	104480	51.9N	10.4E	563		R		
Cork	039550	51.9N	8.5W	162		R		
De Bilt (P Kor)	062600	52.1N	5.2E	4	T 4,4	R	P 1,1	
Debrecen	128820	47.5N	21.8E	112	T 4,1			
Dijon	072800	47.2N	5.0E	227		R		H
Dresden	104880	51.2N	13.8E	226	T 1,3			
Dublin	039690	53.4N	6.3W	85	T 4,1	R		H
Edinburgh (T kor)	031600	55.9N	3.2W	41	T 3,1	R	P 2,1	
Emden	102030	53.4N	7.2E	5		R		
England and Wales	03........	54.3N	2.3W	150		R		
Erfurt	105540	51.0N	11.0E	323		R		
Essen	104100	51.4N	7.0E	161				H
Feldkirch	111050	47.3N	9.8E	439		R		
Frankfurt/Main	106380	50.0N	8.4E	112	T 5,5		P 5,2	
Frankfurt/Feldbg-47	10638..	50.1N	8.7E	109		R		
Frankfurt/Stadt	10638..	50.1N	8.7E	112		R		
Freudenstadt	108150	48.4N	8.3E	797		R		
Fruholmen	010550	71.1N	24.0E	14			P 5,2	
Funchal	085210	32.6N	16.9W	25			P 5,3	
Geisenheim	106280	50.0N	7.9E	108	T 2,1			
Geneve (T kor)	067050	46.3N	6.2E	416	T 3,3	R	P 5,5	
Genova	161210	44.3N	8.3E	55	T 5,1		P 3,4	
Gibraltar	084950	36.2N	5.4W	5	T 5,1	R	P 5,5	H
Godthab	042500	64.2N	51.7W	20	T 5,5			
Goeteborg	025120	57.8N	11.9E	53		R		
Görlitz	104990	51.2N	15.0E	238		R		
Halle	104798	51.6N	12.0E	96		R		
Hamburg	101470	53.6N	10.0E	16	T 5,2			H
Hamburg/St.Pauli	101480	53.3N	9.6E	14		R		
Hannover	103380	52.5N	9.7E	54		R		H
Haparanda	021960	65.8N	24.2E	6	T 3,1	R	P 2,1	
Helsinki (T kor)	029740	60.4N	24.9E	56	T 2,1	R	P 1,1	
Heraklion	167540	35.2N	25.1E	39				H
Hohenpeißenberg	109620	47.8N	11.0E	977	T 1,1	R		H
Hoofddorp	062327	52.2N	4.4E	0		R		
Innsbruck/Univ.	11120..	47.3N	11.4E	577	T 1,1	R		
Jerusalem	401840	31.8N	35.2E	809	T 5,5	R		
Jokkmock	016988	66.4N	19.5E	260			P 2,1	
Jonkoping	025500	57.7N	14.1E	232		R		
Jyvaskylae	029350	62.4N	25.7E	145	T 5,5	R		

Name of station [1]	WMO-No.	Geographical coordinates in °		Altitude in m	Climate element and homogeneity class [2]			
Kajaani	028970	64.3N	27.6E	136		R		
Kalmar	026720	56.6N	16.1E	15			P 3,1	
Karesuando	020800	68.5N	22.5E	327	T 5,2	R	P 2,1	
Karlsruhe	107270	49.0N	8.4E	116	T 5,1	R	P 4,1	H
Karlstad	024180	59.4N	13.5E	55		R		
Kassel	104380	51.3N	9.5E	233	T 5,5			
Kazan	275950	55.8N	49.2E	64	T 5,1	R	P 3,2	
Kerkyra	166410	39.6N	19.9E	4				H
Kiev	333450	50.4N	30.5E	179	T 3,1	R	P 2,1	H
Klagenfurt	112310	46.6N	14.4E	452	T 3,1	R		
Kleve	103010	51.8N	6.1E	45		R		
Kobenhavn	061860	55.7N	12.0E	9	T 4,1	R	P 3,1	
Köln	10513..	51.0N	7.1E	45		R		
Kornat	112057	46.7N	12.9E	1037		R		
Kremsmünster	110120	48.1N	14.1E	388	T 3,3	R		
Kristianstad	026490	56.0N	14.1E	3	T 5,1	R		
La Coruna	080010	43.4N	8.4W	67				H
Lambach	11........	48.1N	13.5E	360		R		
Landau	107237	49.2N	8.1E	150		R		
Lappeenranta	029580	61.1N	28.2E	106		R		
Limoges	074350	45.9N	1.1E	396		R		
Linz	110100	48.3N	14.3E	263	T 2,2			
Lisboa (P kor)	085350	38.7N	9.2W	95	T 5,2	R	P 5,5	
London/Gatwick	037760	51.2N	0.2W	62				H
London/Greenw/Kew	0377....	51.2N	0.2W	62			P 1,1	
Lübeck	101560	53.9N	10.6E	14		R		
Lugano	067700	46.0N	9.0E	276		R		
Luqa	165970	35.8N	14.5E	91	T 3,4	R		
Luxembourg	065890	49.6N	6.2E	379	T 5,1	R	P 5,5	
Lvov	333930	49.8N	24.0E	325		R		
Lyon	074800	45.6N	4.9E	201	T 3,3		P 3,1	H
Maarianhamina	029700	60.1N	19.9E	4	T 3,1			
Madrid	082220	40.4N	3.7W	667	T 5,2	R		H
Malin Head	039800	55.4N	7.3W	25				H
Manchester	033340	53.4N	2.3W	78		R		H
Marseille	076500	43.5N	5.2E	36	T 5,4	R	P 5,1	H
Messina	164200	38.2N	15.6E	51				H
Metten	107767	48.9N	12.9E	313		R		
Milano/Linate	160800	45.5N	9.2E	103			P 5,3	
Moskva	276120	55.8N	37.5E	156	T 4,1	R	P 3,4	
München	108650	48.1N	11.7E	527	T 3,1	R		H

Name of station [1]	WMO-No.	Geographical coordinates in °		Altitude in m	Climate element and homogeneity class [2]			
Nancy	071800	48.7N	6.2E	217		R		H
Nantes	072220	47.2N	1.6W	27	T 1,3	R	P 2,1	H
Napoli	162890	40.9N	14.3E	3				H
Nice	076900	43.7N	7.2E	10				H
Nimes	076450	43.9N	4.4E	62				H
Nuernberg	107630	49.5N	11.1E	312				H
Odessa	338370	46.5N	30.7E	64	T 4,1	R	P 3,1	
Orland	012410	63.7N	9.6E	7				H
Oslo (T kor)	014920	59.9N	10.7E	96	T 2,1	R	P 2,1	
Ostersund	022260	63.2N	14.5E	366		R		
Oulu	028750	64.9N	25.4E	15	T 2,2	R		
Oxford	03.......	51.7N	1.2W	63	T 1,1	R		
Palma de Mallorca	083010	39.4N	2.7E	8	T 4,3	R	P 3,5	
Paris	071500	48.9N	2.5E	65	T 5,1	R	P 3,1	
Perpignan	077470	43.8N	2.9E	48		R		
Pisa	161580	43.7N	10.4E	1				H
Plymouth	038270	50.4N	4.2W	27	T 5,1			H
Ponta Delgada	085120	37.5N	25.4W	36	T 4,3			
Potsdam	103790	52.4N	13.2E	100	T 1,1		P 1,1	H
Praha (T kor)	115180	50.1N	14.3E	369	T 5,5	R	P 4,2	
Regensburg	107760	49.0N	12.1E	366		R		
Roma	162390	41.7N	12.2E	3	T 5,4	R	P 5,3	
Salzburg	111500	47.8N	13.0E	450				H
Säntis	066800	47.3N	9.4E	2500	T 2,2	R		
Schleswig	100350	54.5N	9.6E	48				H
Shannon	039620	52.7N	8.9W	20		R		H
Sibiu	152600	45.8N	24.2E	444	T 4,3	R	P 5,3	
Sonnblick	111460	47.1N	12.9E	3107	T 2,2	R		
Southampton	038650	50.9N	1.4W	3	T 5,1	R		
St.Petersburg	260630	60.0N	30.4E	4	T 5,1	R	P 3,1	H
Stavanger	014150	58.9N	5.6E	9				H
Stensele	021270	65.0N	17.1E	326	T 3,4	R		
Stockholm	024640	59.4N	18.0E	11	T 4,2			
Stornoway	030260	58.2N	6.3W	13				H
Strasbourg	071900	48.5N	7.7E	154	T 4,1	R		H
Stuttgart/Echterd	107380	48.7N	9.2E	311				H
Stuttgart/Schnarrbg	107390	48.8N	9.1E	311		R		
Stykkisholmur(T kor)	040130	65.1N	22.8W	26	T 4,2	R	P 5,2	
Sulina	153600	45.2N	29.7E	9	T 2,2	R	P 2,2	
Szczecin	122050	53.4N	14.6E	3		R		
Tampere	029440	61.5N	23.7E	85	T 5,1	R		

Name of station 1)	WMO-No.	Geographical coordinates in °		Alti-tude in m	Climate element and homogeneity class 2)			
Tbilisi	375490	41.7N	44.9E	490	T 3,3	R	P 3,4	
Thessaloniki	166220	40.5N	23.0E	4	T 5,5			
Thorshavn	060110	62.0N	6.8W	24	T 5,2	R	P 4,1	
Toulouse	076300	43.6N	1.4E	153	T 5,1	R		
Trapani	164290	37.9N	12.5E	14				H
Trier/Petrisberg	106090	49.8N	6.7E	273		R		
Trieste	161100	45.6N	13.7E	20	T 3,3	R	P 3,1	
Trondheim	012710	63.4N	10.4E	58	T 3,4		P 2,1	
Tunis	607150	36.5N	10.1E	4		R		
Uccle	064470	50.8N	4.4E	104	T 3,3	R	P 3,4	H
Uppsala	024620	59.9N	17.6E	15	T 1,1	R	P 4,4	H
Vaasa	029110	63.1N	21.7E	4	T 3,4	R		
Valentia	039530	51.9N	10.3W	14	T 5,1	R	P 2,2	
Vardoe	010980	70.4N	31.1E	10	T 3,1		P 4,1	H
Vestmannaeyj.(T kor)	040480	63.4N	20.3W	122	T 5,3	R	P 4,3	
Visby	025900	57.6N	18.4E	47		R		
Vologda	270370	59.3N	39.9E	118				H
Warszawa	123750	52.2N	21.0E	107	T 5,1	R	P 5,1	
Wien/Hohe Warte	110350	48.3N	16.4E	209	T 3,1	R	P 3,1	H
Wroclaw	124240	51.1N	16.6E	121	T 5,1	R		
Zagreb	131290	45.8N	16.0E	163	T 5,3	R		
Zaragoza	081600	41.7N	1.0W	258				H
Zell am See	111430	47.4N	12.8E	753		R		
Zugspitze	109610	47.4N	11.0E	2962				H
Zürich (T kor)	066600	47.4N	8.6E	569	T 3,1	R	P 5,5	

APPENDIX B

Some aspects of spectral variance

In the following a few selected results of variance (power) spectrum analysis are presented where in all cases the time series under consideration are annual data 1891–1990. We used the autocorrelation spectral analysis (ASA) method as described by Mitchell *et al.* (1966) and many other authors in their textbooks on time series analysis (e.g. Båth, 1974; Box and Jenkins, 1970; Jenkins and Watts, 1968; Olberg and Rákószik, 1984; Schönwiese, 1992; in some of these publications alternative techniques like fast Fourier transform, FFT, or maximum entropy spectral analysis, MESA, are discussed, too). These spectral estimate techniques may also be applied moving in time (Schönwiese, 1987, 1992).

The basic idea of ASA is the Fourier transformation of the autocorrelation (or autocovariance) function

$$r_A = \frac{\sum_{i=1}^{n-\tau} a_i' \cdot a_{i+\tau}'}{\sqrt{\sum_{i=1+\tau}^{n} a_i'^2 \cdot \sum_{i=1}^{n-\tau} a_i'^2}}, \tag{B.1}$$

where $a_i = a_i(t_i)$ is any time series, i=1, ..., n; $\Delta t = t_{i+1} - t_i$; a_i are the deviations from the arithmetic mean and $\tau = k\,\Delta t$ is the time lag (k=0,1, ..., M, M = maximum lag). The generalized theory of the spectrum estimation in the case of discrete data a_i and finite observation period leads to (Mitchell *et al.*, 1966)

$$Sp(0) = \frac{1}{2M}\left[s_A^{\,2}(0) + \sum_{k=1}^{M-1} D(k)\, s_A^{\,2}(k)\right] \quad \text{for } h = 0; \tag{B.2}$$

$$Sp(h) = \frac{1}{M}\left[s_A^{\,2}(0) + \sum_{k=1}^{M-1} D(k)\, s_A^{\,2}(k) \cos\frac{\pi h k}{M}\right] \quad \text{for} \begin{cases} h > 0 \\ h < M \end{cases}; \tag{B.3}$$

$$Sp(M) = \frac{1}{2M}\left[s_A^{\,2}(0) + \sum_{k=1}^{M-1} D(k)\, s_A^{\,2}(k)\, (-1)^k\right] \quad \text{for } h = M. \tag{B.4}$$

where D(k) is a filter function, usually

$$D(k) = \begin{cases} 1/2\,(1+\cos \pi k/M) & \text{if } 0 \le k \le M \\ 0 & \text{if } -\infty \le k \le 0 \text{ and } M < k < +\infty \end{cases}, \tag{B.5}$$

where h = 0, 1, ..., M. Because of frequencies $f_h = h/2M\Delta t$ and period = $1/f$ the extreme spectral estimates relate to $f_{min} = 0$ or $T_{max} = 2M\Delta t$, $f_{max} = 1/2\Delta t$ or $T_{min} = 2\Delta t$. In simple words, a variance spectrum reveals the disintegration of a time series variance s^2 in respect to frequencies f_h or periods $T_n = 1/f_h$, respectively. Particular spectrum peaks point to cyclical variance; the long-term residuum (variance at f_{min}, T_{max}) indicates trends or (relatively) very low-frequency components.

The background spectrum may be white (W, theoretically equal variance contributions to all frequencies, typically for a random process) or, in the case of significant autocorrelations, red (R, variance increases with period T) where, in the latter case, often the Markow type of red noise is assumed (Mitchell *et al.*, 1966):

$$r_A(\tau = k\Delta t) = r_A{}^k(\tau = 1\Delta t);\ k = 2, 3, \ldots, M; \tag{B.6}$$

$$Sp_R\left(\tfrac{k}{M}\right) = Sp_W \frac{1 - r_1{}^2}{1 + r_1{}^2 - 2\,r_1 \cos\left(\tfrac{\pi k}{M}\right)}, \tag{B.7}$$

where $r_1 = r_A(\tau = 1\Delta t)$. Thus, the decision W or R is based on a non-significant or significant value r_1. In the case of statistically significant spectral peaks, these should differ from white noise W or red noise R, respectively, so that certain confidence levels cl are exceeded. Usually a χ^2 test is used, degrees of freedom $\Phi = (2n - M/2) / M$, where the values W or R, respectively are multiplied by χ^2/Φ.

In the figures presented here, concerning temperature, precipitation, pressure and humidity, the variance spectra $s^2(f)$ are plotted in a linear way. W or R, respectively, and the confidence levels cl 80 (80 % confidence or 0.2 error probability), cl 90 (90 % confidence or 0.1 error probability), etc. are indicated by dotted lines. Finally pronounced spectral peaks are specified by their period valued in years.

In most but not all cases the background spectra appear to be white. However, if the considered time interval (here 100 yr) is extended, there is a tendency to red background spectra as demonstrated here in the case of Hohenpeissenberg temperature (see Figures 1.1b). Often the long-term residuum variance is confident or even highly confident (e.g. temperature Bergen and Lisboa, air pressure London), but not in case of the precipitation series. Also in the case of confident spectral peaks there are some similarities concerning temperature, pressure, and humidity: relatively frequent peaks occur at 2.1–2.3 yr (quasi-biennal oscillation, QBO), 3.2–3.4 yr, around 5, 8, and 10–14 yr, whereas in the case of precipitation, in addition to QBO, a roughly 4 yr peak often occurs. The spectra of temperature, Lisboa, and humidity, Hohenpeissenberg, are dubious due to possible data inhomogeneities or data errors. Finally, note that these are just a few examples of time series spectral analysis in climatology. A much more comprehensive collection of results is available, for example, by Schönwiese *et al.* (1990).

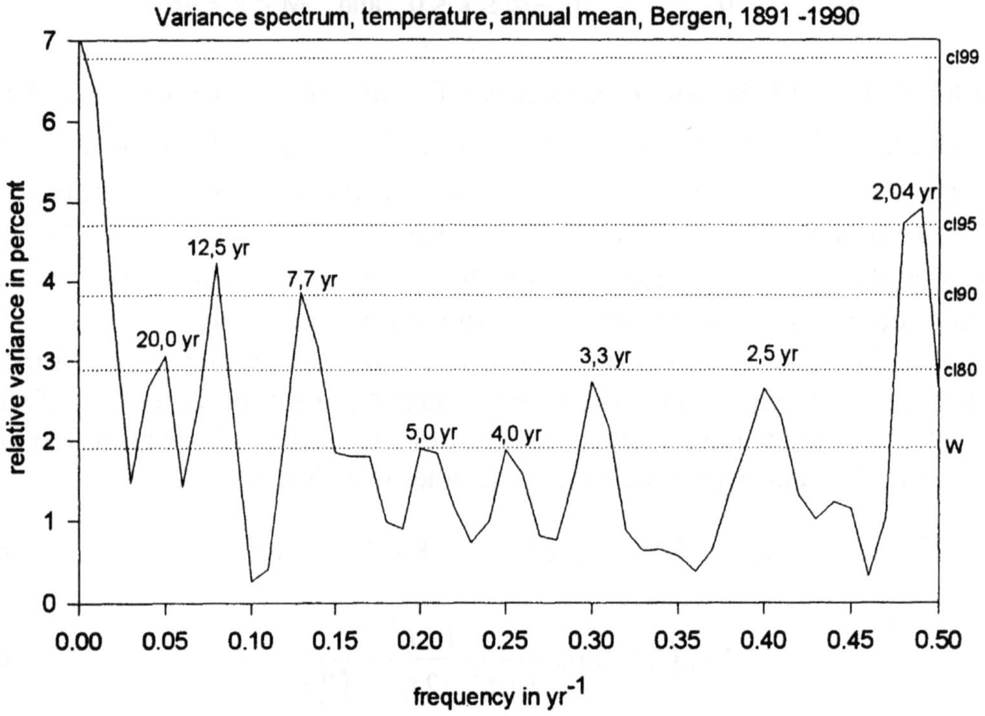
Variance spectrum, temperature, annual mean, Bergen, 1891 -1990
relative variance in percent
frequency in yr-1
cl99
cl95
cl90
cl80
W
2,04 yr
12,5 yr
7,7 yr
20,0 yr
3,3 yr
2,5 yr
5,0 yr
4,0 yr
0.00
0.05
0.10
0.15
0.20
0.25
0.30
0.35
0.40
0.45
0.50

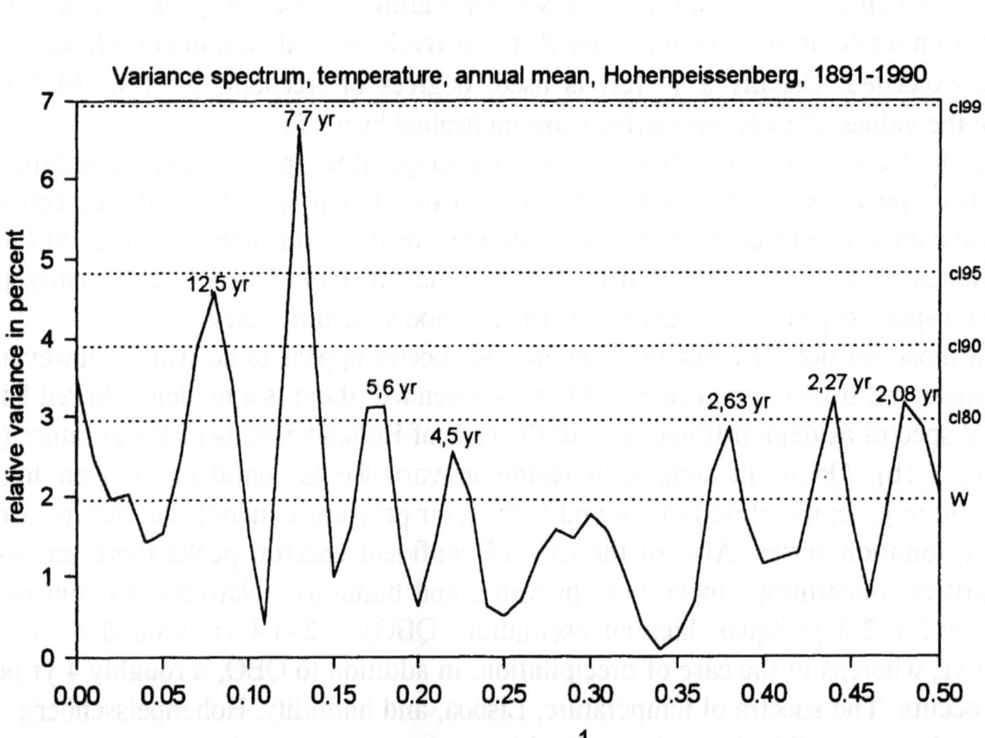
Variance spectrum, temperature, annual mean, Hohenpeissenberg, 1891-1990
relative variance in percent
frequency in yr-1
cl99
cl95
cl90
cl80
W
7,7 yr
12,5 yr
5,6 yr
4,5 yr
2,63 yr
2,27 yr
2,08 yr
0.00
0.05
0.10
0.15
0.20
0.25
0.30
0.35
0.40
0.45
0.50

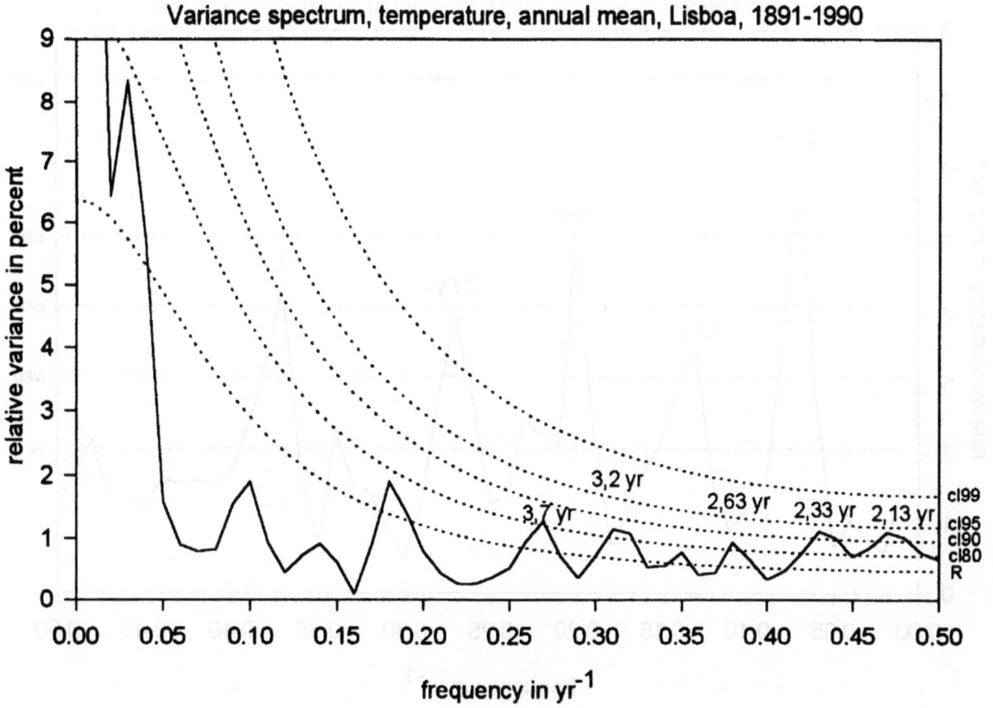
Variance spectrum, temperature, annual mean, Lisboa, 1891-1990
relative variance in percent
frequency in yr-1
3,2 yr
3,7 yr
2,63 yr
2,33 yr
2,13 yr
cl99
cl95
cl90
cl80
R
0.00
0.05
0.10
0.15
0.20
0.25
0.30
0.35
0.40
0.45
0.50

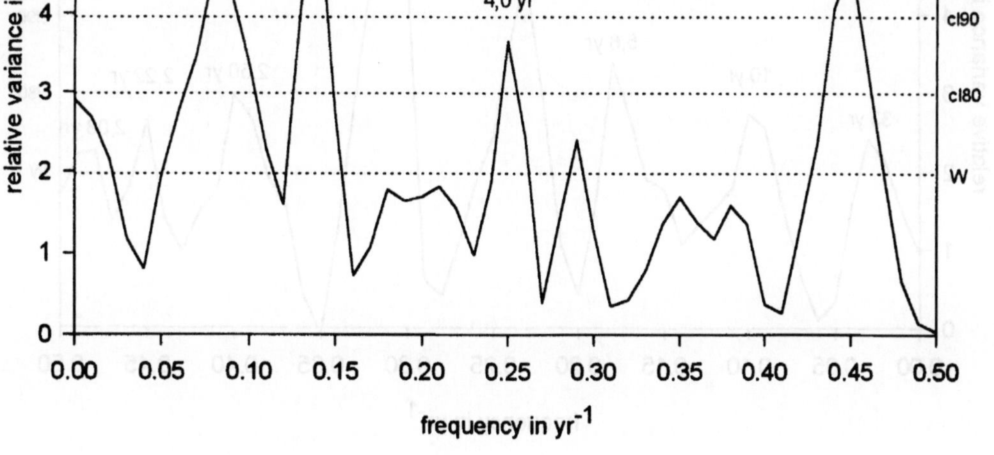
Variance spectrum, temperature, annual mean, Sulina, 1891-1990
relative variance in percent
frequency in yr-1
12,5 yr
7,1 yr
4,0 yr
2,22 yr
cl99
cl95
cl90
cl80
W
0.00
0.05
0.10
0.15
0.20
0.25
0.30
0.35
0.40
0.45
0.50

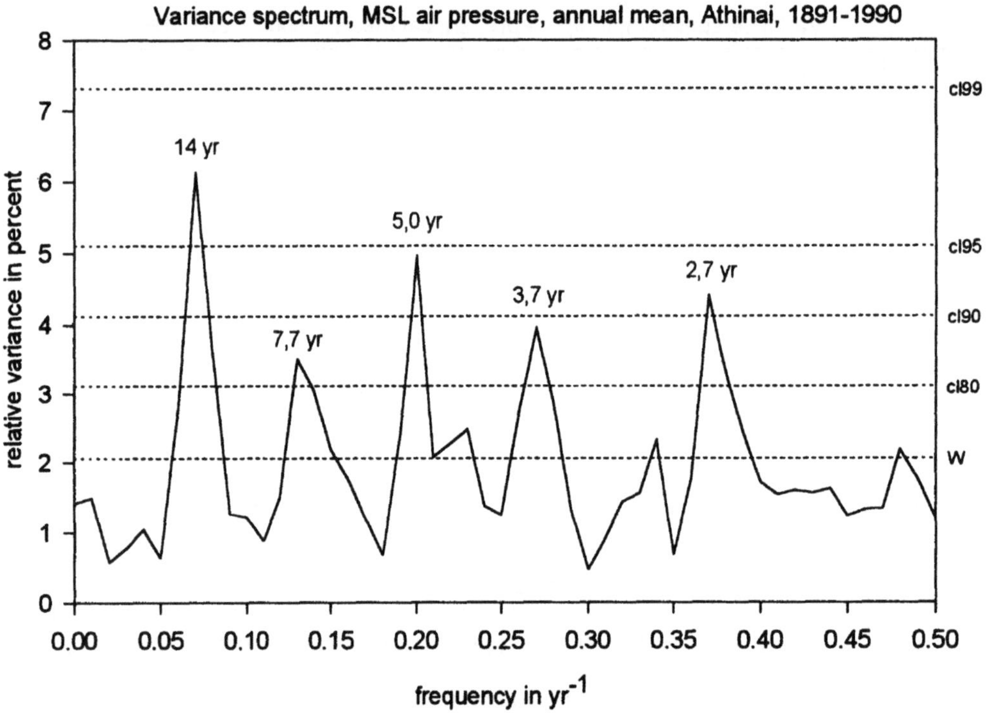
Variance spectrum, MSL air pressure, annual mean, Athinai, 1891-1990
relative variance in percent
frequency in yr-1
14 yr
7,7 yr
5,0 yr
3,7 yr
2,7 yr
cl99
cl95
cl90
cl80
W
0
1
2
3
4
5
6
7
8
0.00
0.05
0.10
0.15
0.20
0.25
0.30
0.35
0.40
0.45
0.50

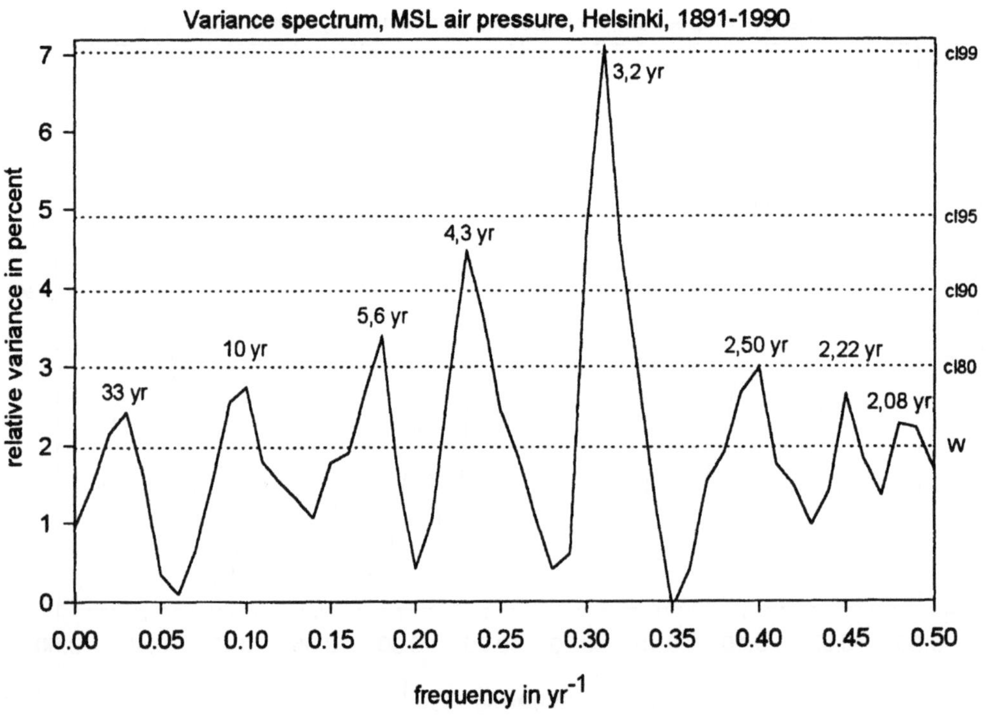
Variance spectrum, MSL air pressure, Helsinki, 1891-1990
relative variance in percent
frequency in yr-1
33 yr
10 yr
5,6 yr
4,3 yr
3,2 yr
2,50 yr
2,22 yr
2,08 yr
cl99
cl95
cl90
cl80
W
0
1
2
3
4
5
6
7
0.00
0.05
0.10
0.15
0.20
0.25
0.30
0.35
0.40
0.45
0.50

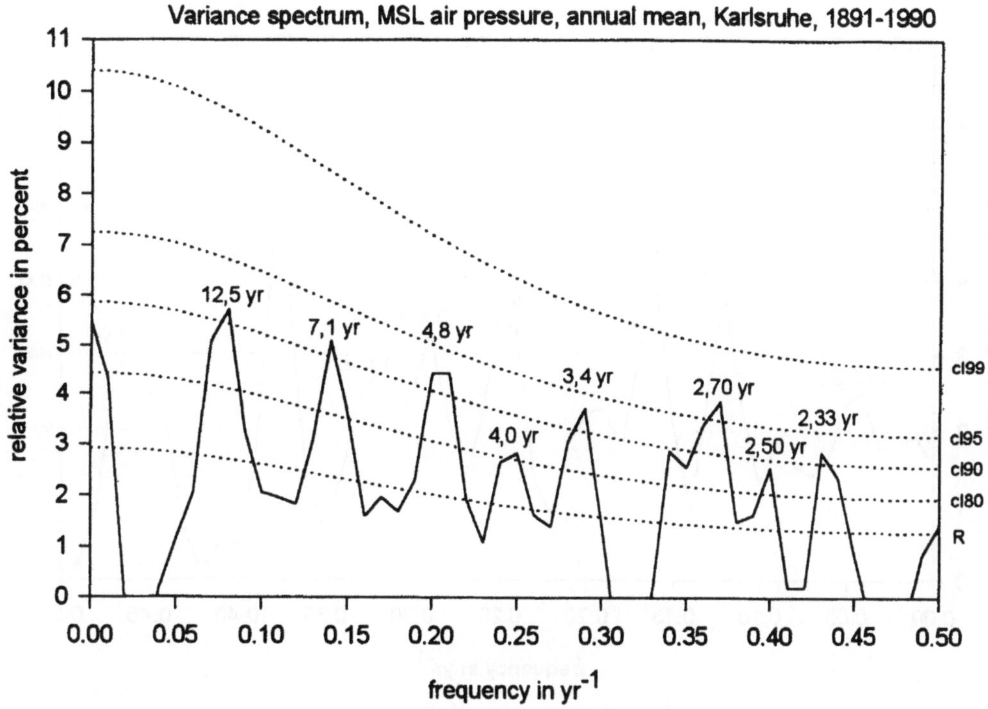
Variance spectrum, MSL air pressure, annual mean, Karlsruhe, 1891-1990
relative variance in percent
frequency in yr-1
12,5 yr
7,1 yr
4,8 yr
3,4 yr
2,70 yr
2,33 yr
4,0 yr
2,50 yr
cl99
cl95
cl90
cl80
R
0.00
0.05
0.10
0.15
0.20
0.25
0.30
0.35
0.40
0.45
0.50

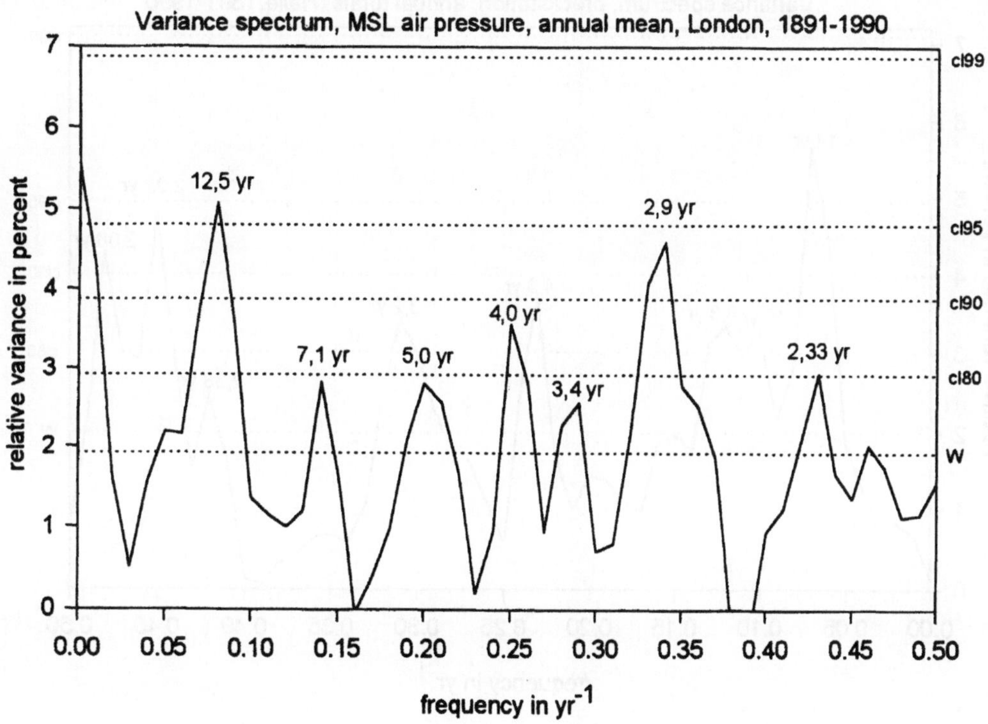
Variance spectrum, MSL air pressure, annual mean, London, 1891-1990
relative variance in percent
frequency in yr-1
12,5 yr
2,9 yr
4,0 yr
7,1 yr
5,0 yr
3,4 yr
2,33 yr
cl99
cl95
cl90
cl80
W
0.00
0.05
0.10
0.15
0.20
0.25
0.30
0.35
0.40
0.45
0.50

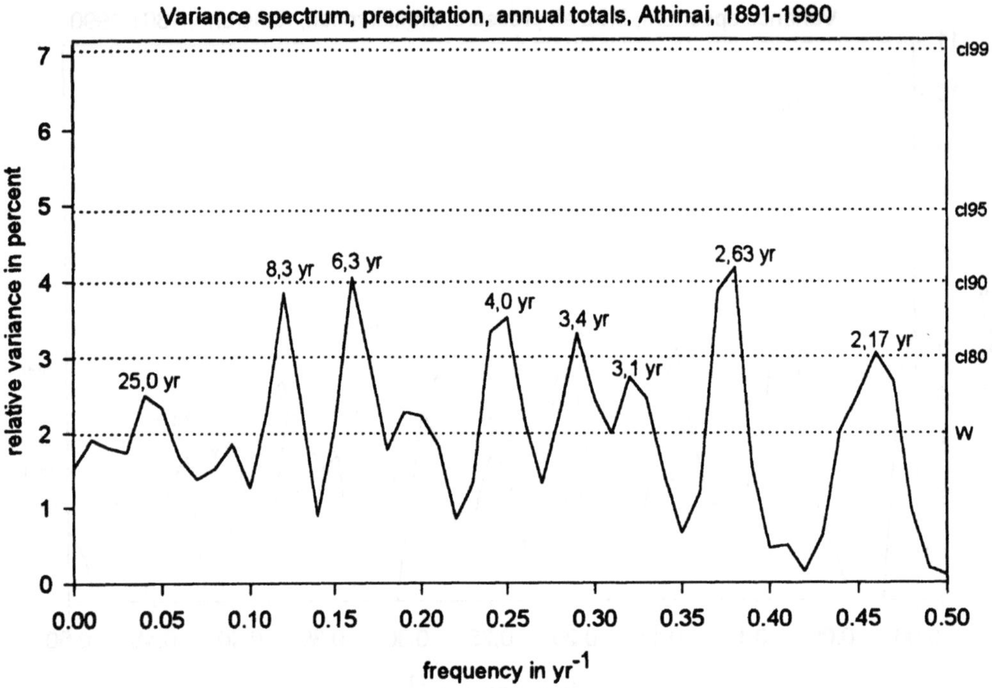
Variance spectrum, precipitation, annual totals, Athinai, 1891-1990
relative variance in percent
frequency in yr-1
cl99
cl95
cl90
cl80
W
25,0 yr
8,3 yr
6,3 yr
4,0 yr
3,4 yr
3,1 yr
2,63 yr
2,17 yr
0.00
0.05
0.10
0.15
0.20
0.25
0.30
0.35
0.40
0.45
0.50
0
1
2
3
4
5
6
7

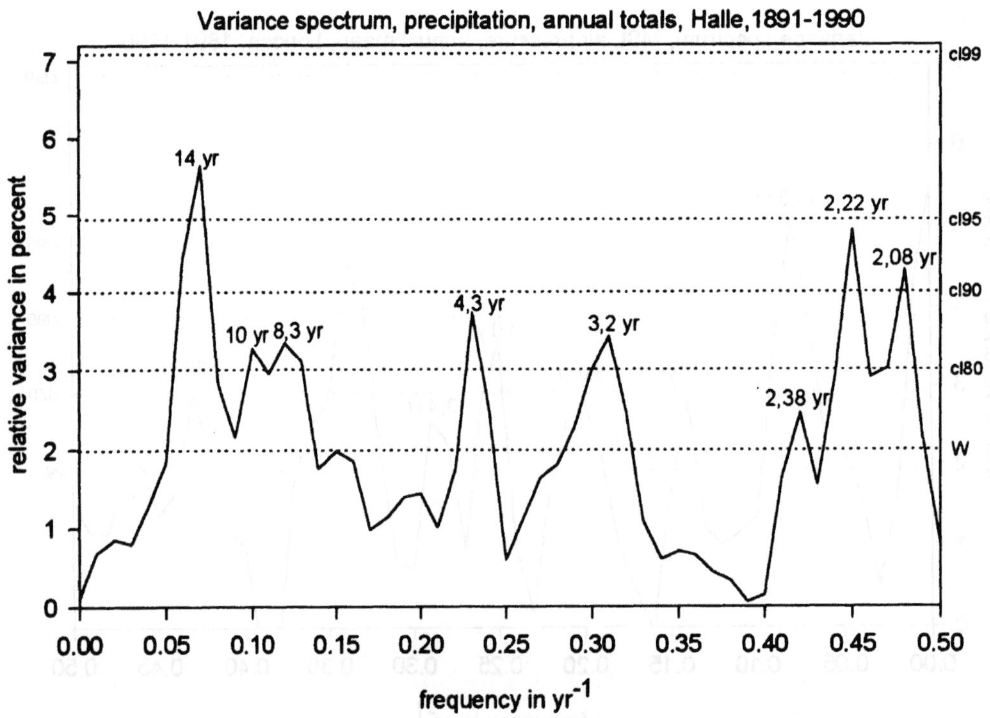
Variance spectrum, precipitation, annual totals, Halle,1891-1990
relative variance in percent
frequency in yr-1
cl99
cl95
cl90
cl80
W
14 yr
10 yr
8,3 yr
4,3 yr
3,2 yr
2,38 yr
2,22 yr
2,08 yr
0.00
0.05
0.10
0.15
0.20
0.25
0.30
0.35
0.40
0.45
0.50
0
1
2
3
4
5
6
7

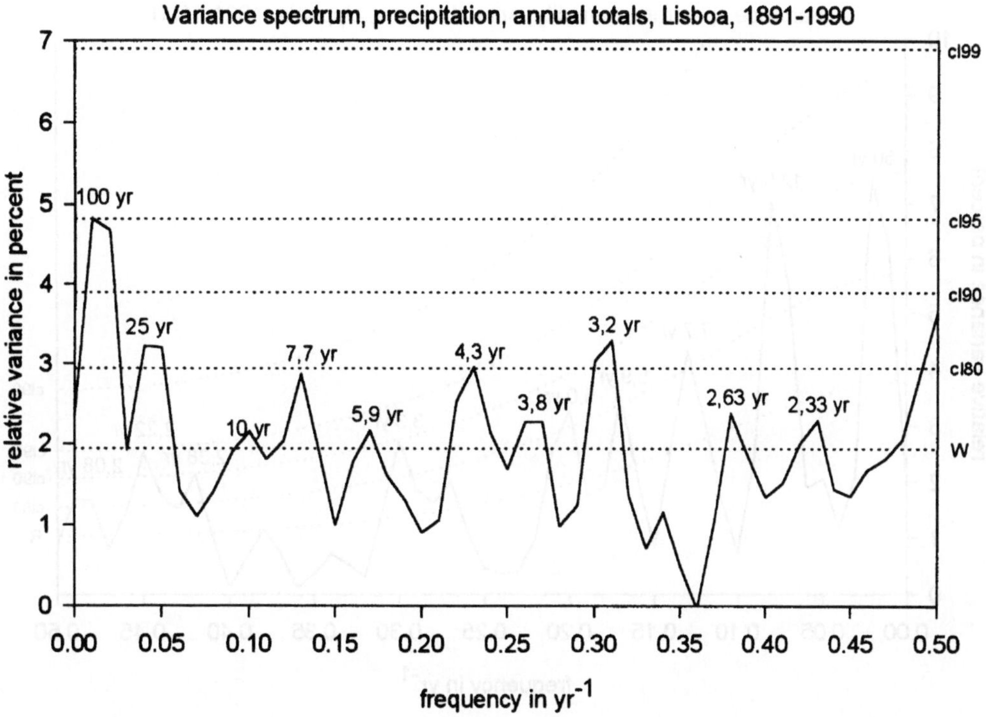
Variance spectrum, precipitation, annual totals, Lisboa, 1891-1990
relative variance in percent
frequency in yr-1
cl99
cl95
cl90
cl80
W
100 yr
25 yr
10 yr
7,7 yr
5,9 yr
4,3 yr
3,8 yr
3,2 yr
2,63 yr
2,33 yr

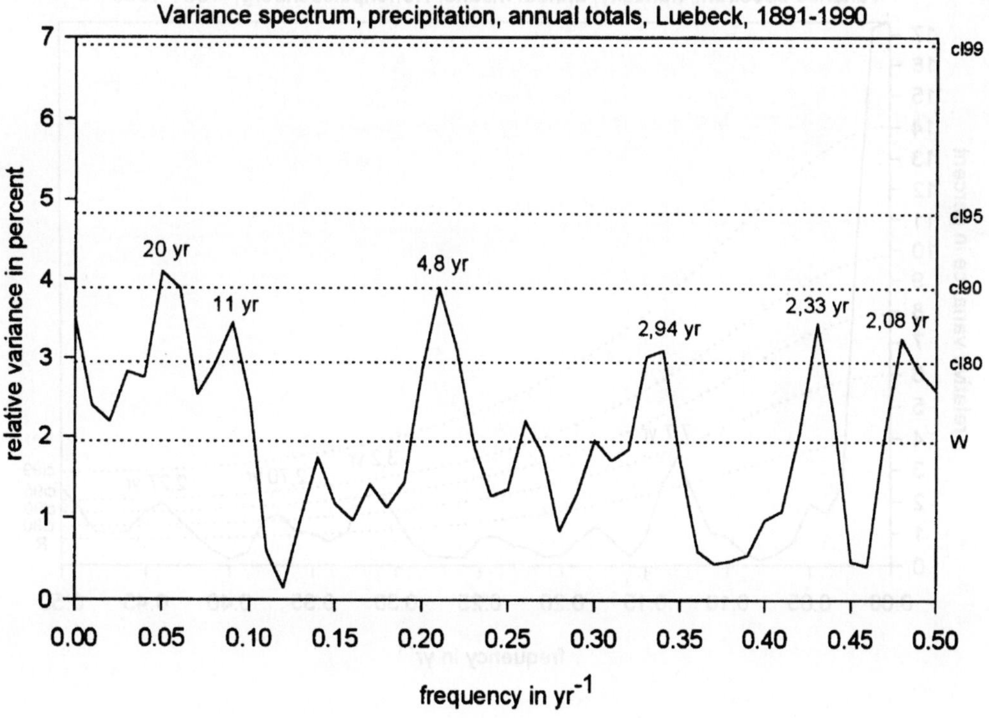
Variance spectrum, precipitation, annual totals, Luebeck, 1891-1990
relative variance in percent
frequency in yr-1
cl99
cl95
cl90
cl80
W
20 yr
11 yr
4,8 yr
2,94 yr
2,33 yr
2,08 yr

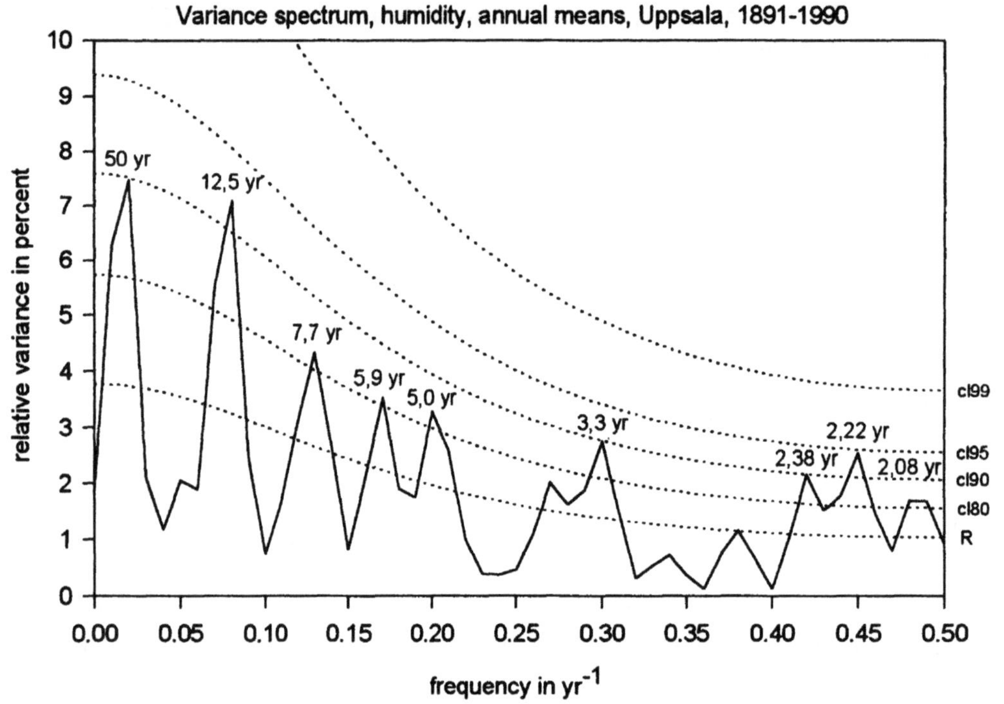

Variance spectrum, humidity, annual means, Uppsala, 1891-1990
relative variance in percent
frequency in yr^{-1}
50 yr
12,5 yr
7,7 yr
5,9 yr
5,0 yr
3,3 yr
2,38 yr
2,22 yr
2,08 yr
cl99
cl95
cl90
cl80
R

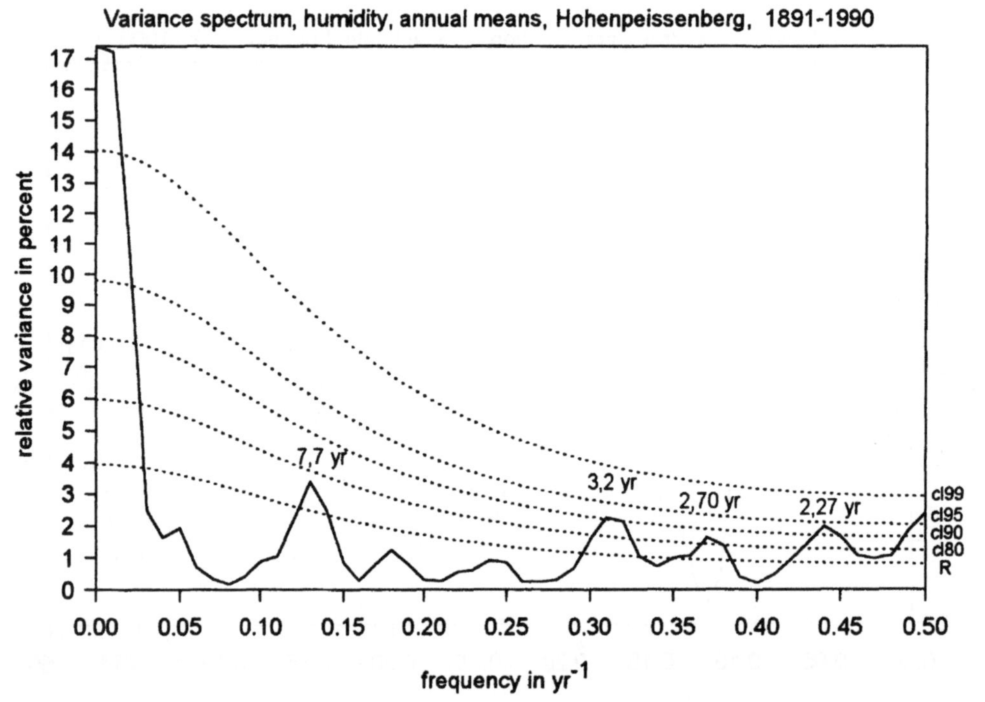

Variance spectrum, humidity, annual means, Hohenpeissenberg, 1891-1990
relative variance in percent
frequency in yr^{-1}
7,7 yr
3,2 yr
2,70 yr
2,27 yr
cl99
cl95
cl90
cl80
R

MIX
Papier aus verantwortungsvollen Quellen
Paper from responsible sources
FSC® C105338

If you have any concerns about our products,
you can contact us on
ProductSafety@springernature.com

In case Publisher is established outside the EU,
the EU authorized representative is:
Springer Nature Customer Service Center GmbH
Europaplatz 3, 69115 Heidelberg, Germany

Printed by Libri Plureos GmbH
in Hamburg, Germany